T0181332

Rural Road Engineering in Developing Countries

Rural Road Engineering in Developing Countries provides a comprehensive coverage of the planning, design, construction and maintenance of rural roads in developing countries and emerging nations. It covers a wide range of technical and non-technical problems that may confront road engineers working in the developing world, focusing on rural roads which provide important links from villages and farms to markets and offer the public access to health, education and other services essential for sustainable development.

Most textbooks on road engineering are based on experience in industrialised countries with temperate climates or deal only with specific issues, with many aspects of the design and construction of roads in developing regions stemming from inappropriate research undertaken in Europe and the USA. These approaches are frequently unsuitable and unsustainable for rural road network environments, particularly in low- to middle-income countries. This book takes on board a more recent research and application focus on rural roads, integrating it for a broad range of readers to access current information on good practice for sustainable road engineering in developing countries.

The book particularly suits transportation engineers, development professionals and graduate students in civil engineering.

Rural Road Engineering in Developing Countries

Jasper Cook and Robert Christopher Petts

CRC Press
Taylor & Francis Group
Boca Raton London New York

CRC Press is an imprint of the
Taylor & Francis Group, an **informa** business

Cover image: Jasper Cook

First edition published 2023
by CRC Press
6000 Broken Sound Parkway NW, Suite 300, Boca Raton, FL 33487-2742

and by CRC Press
4 Park Square, Milton Park, Abingdon, Oxon, OX14 4RN

CRC Press is an imprint of Taylor & Francis Group, LLC

Library of Congress Cataloging-in-Publication Data
Names: Cook, Jasper, author. | Petts, Robert Christopher, author.
Title: Rural road engineering in developing countries / Jasper Cook and
Robert Christopher Petts.
Description: First edition. | Boca Raton : CRC Press, [2023] |
Includes bibliographical references and index.
Identifiers: LCCN 2022056961 | ISBN 9781482259704 (hbk) |
ISBN 9780367860592 (pbk) | ISBN 9780429173271 (ebk)
Subjects: LCSH: Rural roads—Developing countries.
Classification: LCC TE127 .C66 2023 |
DDC 625.709173409172/4—dc23/eng/20230110
LC record available at https://lccn.loc.gov/2022056961

ISBN: 978-1-4822-5970-4 (hbk)
ISBN: 978-0-367-86059-2 (pbk)
ISBN: 978-0-429-17327-1 (ebk)

DOI: 10.1201/9780429173271

Typeset in Sabon
by codeMantra

Contents

5 Appraisal of low volume rural road initiatives 61

6 The rural road environment 75

Foreword

The authors have produced a timely and comprehensive coverage of a vital element of infrastructure, namely rural roads, delivery of which is still very much needed to alleviate poverty across the world. They have many years of experience in dealing with these structures and this is evident in the pages that follow, which guide the reader from the early planning through the design and construction phases onto maintenance, both routine and periodic. In the coverage, they have brought ideas developed through their and other people's experiences. They have given useful pointers to working in difficult terrain and the new concern for climate change.

Rural roads, which have to deal with low volumes of traffic of variable type, sometimes non-motorised, form the main routes of district jurisdictions and between villages, and serve remote places of habitation. There are many thousands, not to say millions of kilometres, of these roads existing, which require regular upkeep, but there remain many thousands of kilometres more to be constructed. This might have been said over half a century ago when road design or maintenance of these routes was beginning to be taken seriously, e.g. by TRL and other research and aid agencies.

The global population was then 3 billion, but the requirement to construct rural roads is still there today, if not more, when the population has grown at unprecedented levels and now has reached 8.5 billion. The UN's Sustainable Development Goals (2015–2030) emphasise 17 objectives and target rural roads as vital to alleviate hunger and support food production in a time of changing climate in some locations. The requirement for rural roads can be expected to last for many decades yet.

The authors can be congratulated on using their considerable experience in putting together this very useful book for practitioners in the field, as well as aid agencies. The book is also applicable as a teaching or knowledge transfer tool.

Prof J S Younger OBE DEng PhD MS BSc CEng FICE
Hon Research Fellow, University of Glasgow
Int'l Chancellor, President University, Indonesia

Authors

Jasper Cook is a Chartered Geologist with an undergraduate degree in Geology, an MSc in Engineering Geology and a PhD in Civil Engineering. He has over 50 years of experience as an engineering geologist/geotechnical engineer/climate resilience specialist primarily in the fields of infrastructure development, capacity building and research management. Over 30 years of this time has been spent on projects with emerging nations in Africa and Asia, most notably with UKAID-DFID, World Bank and Asian Development Bank on rural road network development programmes. He is currently advising the World Bank on climate resilience and road engineering issues for multiple transport projects in Sub-Saharan Africa, South Asia and South East Asia. He is the author of a wide range of peer-reviewed papers on aspects of engineering geology, geotechnics and road engineering.

Robert Christopher Petts has more than 50 years of professional experience working in over 40 African, Middle Eastern, Asian, Pacific and European countries, of which 44 years have been involved with developing countries and transport infrastructure for inter-urban routes and rural communities. He has been involved with the development of systems, technologies and techniques appropriate for a range of limited resource environments, making the best use of local resources such as labour, materials, enterprises and locally made equipment. He has advised various countries on development of government policy and strategy. He played a leading role in drafting rural road pavement, structures and maintenance sections within the Ethiopian and South Sudan Rural Road Manuals. He is the author of a number of sector reference documents, such as the WRA (PIARC) *International Road Maintenance Handbook* and *The Handbook of Intermediate Equipment for Roadworks in Developing and Emerging Regions*. He is co-author of the international *Low Volume Rural Road Surfacing and Pavements – A Guide to Good Practice* and the *gTKP Small Structures for Rural Roads Guideline*. He is an Honorary Senior Research Fellow for the University of Birmingham and advisor to the London University School of Oriental and African Studies.

Abbreviations and Acronyms

4	4-Wheel Drive
AADT	Annual Average Daily Traffic
AADT	Annual Average Daily Traffic
AASHTO	American Association of State Highway and Transportation Officials
ADB	Asian Development Bank
ADT	Average Daily Traffic
AfCAP	Africa Community Access Partnership
AfDB	African Development Bank
AIV	Aggregate Impact Value
ALD	Average Least Dimension
ARRB	Australian Road Research Board
AsCAP	Asia Community Access Partnership
ASTM	American Society for Testing and Materials
BMMS	Bridge Maintenance Management System
BoQ	Bill of Quantities
BS	British Standard
CBR	California Bearing Ratio
CO2	Carbon Dioxide
CPT	Cone Penetration Test
CRRN	Core Rural Road Network
CS	Cape Seal
DBM	Dry-Bound Macadam
DBM	Dry Bound Macadam
DBST	Double Bituminous Surface Treatment
DCP	Dynamic Cone Penetrometer
EHS	Environmental Health and Safety
EIA	Environmental Impact Assessment
EMP	Environmental Management Plan
ENS	Engineered Natural Surface
EOD	Environmentally Optimised Design
EPC	Engineer Procure and Construct
ESA	Equivalent Standard Axles

FED	Final Engineering Design
FHWA	Federal Highway Administration (USA)
FMS	Flood Management System
FS	Feasibility Study
GIS	Geographical Information System
GM	Grading Modulus
GPS	Global Positioning System
gTKP	Global Transport Knowledge Partnership
GWC	Gravel Wearing Course
HFL	High Flood Level
hp	horses power
HPS	Hand-Packed Stone
HVT	High Volume Transport
IDF	Intensity Duration Frequency
IEE	Initial Environmental Examination
IFC	International Finance Corporation
IFI	International Finance Institute
iRAP	International Road Assessment Programme
IRR	Internal Rate of Return
JICA	Japan International Cooperation Agency
KfW	Kreditanstellung für Wiederaufbau (German Development Bank)
kW	kilo Watt
LAA	Los Angeles Abrasion
LCB	Local Competitive Bidding
LIC	Lower Income Country
LMIC	Low to Middle Income Country
LVRR	Low Volume Rural Road
LVT	Low Volume Transport
LWC	Low-Water Crossing
masl	meter or metre above sea level
MDB	Millennium Development Bank
MDD	Maximum Dry Density
MDG	Millennium Development Goal
mesa	million equivalent standard axles
MPa	Mega Pascal
MTEF	Medium Term Expenditure Framework
NCB	National Competitive Bidding
NGO	Non-Government Organisation
NMT	Non-Motorised Traffic
NPV	Net Present Value
NRC	Non-Reinforced Concrete
NSEDP	National Socio-Economic Development Plan
OMC	Optimum Moisture Content
OPBC	Output Performance-Based Contract

PCU	Passenger Carrier Unit
PED	Preliminary Engineering Design
PFS	Pre-Feasibility
PMMS	Pavement Maintenance Management System
PMU	Project Management Unit
QA	Quality Assurance
QC	Quality Control
RAI	Rural Access Index
RAMS	Road Asset Management System
RDM	Road Design Manual
ReCAP	Research for Community Access Partnership
RED	Road Economic Development Model
RMMS	Routine Maintenance Management System
RoW	Right of Way
SADC	Southern African Development Committee
SANRAL	South African National Roads Agency
SBL	Sand Bedding Layer
SDG	Sustainable Development Goal
SE	Super-Elevation
SEACAP	South East Asian Community Access Programme
SlS	Slurry Seal
SME	Small to Medium Enterprise
SMS	Safety Management System
SOS	Single Otta Seal
SPT	Standard Penetration Testing
SS	Sand Seal
SuM4ALL	Sustainable Mobility for All
TL	Team Leader
ToR	Terms of Reference
TRL	Transport Research Laboratory
UCS	Unconfined Compression Test
UK	United Kingdom (of Great Britain and Northern Ireland)
UKAid	United Kingdom Aid (Department for International Development, UKDFID)
UN	United Nations
USAid	United States Agency for International Development
VAR	Vent-Area Ratio
VOC	Vehicle Operating Costs
WB	World Bank
WBM	Water-Bound Macadam
WHO	World Health Organisation
WLAC	Whole Life Asset Cost
WLC	Whole-Life Cost
WLRC	Whole Life Road Cycle

Introduction

Rural road networks are recognised as being essential to the sustainable development of rural economies. Rural roads provide important links from homes, villages and farms to markets and offer the public access to health, education and other essential services. These predominantly low volume roads also provide important links between rural communities, main road networks and thence to urban centres. At their lowest level, they provide the 'First Kilometre' link from agricultural areas for farmers to access markets and the 'Last Kilometre' for the provision of health and education services from central or regional governments.

There is some ongoing debate as to the definition of the upper level of *Low Volume Rural Roads*. This book has followed general current practice for Low- and Middle-Income Countries (LMICs) in developing regions by setting the upper traffic limit at around 300 motorised vehicles or the approximately 1–3 million standard axles (mesa).

This book aims to synthesise in a clear manner the key elements of current knowledge and experience to facilitate the application of best practice in improving rural transport infrastructure in an affordable, sustainable and manageable way for Low- and Middle-Income Countries (LMICs). The contents are, to a large extent, built around a core of practical and research experience derived from the authors' work with the development of rural infrastructure in Asia and Africa. It is, however, very applicable to LMICs in other regions.

The structure and contents of this book are outlined below.

Chapter	Key content
1. Roads, Development and Related Policy	A review of the current appreciations of roads, development, poverty alleviation and goals. It includes key issues around: roads in the context of sustainable development, poverty alleviation and the role of policy.
2. The Whole-Life Road Cycle	An Introduction to the concept of the whole-life road cycle. It describes importance of the road life cycle in providing a framework for cost-effective road or road network management from planning through to upgrade and rehabilitation.
3. Road Classification and Function	Roads must be designed to suit their identified function (or task) and the nature of the traffic (the people as well as the vehicles), which must pass safely along them. This chapter describes the key linkages between road function, classification and design as well introducing basic issues of rural road safety.
4. Roads and the Green Environment	Roads can dramatically affect the landscape, destroying or impacting farmland, forests and other natural areas, as well as human and wildlife habitats. This chapter outlines these potential impacts and the processes by which they should be assessed and mitigated.
5. Appraisal of low volume Road Initiatives	Description and guidance on the steps to be taken in appraising and prioritising proposed road projects in terms of benefits, impacts, socio-economic issues and assessing their economic validity.
6. The Rural Road Engineering Environment	It is recognised that the life-time performance of rural roads is influenced significantly by the impacts within what is termed the 'Road Environment', comprising engineering, operational and resource factors. This chapter introduces the concept of the Road Environment and describes the various impact factors associated with it.
7. Site Investigation	Guidance on procedures for collecting and managing site information to be used in the planning and design of rural roads within tropical and sub-tropical environment. Apart from geotechnical and engineering information, this chapter also deals with advice on gathering socio-economic and current and future climate information. Levels of required investigation detail are related to stage of projects within the Project Cycle.
8. Natural Construction Materials	A description of construction materials issues for each component of the road works and the desirability to optimise the use of locally available resources and the challenges of scarce resources. The need to take into account the distinctive characteristics of tropically weathered soil–rock materials is emphasised.
9. Geometric Design	Guidance on rural road geometric design, including influential factors, and aspects of traffic characteristics, cross section, alignment and safety.
10. Earthworks	The sustainable design of rural road earthworks including cut-slopes, embankment and roadside natural slopes. Options for slope stabilisation and protection are discussed together with the importance of lower cost bio-engineering solutions.

(Continued)

Chapter	Key content
11. Hydrology and Drainage	Protecting a road from then impacts surface and groundwater is a key aspect of the design of a road, and an understanding of hydrology is an essential component of the drainage design process. This chapter provides guidance on hydrology and its role in drainage design.
12. Cross-Drainage Small Structures	Guidance on the selection and design of small hydraulic structures such small bridges, causeways, drifts and culverts. The design process is set within the requirement to take account of future climate in terms of changes in rainfall patterns and occurrence of storm events.
13. Low Volume Rural Road Pavement Design Principles	The principles of Low Volume Rural Road pavement design. A summary of the general paving options available and guidance on their selection based on an assessment of key Road Environment factors.
14. BituminousSealed Pavements	Provides guidance on bituminous surfacing and associated structural pavement design based on recent work in Asia and Africa on Low Volume roads. Different approaches to sealed pavement design are summarised and assessed.
15. Non-Bituminous and Concrete Pavement Options	Provides design guidance on a range of non-bituminous pavement options ranging from low-cost engineered natural surface (earth roads), through gravel surfaced roads to brick or block roads, stone surfacing to various concrete paving options.
16. Spot Improvement	Outlines the principles of Spot Improvement and provides guidance on its application in providing effective access in situations of constrained budget. It is concerned with varying pavement selection and design in sensible lengths along an alignment in response to varying Road Environment impact factors.
17. Climate Resilience	Outlines climate change principles and gives guidance on assessment of climate change threats and consequent impact on road assets. A description is presented of the sequence for assessing and prioritising climate risk, identifying engineering and non-engineering resilience options followed by their constriction, monitoring and maintenance.
18. Procurement and Documentation	Outlines good practice in procurement documentation and contract selection. It provides general guidance on the procurement process within the context of the sometime challenging requirements of the rural infrastructure sector. Although the principal focus is on the procurement for road construction or rehabilitation, the general principles also apply to other works such as consultancy services, ground investigation contracts and maintenance contracts.
19. Construction Supervision and Quality Management	Issues relevant to good-practice construction for LVRRs, including construction planning, Quality Plans and as-built reporting. This chapter outlines the use of Technical Audits and summarises their objectives, structure and application.

(Continued)

Chapter	Key content
20. Road Maintenance Strategy	A review of the vital issue of asset management and maintenance and an outline of the basic aims and principles with the context of rural road networks. Options to assess and improve asset management performance are outlined and discussed.
21. Road Maintenance Procedures	Provides guidance on maintenance activities in terms of their operational frequency, routine maintenance, periodic maintenance and urgent/disaster maintenance. The general activities to be carried out are be summarised as regards rural road types.
22. Appropriate Technology	Reviews the general technologies used in constructing and maintaining roads. It outlines key criteria in the selection of the appropriate application within the broad groupings of labour based, intermediate technology and heavy plant based operations.

Chapter 1

Roads, development and related policy

INTRODUCTION

The typical situation in most low-income developing countries is one of a large rural population with agricultural-based economies where the imperative is to provide these rural communities with safe and sustainable access to basic services and opportunities for improving their livelihoods. It is an established principle that effective transportation plays a crucial role in rural socio-economic development and in reducing poverty; as stated by the SuM4All initiative in 2019, 'The contribution of transport to economic development and human capital is undeniable' (SuM4All 2019). Low volume rural roads (LVRRs) can therefore be considered a vital part of a country's road transport network, along with the other volume elements of the transport network such as national highways, urban roads, and higher volume inter-urban roads. However, after more than a hundred years of motor transport, many developing regions are still characterised by a low density of all seasonally serviceable roads (as shown in Figure 1.1).

In addition, in Latin America, parts of Asia and Sub-Saharan Africa, the classified road networks are also still predominantly unsealed (World Bank 2020). This compared starkly to 70%–100% of networks being paved in the high-income countries, as illustrated in Figure 1.2.

This chapter comprises a review and update to the current appreciations of roads, development, poverty alleviation and goals. It includes key issues around:

- Roads in the context of sustainable development and poverty alleviation.
- The role of policy in rural road network development.
- Rural road network strategic planning.

DOI: 10.1201/9780429173271-1

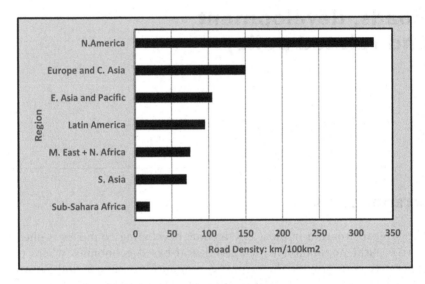

Figure 1.1 World variation in road density (km/100 km²).

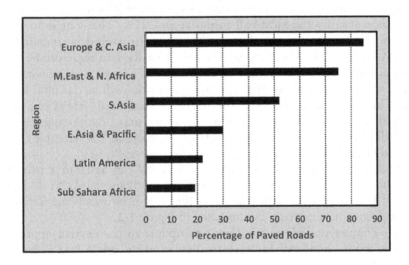

Figure 1.2 World variation in percentage of paved roads.

ROADS AND RURAL DEVELOPMENT

Key issues

The importance of rural road developments has been internationally recognised in two key statements (Sum4All 2019).

Firstly, the *Ashgabat Statement (2016)*; this was endorsed by more than 50 member countries attending the first ever UN Secretary General's Global

Conference on Sustainable Transport held in Ashgabat, Turkmenistan. Regarding rural access, countries reaffirmed their commitment to:

> support efforts to provide communities in rural areas in developing countries with access to major roads, rail lines, and public transport options that enable access to economic and social activities and opportunities in cities and towns and that unleash productivity and competitiveness of rural entrepreneurs and smallholder farmers.

Secondly, *the Vientiane Declaration on Sustainable Rural Transport (2017)*. This voluntary declaration was adopted by representatives of 23 member countries and 14 observer countries at the tenth Regional Environmentally Sustainable Transport Forum in Asia, in Vientiane, Lao PDR. The declaration demonstrated a commitment by government authorities, development agencies, civil society, academia, the private sector, and other relevant stakeholders to promote inclusive, affordable, accessible and sustainable rural transport infrastructure and services, in order to facilitate improved access to basic utilities and services including health and education by the rural poor and vulnerable groups. The declaration expresses a commitment to 'assign due priority to rural transport projects and initiate the development of national strategies and policy frameworks to improve rural transport connectivity to wider local, national and regional transport networks'.

To further underscore the critical role of rural transport in development, a set of key messages was developed by the UKAID funded Research for Community Access Partnership (ReCAP), Table 1.1 outlines these messages (SLoCaT 2017).

Table 1.1 Key Rural Transport Messages

No.	Message	The rationale
1	Improved rural transport drives sustainable rural development and national growth.	Road infrastructure promotes connectivity and social cohesion, drives commercial activities as well as accessibility to social and economic facilities.
2	Better rural transport is the key for food security and zero hunger.	Improving rural access can lead to lower costs for farm inputs and lower transport costs for marketed outputs, thus increasing agricultural production.
3	Poor rural transport condemns the poor to stay disconnected and poor.	Access to markets and employment opportunities through better rural transport infrastructure and services is an essential pre-condition to generating rural income and thus reducing poverty.
4	Additional money and commitment is needed to build and maintain rural road networks.	Funding sources need to be expanded and new funding sources need to be developed, piloted, and implemented not only for building but also for maintaining the asset.
5	Better rural transport calls for local solutions for local challenges.	Rural access challenges require local resource-based solutions that are compatible with the local road environment conditions.

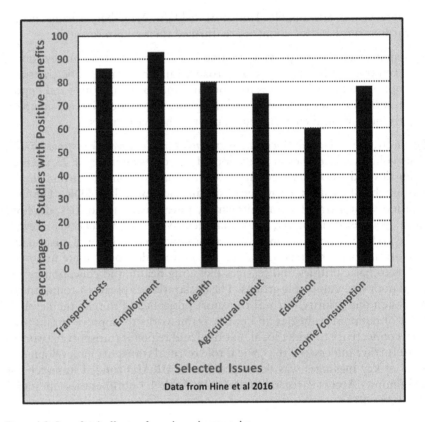

Figure 1.3 Beneficial effects of rural road networks.

A general systematic review by Hine et al. (2016) established that the expansion of rural road networks has a positive impact on poverty reduction. The review presented strong evidence that, over the medium to long term, development of a rural road network leads to an increase in employment, income and consumption, and expansion of the agricultural sector. There is also evidence to suggest that the health impacts are generally positive, although, on the downside, increased connectivity is also shown to lead to an increase in communicable diseases. Figure 1.3 adapted from this study shows the percentage of studies showing a beneficial effect for some of the key characteristics investigated.

The sustainable development goals

The Sustainable Development Goals (SDGs) are a collection of 17 interlinked global goals designed to be a 'blueprint to achieve a better and more sustainable future for all' (UN General Assembly 2015). Following-on from the Millennium Development Goals, which were superseded in 2015. The

Figure 1.4 The 17 SDGs (*Source:* https://www.un.org/sustainabledevelopment/news/communications-material/).

SDGs were endorsed by the United Nations General Assembly and are intended to be achieved by the year 2030. Figure 1.4 presents the 17 SDGs.

Although there is no dedicated SDG target on rural transport, there are numerous linkages between rural access and the SDGs (Cook et al. 2017). Successful scaled up implementation of rural transport will contribute to realizing:

- SDG 1 to alleviate poverty.
- SDG 2 to achieve zero hunger and ensure food security.
- SDG 3 to ensure health and well-being.
- SDG 4 to provide access to education.
- SDG 5 to empower women in rural areas;
- SDG 6 to facilitate access to clean water and sanitation.
- SDG 8 to promote inclusive growth and economic opportunities.
- SDG 9 and SDG 11 to contribute to sustainable infrastructure and communities for all.
- SDG 13 to increase climate resilience and adaptation in rural areas.

Details on reporting of the SDGs are contained in the UNDG publication Guidelines to Support Country Reporting on the Sustainable Development Goals (UNDG 2017).

The rural access index

In addition to indirect linkages to SDGs and associated targets, there is a direct linkage to rural access in SDG indicator 9.1.1. Indicator SDG 9.1.1 is also referred to as the Rural Access Index (RAI) The RAI is currently defined

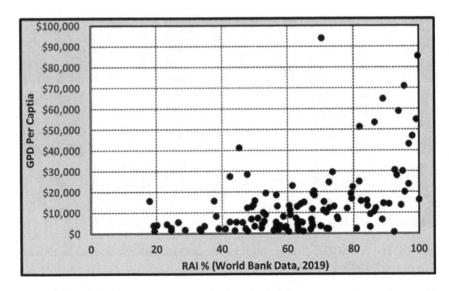

Figure 1.5 RAI versus GDP.

as the percentage of people within a country, region or district who have access to an all-season road within an approximate walking distance of 2 kilometres (km) (World Bank 2019). The reasonable assumption is that a 2 km threshold is a reasonable extent for people's normal economic and social purposes. Figure 1.5 indicates a broad correlation between RAI (SDG 9.1.1) and GDP.

An 'all-season road' is a road that meets its assigned level of service all year round in terms of the prevailing means of rural transport (for example, a pick-up or a truck that does not have four-wheel-drive). Predictable interruptions of short duration during inclement weather (e.g. heavy rainfall) are accepted, particularly on low volume roads (Roberts et al. 2006). A road that it is likely to be impassable to the prevailing means of rural transport for a total of 7 days or more per year is not regarded as all-season (ReCAP 2019a).

The RAI is simple enough to understand and use as an indicator for strategic planning not only in the transport sector but also in a broader rural development context. RAI is most commonly used at a national level, although some countries have begun to use it at regional or district level. It also has potential applications as a road project-level indicator as to the socioeconomic impact (ReCAP 2019b).

RURAL ROAD POLICY

The role of policy

An effective policy framework can help to ensure that decisions made by different organisations and bodies are achieving a common overall aim and are consistent with each other. Policies can be a powerful driver to facilitate

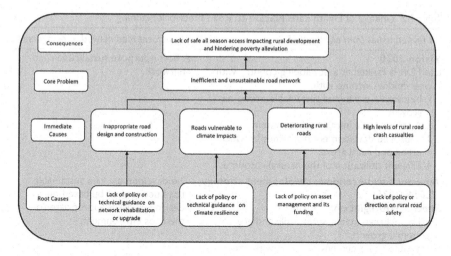

Figure 1.6 The consequences of a lack of policies in the rural road sector.

and sustain change. In the transport sector, policy provides a framework within which decisions can be taken about all aspects of road network management. The policy framework helps to ensure that cross-sectorial decisions made by different government organizations and bodies are achieving a common overall aim and are consistent with each other. At its simplest, policy may be considered as the 'What and Why' of rural road network management (Howe 1996).

A well-structured policy is particularly helpful in the rural road sector where budgets are constrained and where it facilitates objectives being more transparently set and prioritized, which reflect strategic road user requirements more accurately. Figure 1.6 outlines how a lack of relevant policies inhibits sustainable development.

Policy should adopt a multi-tiered, multi-modal approach that supports both attainment of universal rural access, and further upgrading to higher access tiers based on affordability and feasibility (SuM4All 2019). Policies can be set at a range of levels; from regional agreements through national right down to village or community level. At the highest government level policy will set national development targets, frequently within 5–10 year rolling programmes. National development policy should drive the formation of clear transport policies, including those relevant to rural network development.

Table 1.2 provides an example of the links between an overarching national policy and consequent rural road network implications, taken from the Lao PDR eighth National Socio-Economic Development Plan (NSEDP), Government of Lao (2015) and Figure 1.7 indicates typical target levels of policy within an overall national policy that has links to international agreements and protocols.

Table 1.2 Linkage of Eighth Laos NSEDP to Road Network Planning

CR related issues from eighth NSEDP	Consequent road network requirement
Vision 2030 Lao PDR is ranked as a developing country with upper-middle income with: 1. Innovative, green and sustainable economic growth. 2. A strong basic infrastructure system. 3. Improved development disparities between urban and rural areas. 4. Efficient utilization of the natural resources.	A fit-for-purpose sustainable road network.
Ten-year socio-economic development strategy (2016–2025: ***Specific strategies on:*** Quality, inclusive, stable, sustainable and green economic growth. Sustainable and green environment with effective and efficient use of the natural resources.	The development of a sustainable road network to be based on design standards appropriate to the Laos environment.
Outcomes of the eighth NSEDP (2016–2020) **Outcome 3** Natural resources and the environment are effectively protected and utilized according to green growth and sustainable principles; there is readiness to cope with natural disasters and the effects of climate change and for reconstruction following natural disasters.	The road network is designed, constructed and maintained in an environmentally appropriate and sustainable manner. Road designs must take account of potential natural hazards, including current and future climate.

Policy implementation

The following are some general points to be considered from policy delivery and ease-of-implementation viewpoints:

1. Establish a comprehensive and pragmatic policy that is aimed at national rural access targets within a defined time frame, preferably around 10 years.
2. Adopt a multi-tiered, multi-modal approach that supports both early attainment of rural access targets, and then further upgrading to higher access and mobility levels.
3. Agree policy targets with key stakeholders on cross-sector planning, financing and project management.
4. Include measures to expand and improve rural transport services and rural logistics.
5. Establish a national programme to implement the policies through a partnership involving central and local governments, communities and the private sector.
6. Establish technical standards for each of the multiple tiers of rural access, ensuring protection against water penetration, screening for

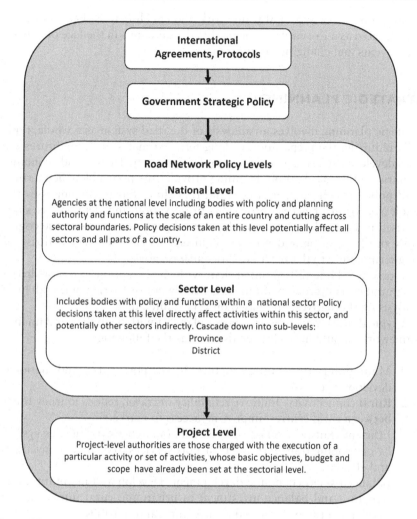

Figure 1.7 Levels of policy and planning.

climate vulnerability, use of local materials and resources where feasible, and incorporation of traffic safety.

7. Establish reliable approaches to asset management, with communities performing routine maintenance where feasible and contactors performing periodic maintenance selected using a Road Asset Management System (RAMS), and implement measures to control axle loads.

8. Prepare procurement rules, procedures, standard bidding documents and contract documents for the programme, appropriate for the local operational environment, supported by an e-procurement platform.

Once policies have been developed and agreed that they should be accessible to stakeholders and practitioners through dissemination and awareness creation initiatives as a vital framework and enabling tool. They will need to be treated as a live entity, to be modified and refined in the face of experience, events and changing circumstances.

STRATEGIC PLANNING

Strategic planning involves an analysis of the road system as a whole, typically requiring the preparation of long-term estimates of expenditures for road development and conservation under various budgetary and economic scenarios (Overgaard 2004). Planning provides the means by which a rural road policy is delivered. It is essential that this strategy encompasses the whole spectrum of access provision, preservation and delivery, as summarised in Figure 1.8. A focus only on road provision, without due cognisance of preservation and delivery requirements runs the very real risk of long-term failure to deliver fitness-for-purpose goals.

In practice, Hine (2014) noted that, with the exception of donor-driven programmes, rural road planning is generally poorly carried out, with little analysis of alternatives, and based on very limited data.

A typical rural road network master plan, set within a policy and finance framework, should have clear goals, such as the following:

1. Maintaining rural infrastructure to support rural and national development.
2. Rural connectivity: Improve reliability of travel, reduce journey times between rural centres, improving access to markets.
3. Efficient transport services: Create a comprehensive logistics system by developing transport and transport services to be more convenient and efficient.
4. Balanced socio-economic development: Develop and maintain infrastructure and balance investment in urban and rural infrastructure development in order to reduce urban/rural disparities.
5. Resource use: Optimise the use of local resources, minimise unnecessary imports of goods and services, and minimise waste or inappropriate use.

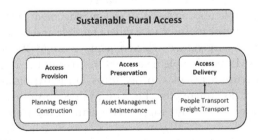

Figure 1.8 The rural access continuum.

6. Climate resilience: Improve the quality and resilience of infrastructure development for climate resilience to ensure that public works and transport networks are built and maintained to standards that provide safe and reliable facilities in all weathers.
7. Environmental improvement: Implementation of a strategy reducing environmental pollution, toxic emissions and waste. Reduce the impact of natural disasters.
8. Safety: Reduce road traffic fatalities and injuries.

The following is a typical list of key headings for a rural network development Master Plan

- Definition of objectives and scope of the master pan.
- Assessment of socio-economic and natural environments.
- Definition transport demand to be met by plan.
- Options for alternative transport development scenarios.
- Materials, human, enterprise, and technology resource options.
- Analysis and prioritisation of master plan options.
- Define investment requirements.
- Roll-out of Transport Master Plan(s).

REFERENCES

Cook, J. R., R. C. Petts, C. Visser and A. You. 2017. The contribution of rural transport to achieve the Sustainable Development Goals. Research Community for Access Partnership (ReCAP) Paper, ref. KMN2089A, for UKAID-DFID. https://www.research4cap.org/index.php/resources/rural-access-library.

Government of Lao PDR. 2015. *8th National Socio-economic Development Plan* (NSEDP).

Hine, J. 2014. Good policies and practices on rural transport in Africa; Planning Infrastructure & Services. SSATP, The World Bank Group, Washington. https://documents1.worldbank.org/curated/en/127531467999362783/pdf/937790NWP0Box30l0Transport0Planning.pdf.

Hine J., M. Abedin, R. J. Stevens, T. Airey and T. Anderson. 2016. *Does the extension of the rural road network have a positive impact on poverty reduction and resilience for the rural areas served? If so how, and if not why not? A systematic review.* London: EPPI-Centre, Social Science Research Unit, UCL Institute of Education, University College London. https://www.gov.uk/research-for-development-outputs/does-the-extension-of-the-rural-road-network-have-a-positive-impact-on-poverty-reduction-and-resilience-for-the-rural-areas-served.

Howe, J. 1996. Transport for the poor or poor transport? A general review of rural transport policy with emphasis on low-income areas. IHE working paper IP-2, Delft: International Institute for Infrastructural, Hydraulic and Environmental Engineering.

Overgaard, K. R. 2004. Chapter 6 Planning methods. In *Road engineering for development*. Second edition, Robinson R. and Thagesen B., Eds. Taylor and Francis. ISBN 0-203-30198-6.

RecAP. 2019a. *Consolidation, revision and pilot application of the Rural Access Index (RAI): TG2 Final Report*. TRL Ltd. ReCAP GEN2033D. London: ReCAP for DFID. https://www.research4cap.org/index.php/resources/rural-access-library.

ReCAP. 2019b. *Measuring rural access using new technologies: supplemental guidelines*. TRL Ltd. ReCAP Ref. GEN2033D. London: ReCAP for DFID. https://www.research4cap.org/index.php/resources/rural-access-library.

Roberts, P., K. C. Shyam and C. Rastogi. 2006. Rural access index: A key development indicator. Transport paper TP-10. World Bank, Washington DC. https://documents1.worldbank.org/curated/en/721501468330324068/pdf/Rural-access-index-a-key-develpment-indicator.pdf.

SLoCaT. 2017. Promotion of sustainable rural access in the implementation of the 2030 global agenda on Sustainable Development: Key messages consultation analysis. Partnership on sustainable, low carbon transport. London: ReCAP for DFID-UKAID. http://research4cap.org/Library

SuM4All. 2019. Global roadmap of action toward sustainable mobility: Universal rural access. Washington DC. https://thedocs.worldbank.org/en/doc/662991571411009206-0090022019/original/UniversalRuralAccess GlobalRoadmapofAction.pdf.

UN Development Group. 2017. Guidelines to support country reporting on the Sustainable Development Goals. https://www.un.org/development/desa/publications/sdg-report-2017.html.

UN General Assembly. 2015. Transforming our world: The 2030 agenda for sustainable development. A/RES/70/1. https://sustainabledevelopment.un.org/content/documents/21252030%20Agenda%20for%20Sustainable%20Development%20web.pdf.

World Bank. 2019. World measuring rural access: Update 2017/18. Report no: ACS26526. https://openknowledge.worldbank.org/bitstream/handle/10986/32475/World-Measuring-Rural-Access-Update-2017-18.pdf?sequence=1&isAllowed=y.

World Bank. 2020. To Pave or Not to Pave. Developing a Framework for Systematic Decision Making in the Choice of Paving Technologies for Rural Roads. Mobility and Transport Connectivity Series. https://openknowledge.worldbank.org/handle/10986/35163.

Chapter 2

The whole-life road cycle

INTRODUCTION

Most of the features of a rural road project can be related to the need to take a series of decisions and action linked to the logical progression of the road project from inception to delivery of design life performance. A Whole-Life-Road Cycle (WLRC) can thus be considered as a framework around which the rural road and road network management process can be structured. Cook et al. (2013) advocate the use of a WLRC approach to allow decisions to be taken not only in a logical sequence but also within a wider strategic and holistic framework. In the initial phases of a road project leading to construction, the resources and time needed and consequently the cost for obtaining the information and carrying out the necessary work increases VARY substantially at each stage of the process leading to construction completion. The maintenance phase will require different information and decision processes. Possible future rehabilitation or upgrade will also require adapted information and decision-making processes.

The increasing cost of each successive stage of the process means that it is very important to review the process at each stage and to make a clear decision whether the next stage of the process is justified. Major decisions on whether or not to continue with project preparation are made at the end of each stage. As the cost of each successive stage leading to construction is many times greater than that of the previous stage, the importance of the decision and the amount of information needed to make it also increase at each stage.

This chapter aims to place rural road activities and decisions within the context of the WLRC, and provide readers with a clear pathway to understanding and implementing effective rural road management as a rational 'joined-up' process.

STRUCTURE OF THE WHOLE-LIFE ROAD CYCLE

In its simplest form, the WLRC comprises a series of linked phases as shown in Figure 2.1 (Cook et al. 2013).

DOI: 10.1201/9780429173271-2

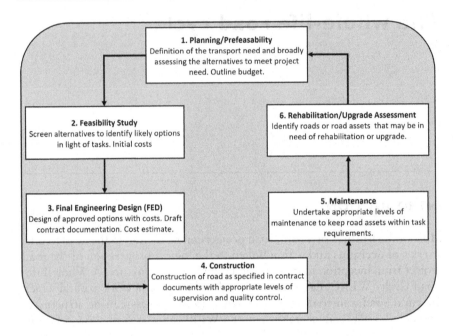

Figure 2.1 The whole-life cycle for roads.

There are potentially many possible variations to this basic diagram. It is quite common, particularly in larger rural programmes, to include a Pre-Feasibility Stage (PFS) between Planning and Feasibility. This PFS is normally inserted to allow further discussion and screening of potential road or network options, perhaps within a wider national or regional context.

The WLRC may also be used as a guiding framework for the application of specific topics; for example, it is used as a guide for the planned integration of climate change adaptations into road programmes (Chapter 17) and as an input to appropriate selection of paving options (Chapter 12).

A life-cycle approach can also be a strategic aid to overall Road Asset Management in providing a framework around which to plan and monitor rolling programmes of maintenance, rehabilitation and upgrade.

The preparation of a major road project may take several years, from its initial planning conception to its construction. During this time, the national economic or political situations may change, and hence a further benefit of a staged decision-making process is that it provides the necessary review points at the end of each stage at which any changes in the external situation can be reflected in the approach used for the next stage.

The WLRC encourages a focus on collecting, collating and screening data so that key, well-informed decisions can be made to support moves between cycle stages as well as defining internal actions for each stage. It is also important to note that this assessment and decision process facilitates decisions such as 'rejection', 'pause', or 'reject and reconsider' to be considered at each stage. This may be particularly important in the early Planning–Feasibility–Design stages, where valid challenges to important socio-economic, engineering or financial assumptions can result in costly mistakes being avoided.

LIFE-CYCLE STAGE DETAIL

Planning

The input into this initial stage normally comes from a combination of existing road upgrade assessments and recommendations or directives initiated by a national policy; for example, a Rural Development Programme or Rural Poverty Alleviation agenda.

This is the stage at which the overall project and its strategic objects are refined, potential budgets are outlined, and strategic financial and broad engineering risks are identified. This process takes into account government policies and strategic development targets that impact on road development. Proposed projects at this stage are, therefore, examined in a very wide socio-economic and policy-orientated context. There will normally be an initial assessment of the project against the previously defined criteria. Projects that do not meet selection criteria are screened out or modified.

During the planning stage, the general road tasks to be met are defined and evaluated. The identification of whether, or not, the likely road falls within the Low Volume Roads envelope is assessed.

Road project decisions impact on local communities as well as other sectors, such as agriculture, water, health, environmental and education, alongside commercial activities such as local transporters, suppliers and traders. Consideration of these impacts and initial consultation with other ministries and stakeholders at this early stage will help mobilise support and maximise the beneficial impacts of the road works.

The following is a checklist of key decisions or criteria for continuing with the proposed project in some form:

- A rural road project meets the identified project requirement.
- The project modes are identified: new construction, rehabilitation, periodic maintenance, upgrade or Spot Improvement.
- The project lies within general policy or transport master-planning guidelines.
- The project is compatible with potentially available funding sources.
- The project should proceed to the next stage or is a Pre-Feasibility Study (PFS) required.

Actions to inform the key decisions are summarised in Table 2.1. These actions would be most likely be undertaken either by Road Agency personnel or by a retained Consultant.

Table 2.1 Key Actions at the Planning/Pre-feasibility Stage

Pre-feasibility action	Comment
Initial desk study.	Collate available information on policy, masterplans and project background. Identify general physical and climate setting, and sources of more detailed information.
Road task, classification and general standards.	Define the purpose of the project and identify the appropriate classification and general standards.
Initial visits to project area.	Initial orientation visit to project area to identify key features and potential meetings with local stakeholders; see Chapter 7.
Construction materials	Collate local materials resources data.
Technology and local resources review.	Collate data on local enterprises and skills capacity.
Climate character and level of resilience identified.	The broad regional climate characteristics are captured; see Chapter 17.
Identify requirement for major structures.	Location of any major and potentially costly structure.
Maintenance regime.	Review current asset management capability.
Identify any potential high-risk issues.	Early definition of major high-risk or high-cost challenges.
Conditions for approval.	Define any adjustments required to the basic project assumptions and outline.

Feasibility

The Feasibility Stage (FS) of a rural road project assumes that road transport or access need has been identified and that the solution to the need falls within the rural road envelope. Input to the FS normally comes from recommendations in a Planning or Pre-Feasibility report or outline. This is the stage where more detailed economic and engineering assessments are made, and the main engineering problems and other issues affecting the route or network are identified.

Where Pre-Feasibility (PFS) stages are required, they normally concentrate on assessing alternate strategic options such an alignment, road function or network extent options within likely budgets.

The FS assesses alignment, paving, earthwork, drainage and bridge/structure options, and identifies those most likely to provide a sustainable solution within the governing Road Environment and within the anticipated available budgets. This is generally seen as a critical stage by road authorities and external funders and donors such as the World Bank, Asian Development Bank (ADB), African Development Bank (AfDB) or Japan International Cooperation Agency (JICA). Relevant Ministry planners and road consultants are normally involved at this stage.

As part of the FS, it is important to identify and investigate the major technical, environmental, financial, economic and social constraints in order to obtain a broad appreciation of the viability of the competing options. For rural roads, one of the most important aspects of the Feasibility Study is communication with the local stakeholder groups who will be affected by the road.

An assessment of available resources is generally required both to confirm the feasibility of a proposed road and to identify sustainable and appropriate strategic design options. At FS, sufficient data is required to identify the most suitable options appropriate to the specific road requirements. Information is generally required that is sufficient to obtain likely costs to an accuracy of at least ±25%.

Such is the importance of the FS that on larger projects, as noted above, it may be split into sub-stages or sub-cycles in dealing with specific issues. There are some issues that form a crucial part of the FS, for example:

1. The appropriate use of locally available construction materials and skill resources. Particularly important is early knowledge of any potential problems with existing sources. If there are no existing material sources, then more detailed materials exploration investigations need to be defined for further design stage investigations.
2. The definition of the climate change impacts, their mitigation and the budgetary consequences.
3. Environmental impacts; for example, the identification of constraints such as National Parks, or Protected Zones.

The overall checklist of key decisions or criteria for proceeding with the proposed project beyond FS is as follows:

- Local resources review including materials, enterprises, skills and technology options.
- Road pavement options.
- The project is technically feasible.
- Road long lists or network options are screened and prioritised.
- Proposed solutions are compatible with the defined road tasks.
- There are no unsurmountable major engineering, environmental or social issues.
- The existing asset management and maintenance regime has been pragmatically assessed.
- The costed preliminary designs are within the defined budget.
- The project should move on to the next stage or be amended.

The output from the FS is a decision on whether the project should progress and what, if any, modifications are required. If the project is to proceed, then the FS report should include costed preliminary designs and Terms of Reference (ToR) for the next stage.

Actions to inform the key decisions are summarised in Table 2.2. These actions could be undertaken by Road Agency personnel but are commonly

Table 2.2 Key Actions for the Feasibility Stage

Feasibility action	Comment
Desk study to review available information relevant to preliminary design.	Enhance the planning desk study to consider specific engineering, environmental, climatic and socio-economic issues; see Chapter 7.
Walkover surveys of alignment(s).	Identify general character, areas of potential hazard and climate vulnerability.
Initial field investigations.	Designed to identify and define key geotechnical and materials issues, as Chapter 7.
Define the key aspects of the Road Environment.	Preliminary data for design, as listed in Chapter 6.
Identify locations for stream and river crossing.	Climate issues are of key importance at this stage.
Spot Improvement.	Collate data on whether Spot Improvement of roads (Chapter 16), road network or individual road assets could meet the defined requirement.
Undertake an assessment of traffic in terms of vehicle numbers and type.	Traffic counts and axle loads assessments required for preliminary geometric design (Chapter 9), pavement structure and safety issues in FS report.

done by a retained Consultant. There may be input into the structure and analysis of the FS from any external funding sources, such as a World Bank, AfDB or ADB.

Final Engineering Design (FED)

The FED stage requires sufficient data for preparation of the contract documents, including technical specifications, drawings and Bills of Quantity. Final detailed cost estimation is also likely to be required. The FED stage requires more investigation and considerably more data than the previous stages. The process of project design should normally be completed with sufficient accuracy to minimise the risk of significant changes being required after the works contract has been awarded. The exception to this can occur in some complex projects where some aspects of detailed design remain to be completed or modified during the construction phase, for example, for roads in difficult terrain where the detailed final Ground Model may only become clear during construction and an agreed Observational Method is being followed (Peck 1969).

For the specific case of the pavement and surfacing elements, the FED stage may form part of a two-phase pavement design procedure, following on from the identification of a short list of options identified at FS. This FED will include the design and specification of the pavement structural layers and any overlying surfaces together with associated shoulders and pavement drainage.

Feasibility assumptions on traffic patterns should be cross-checked and, if required, additional surveys should be undertaken aimed specifically at obtaining data for each vehicle category and axle loading for the pavement layer design. Risks of likely axle-overloading should be pragmatically assessed.

Sources of material should at this stage be defined in terms of location, quality and quantity, such that it is clearly established that the road or roads can be built to the required specification with the available materials. Source, haulage, processing, and placement costs need to be investigated and any inflation factors considered.

Climatic patterns and the incidence of severe climatic events should be confirmed and assessed within the levels of required climate resilience.

A checklist of key decisions or criteria for proceeding with the proposed project is as follows:

- The project has been designed in detail and costed satisfactorily.
- The technologies selected are compatible with local enterprise and skills availabilities.
- The 'Engineer's Estimate' of total construction and associate costs are within the budget.
- The proposed designs are technically feasible within the available contracting resources.
- The project can be constructed within existing social and environmental requirements.
- The project adheres to national policies on climate mitigation and adaption.
- A practical implementation programme exists.
- Asset management and maintenance funding sources are confirmed.
- There are no objections from external funding sources.
- The project is cleared to proceed to procurement.

Actions to inform the key decisions are summarised in Table 2.3. These actions could be undertaken by Road Agency personnel but are commonly done by a retained Design Consultant. There may be input from any external funding sources, such as a World Bank, AfDB or ADB. Data are generally required that are sufficient to obtain likely costs to an accuracy of better than about ±10%.

Construction

The aim of this phase of the road cycle is the satisfactory construction of road (or roads) as specified, and costed in the contract documents with appropriate levels of supervision and Quality Control. This phase should also include an as-built survey as part of the completion certification and as a link to the future maintenance and long-term management phases,

Table 2.3 Key Actions for the FED Stage

FED actions	Comments
Undertake detailed ground investigations.	Following recommendations from FS reports.
Undertake detailed materials investigations.	Ensuring that materials are available in sufficient quality and quantity; see Chapter 8.
Undertake detailed assessment of local contracting and skills capacity.	Select technology options compatible with local capacities and identify any enhancement needs; see Chapter 22.
Final Engineering Designs pavement.	Detailed phase of pavement designs; see Chapters 14 and 15.
Final Engineering Designs earthworks.	Embankment and cut-slope designs including erosion and landslide protection; see Chapter 10.
Final Engineering Designs drainage.	Pavement, earthwork and alignment drainage issues.
Final Engineering Designs structures.	Detailed culverts, causeway and bridge designs; see Chapter 12.
Final climate resilience applications.	Ensure future climate criteria are used for detailed design of earthworks, vertical alignment and hydrology input to bridge and culvert designs; see Chapters 11 and 17.
Asset management requirements defined.	Prepare detailed maintenance requirements and draft proposals for any required enhancement of funding or capacity.
Agree procurement arrangements and draft documents.	Draft appropriate technical specifications, ToR, BoQ and general contract documents appropriate to contract model; see Chapter 18.

although in increasing popular models the construction period may roll on to include a 5–10-year maintenance period.

The construction process itself is seldom as well controlled as expected or desired. Sources of variability in quality arise in all aspects of the construction process, and some are inherently more serious than others. This is further discussed in Chapter 18.

The following is a checklist of key decisions or criteria for successful construction:

- Contractors have been satisfactorily procured within financial and technical limits.
- Construction works are carried out as specified.
- Contract variations are within those anticipated for the contract model.
- Design amendments are within those anticipated for the ground and design models.
- Quality Assurance arrangements have been implemented satisfactorily.
- Close liaison with the relevant road authority is achieved throughout the construction phase.
- As-built drawings and completion reports are satisfactory.

Table 2.4 Key Actions at the Construction Stage

Construction actions	Comments
Procurement	Procurement of an appropriate contractor or contracting joint venture.
Construction	Construction as per finalised engineering drawings, Bills of Quantity and technical specifications within an approved Quality Control framework and programme.
Supervision arrangements	Quality Management arrangements put in place. This requires in traditional contracts clear assignment of responsibilities for Quality Control (Contractor) and Quality Assurance (Supervising organisation) through an approved Quality Plan; may also include Technical Audit where appropriate. Supervision of the Environmental Management Plan is also an important issue.
Construction planning	Draft and implement an appropriate construction plan that takes into account climate impacts on key construction items such as drainage, earthworks, surfacing and structures.
Design variations	Additional investigations may be required, particularly if there is an 'observation and amend' aspect to the earthworks design. Additional proving of materials sources may be required.
As-built reporting	Should include accurate as-built drawings and collation of all quality and Technical Audit reports.
Maintenance recommendations	Depending on the contract model, there may be a requirement to include recommendations on hand-over condition and identification of specific maintenance requirement – for example, ongoing bioengineering supervision.

Actions to inform the key decisions are summarised in Table 2.4. These actions could be undertaken by Road Agency personnel but are more commonly done by a retained Supervision Consultant. There may be additional overview supervision from any external funding sources, primarily concerned with satisfactory stage payments being in line with achieving quality and quantity indicators.

Maintenance

Maintenance is the range of activities necessary to keep a road and associated structures in an acceptable condition for road users during the defined design life. Despite the recognised importance of maintenance in sustaining roads and road networks, it continues to be generally poorly delivered, particularly at rural road level and is a clear weak link in the overall WLRC. Chapters 21 and 22 deal with the topics of asset management and maintenance delivery in some detail, whilst the following is a basic checklist of key decisions or criteria for successful implementation of maintenance at road level:

• An appropriate maintenance model is in place along with associated ongoing funding.

Table 2.5 Key Actions for the Road Maintenance Stage.

Maintenance actions	Comment
Maintenance model selection	An appropriate maintenance model is identified that is compatible with the road type, condition and general environment; see Chapter 21. Should follow on from any construction stage recommendations.
Maintenance planning	Prioritising of roads or section of road for maintenance based on condition surveys. Ensure programme requirements are compatible with available finance; see Chapter 21.
Capacity enhancement	Implement defined requirements for the asset management and maintenance funding and capacity.
Supervision arrangements	Appropriate supervision and evaluation procedures should be in place; see Chapter 22.
Implementation	Maintenance procedures should follow relevant routine and periodic guidelines.
Monitoring and feedback	Monitoring of effectiveness of maintenance and identification of roads or sections of road where and rehabilitation, climate resilience, or Spot Improvement is required; see Chapters 16, 17 and 22.

- A prioritised rolling programme of maintenance (routine, periodic and contingency) exists.
- An appropriate contractor, local organisation or group has been identified to undertake the works.
- Quality Assurance arrangements are in place based on relevant manuals or guides.
- Identified enhancement initiatives required for the asset management and maintenance funding, capacity and arrangements are implemented.
- Procedures in place to identify when rehabilitation or Spot Improvement is required rather than maintenance.

Actions to inform the key decisions are summarised in Table 2.5. These actions could be undertaken by an in-house Road Agency unit, by a retained Supervision Consultant or a Maintenance Contractor. There may be additional overview supervision from funding sources; for example, road fund managers or Multilateral Development Banks (MDBs).

Rehabilitation or upgrade

This stage of the WLRC is primarily concerned with identifying roads or sections that may need rehabilitation or upgrade to meet changes in task or because of significant degradation that has left roads in a non-maintainable condition. In this context, rehabilitation is defined as the actions to bring a designated road or group of roads back to their original as-built condition, whilst upgrade aims to raise their standard above that level due to changes

Table 2.6 Key Actions for the Rehabilitation/Upgrade Stage

Upgrade actions	Reference
Condition assessments	Review road asset condition assessments and effectiveness of maintenance inputs in ensuring required levels of performance within design life.
Road task or purpose assessment	Review traffic and travel destination changes through surveys.
Link to overall asset management programme	Assess whether and upgrade (change of road class) or rehabilitation to original constructed condition meets requirements.
Prioritisation	Use relevant filter criteria to give a first outline prioritisation for next stage.
Extent of upgrade or rehabilitation	Review condition assessment and task data in decision on whether Spot Improvement would be suitable.

in traffic or other factors. This phase is of particular interest to national or regional strategic transport and infrastructure planners.

The checklist of key decisions or criteria for moving into the rehabilitation or upgrade stage of road or road network is as follows:

- Road task or purpose may have changed sufficiently to justify an upgrade.
- Road conditions have deteriorated sufficiently to require a road rehabilitation.
- Road rehabilitation or upgrade could be implemented by either whole road or Spot Improvement strategies.
- Collate policy and transport master plan data for input to a Planning Stage.
- Confirm whether or not the identified road or roads can be moved into a Planning Stage.

Actions to inform the key decisions are summarised in Table 2.6. These actions could be undertaken by Road Agency management personnel by a retained Supervision Consultant.

REFERENCES

Cook, J. R., R. C. Petts and J. Rolt. 2013. Low volume rural road surfacing and pavements: A guide to good practice. Research report for AfCAP and UKAID-DFID. https://www.research4cap.org/ral/Cook-etal-Global-2013-LVPGuideline-AFCAP-v130625.pdf.

Peck, R. B. 1969. Advantages and limitations of the observational method in applied soil mechanics. Ninth Rankine Lecture. *Geotechnique* 19(2):171–187.

Chapter 3

Road classification and function

INTRODUCTION

Effective rural road design requires a technical framework within which to work in order to maximise the cost-effectiveness of road network investment. Classification is a key element of this framework alongside appropriate standards and specifications (Dingen and Cook 2018). Figure 3.1 illustrates the links within this framework.

Roads should be designed to suit their identified function (or task) and the nature of the traffic (the people as well as the vehicles) which will pass along them. The types of traffic using rural roads in developing countries vary significantly and include both motorised and non-motorised traffic (NMT) involving a wide spectrum of road users from pedestrians and animal-drawn vehicles to large commercial vehicles. It is essential to define road task for the inter-related reasons of geometry, safety, and pavement suitability. Safe achievement of road task should be a key driver of an effective design process.

Experience indicates that a relevant road classification allows for a consistent treatment of all similar roads within the road infrastructure system in terms of their design, construction, maintenance requirements, users expectations, and safety with the clear aim that the roads within a rural network can be designed to be 'fit-for-purpose'.

This chapter describes the key linkages between road function, classification and design, as well as introduces basic issues of rural road safety.

APPROPRIATE CLASSIFICATION

Road function

Rural roads can have widely differing aims in terms of the primary function and service level. A broad distinction is generally made between roads with a primary mobility function and those with a primary access function. Low Volume Rural Roads (LVRRs) would generally fall into the access function group (Table 3.1).

DOI: 10.1201/9780429173271-3

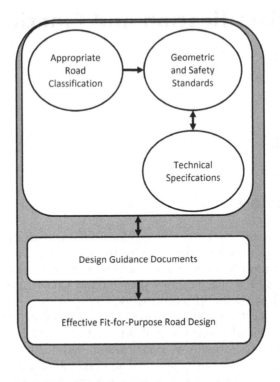

Figure 3.1 The road classification framework.

Table 3.1 Access and Mobility

Primary function	Definition	Typical examples
Access	Reliable all-season access to rural services, communities and markets for the prevailing vehicle types.	Earth and gravel surfaced 3.5 m wide carriageway road with spot sealing in vulnerable areas and through villages.
Mobility	All-weather, high-quality delivery of a transport service suitable for a wide range of vehicle types.	Bituminous sealed or concrete paved road with 5–6 m carriageway with sealed or unsealed shoulders.

Roads with a primary mobility function require relatively consistent cross section and design speed throughout the alignment and would normally be designed in accordance with conventional highway design principles. In contrast, access-classified roads with light traffic may take advantage of cost-saving developments in LVRR design (Cook et al. 2013).

For the purposes of design and the evaluation of benefits, the volume of current traffic needs to be classified in terms of vehicle type. It is the norm in economically developed regions for road design and traffic control measures to be geared to meeting the demand of motorised traffic in interaction

Figure 3.2 Typical rural road vehicles. (a) Light truck, Laos. (b) Medium truck, Myanmar.
(c). Light agricultural tractor and trailer. (d) NMT and light truck, Myanmar.

with a limited number of pedestrians and cyclists. This is not necessarily a valid approach in Low and Middle-Income Countries (LMICs) where large amounts of pedestrians, cyclists, motorcyclists and NMT may change the nature of the traffic flow significantly. The capacity of a road may vary substantially, depending on the volumes of slow-moving vehicles. Figure 3.2 illustrates typical rural vehicle variation in South East Asia.

Variation in the balance of vehicle types across different world regions is exemplified by Figure 3.3 showing variation in the numbers of 2–3 wheel vehicles as a percentage of all motorised traffic.

The role of classification, standards and specifications

Classification facilitates the division of road networks into manageable groups that allow for broadly similar good practice design options that are neither under-designed nor over-conservative and costly. Classification provides the country-specific context for broad programme design, whilst standards should guide the designs toward the needs of a classification.

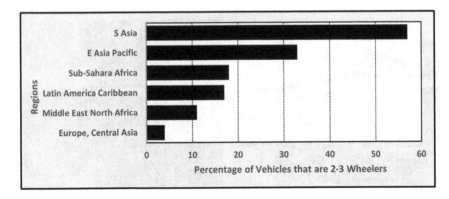

Figure 3.3 Regional variation in numbers of 2–3 wheel vehicles.

These should be supported by technical specifications that define how the roads and associated structures and earthworks must be built to comply with those standards.

The aim of the classification–standards–specification framework is to ensure that roads within a rural road network adhere to key strategic principles and be consistent with any rural development strategy.

It is to be expected that there is a strong correlation between traffic level, traffic growth rates and the administrative function of a road, and therefore an administrative classification is commonly seen as a suitable option. However, although traffic levels often increase in line with the administrative classification hierarchy, this is not always true and, furthermore, the traffic levels are likely to differ considerably between different areas and different regions of one country, depending on local economic and social activities (Giummarra 2001). The design of the road should reflect this.

Although an administrative classification may be necessary to enable ownership, responsibilities, resources and management to be assigned, it should not be the primary basis for an engineering design. Experience indicates that road classifications should be based on road task as well as administrative or political considerations. Table 3.2 is a typical LMIC road classification and Figure 3.4 illustrates this in network terms.

A task-based road classification allows for a consistent treatment of all similar roads within the infrastructure system in terms of their design, construction, maintenance requirements, users expectations and safety. A classification system is necessary for effective management and delegation of responsibilities for different parts of the road network and facilitates the following:

- Establishment of road design criteria.
- Development of road management systems.
- Planning of road construction and maintenance.
- Rational allocation of resources.
- Guidance to the general public.

Table 3.2 Typical Generic Road Network Classification

Class	Description and primary functions (tasks)	Likely traffic designation
A	International highway: Connects international borders to cities or national road network.	HVT
B	National road: Connects towns and cities and to lower-volume road network.	HVT
C	Regional road (feeder road): Generally, connects directly to the higher-level road network or to towns and cities.	LVRR or HVT
D	District: Generally, connects village areas or class E roads to class A rural roads, although they may connect smaller villages directly to higher-level roads or towns.	LVRR
E	Village: Village to village or village to agricultural area.	LVRR
S	Special purpose: Variable, e.g., quarry road, mine or dam access road, legal logging access. Likely to have specific vehicle/axle loading requirements.	LVRR, modified as required
U	Unclassified: Not within formal road authority jurisdiction but may be providing basic village or agricultural access.	Trail or LVRR

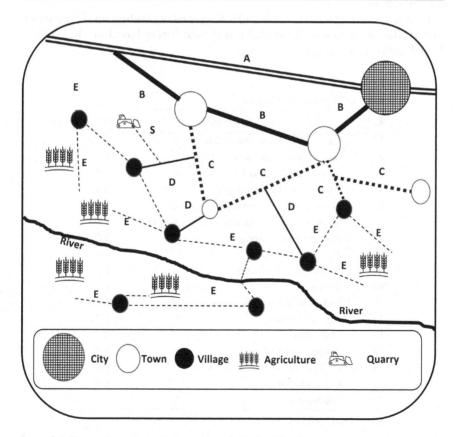

Figure 3.4 Illustration of a typical rural network classification.

RURAL ROAD TRAFFIC

Traffic types

In order to quantify traffic for normal capacity design, the concept of equivalent Passenger Car Units (PCUs) is often used. PCU assessment is related to but differs from the AADT values often used as a basis for pavement selection and design options. PCUs are concerned with the combined space and time that different vehicles occupy. For example, a typical 10-tonne truck requires about three times as much road space as a typical car; hence it is equivalent to 3 PCUs. A motorcycle requires less than half the space of a car and is often assessed as equivalent to 0.5 PCU. Vehicles that are slow-moving cause congestion problems because of their speed rather than because of their size. In effect, they can be considered to occupy more road space than would be expected from their size alone. An animal-drawn cart, although physically smaller than a 10-tonne truck, is very slow moving, and its total impact in terms of size and road space occupation time is high and, therefore, may be assigned a relatively high PCU.

Differing countries have different PCU factors depending on their particular traffic environment; Table 3.3 is a typical listing based on the current Nepal national standard.

Table 3.3 Typical PCU Values

Ref.	Vehicle type	PCU
1	Car, light van, jeep and pick up	1.0
2	Light truck up to 2.5 tonnes gross	1.5
3	>2-axle truck up to 10 tonnes gross	3.0
4	Truck up to 15 tonnes gross	4.0
5	2-axle tractor towed trailer – standard	3.0
6	Single-axle tractor towed trailer – standard	1.5
7	Bus up to 40 passengers	3.0
8	Bus over 40 passengers	4.0
9	Motorcycle or scooter	0.5
10	Bicycle	0.5
11	Rickshaw and tricycle carrying goods	1.0
12	Auto rickshaw	0.75
13	Hand cart	2.0
14	Bullock cart with tyre	6.0
15	Bullock cart with wooden wheel	8.0
16	Horse-drawn cart	6.0
17	Pedestrian	0.2

Source: DoLI, Nepal (2012).

Rural road traffic assessment

The deterioration of pavements caused by traffic results from both the magnitude of the individual axle loads and the number of times that these loads are applied. For pavement design and maintenance purposes, it is therefore necessary to consider not only the total number of vehicles that will use the road but also the axle loads of these vehicles. In many developing countries, the issue of axle-overloading is a major factor in premature pavement damage and increased maintenance or repair costs. Pragmatic forecasts of axles loading allied to realistic assessment of the enforcement regime of legal limits are needed to avoid underestimating future pavement damage and whole-life costs.

The assessment of traffic is undertaken by a combination of traffic count and axle load evaluation. Traffic assessments vary from simple estimations to detailed traffic counts combined with axle load surveys (procedures for traffic assessment and analysis are outlined in Chapter 7). Roads need to provide good service throughout their design life and, therefore, the traffic level to be used in the design process must consider traffic growth (TRL 1993). Traffic as a basis for pavement design is discussed further in Chapters 13 to 15.

RURAL ROAD SAFETY

Road safety in LMICs

Approximately 1.4 million people are killed on the world's roads each year, with Low and Middle-Income Countries (LMICs) suffering around 90% of this and thus are facing a major challenge in road safety. Apart from the human cost, this level of fatalities and accompanying injuries has a significant impact on the socio-economic development of LMICs (Heydari et al. 2019).

Fatalities per 10,000 vehicles is a common indicator used to demonstrate road safety in relative terms. Figure 3.5 indicates regional variation, and

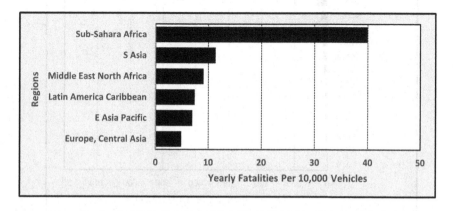

Figure 3.5 Variation of fatality rates between regions.

Figure 3.6 demonstrates the variation within a selection of LMICs in comparison with some typical economically developed countries. These figures are not disaggregated between rural and urban fatalities but do show the far higher figures for low-income countries (with a high rural population percentage), particularly in Sub-Saharan Africa (World Bank 2019; WHO 2018; ADB 2003).

The variation in rates of death observed across regions and countries can be related to differences in the mix of road users and to the road safety environment. The most vulnerable road users (pedestrians, cyclists and motorcyclists)

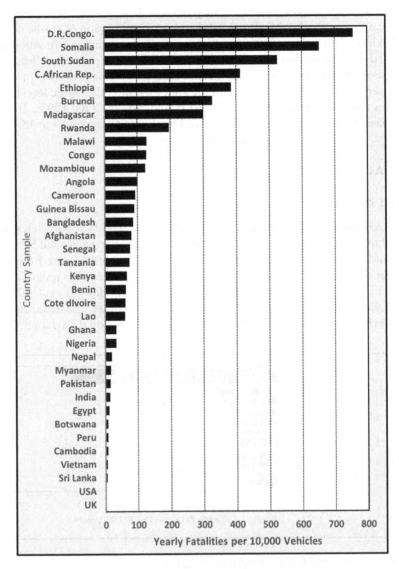

Figure 3.6 Variation of fatality rates between countries, 2016.

are reported as being more than half of all global traffic deaths (WHO 2018). In Sub-Saharan Africa, pedestrian and cyclist fatalities are reported as being 44%, whereas in Southeast Asia the highest percentage of deaths are among riders of motorised two- and three-wheelers (44%).

Improving rural road safety

The Global Plan for the Decade of Action for Road Safety (2011–2020) set out six pillars of road safety:

1. Road Safety Management.
2. Safe Roads.
3. Safe Speed.
4. Safe Vehicles.
5. Safe Road Users.
6. Post-crash Response.

The importance of a holistic approach to rural road safety involving all the above pillars is clear, and that well-designed and well-constructed and maintained roads and roadsides reduce crash risk and injury severity levels. In this context, the International Road Assessment Programme (iRAP 2019) has developed road star ratings that give a simple and objective measure on the level of safety, which is 'built-in' to the road, with five-star roads being the safest and one-star roads the least safe (Table 3.4). Under the road

Table 3.4 Typical Road Safety Star Rating Criteria

Rating	Pedestrian	Bicyclist	2–3 Wheeler	4 Wheeler +
1 star	• No pedestrian way • Poor visibility • 60 km/h traffic	• No cycle path • Poor visibility • Poor road surface • 70 km/h traffic	• No dedicated lane • Trees close to road • Winding alignment • 90 km/h traffic	• Narrow alignment • Winding alignment • Trees close to road • 100 km/h traffic
3 star	• Pedestrian way • Pedestrian refuges • Village lighting • 50 km/h traffic	• Cycle path on carriageway • Village lighting • Good road surface • 60 km/h traffic	• On-carriageway lane • Good road surface • >5 m clear zone • 90 km/h traffic	• Wide carriageway • >5 m clear cone • 100 km/h traffic
5 star	• Pedestrian way • Pedestrian refuges • Village lighting • 40 km/h traffic	• Separated cycle path • Village lighting • Good road surface • Crossing over major roads • 60 km/h traffic	• Separated lane • Straight alignment • No roadside hazards • 80 km/h traffic	• Separated lanes • Road side barriers • Straight alignment • 100 km/h traffic

star procedures, a road safety audit should be undertaken at the planning stage of a new project or before upgrading an existing project. This then forms the basis for a 'Safe Road' approach to the design.

Road infrastructure is strongly linked to levels of fatality and serious injury in road crashes, and research has shown that improvements to the road infrastructure are critical to improving overall road safety. The aim is to create a safe Road Environment, the lack of a dedicated foot or cycle path, for example, presents a major risk for death and injury to pedestrians.

Experience has shown that simply adopting 'international' design standards from developed countries will not necessarily result in acceptable levels of safety on rural roads and may lead to an inappropriate prioritisation of the available resources. Historic road design practice and standards have focussed on meeting the capacity and travel time needs of motorised vehicles, whereas the specific needs of NMT have often been as secondary considerations. Important LVVR safety issues are the completely different mix of traffic, including relatively old, slow-moving and often overloaded vehicles, large numbers of motorcycles and bicycles, poor driver training, and poor enforcement of vehicle and behaviour regulations.

Road design and rural road safety

The following are some of the commonest engineering road design issues that can be addressed at the design stage of rural roads or road networks (Hills et al. 1996; TRL 1994):

1. Inadequate initial Planning or Feasibility Stage assessment for road safety due in part to the non-inclusion of pedestrian and non-motorised traffic in traffic surveys, and consequent failure to take proper account of the operational environment and road purpose.
2. The provision of relatively steep cambers, up to 5%–7% on earth and gravel roads, in order to shed water off the road. This camber may be dangerous for cyclists and motorcyclists who often carry very large/heavy loads and are unable to easily manoeuvre out of the way of approaching traffic. Lower cambers (3%–4%) should be considered for shoulders used by motorcyclists or NMT.
3. The road alignment (Right-of-Way) outside the longitudinal drainage ditch is seldom cleared by more than a few metres, impacting on visibility for vehicle drivers and for pedestrians, particularly young children. Uncontrolled development within the Right-of-Way.
4. Poor road condition, such as a potholed, loose or slippery surface, causing difficulties for cyclists and for motorcyclists to lose control.
5. Relatively fast-moving motorised traffic competing for limited road space with much slower-moving NMT and pedestrians. This need to

cater for all road users has implications for carriageway width, shoulder design, side slopes and side drains.

6. Providing a clear and consistent message to the driver: Roads should be easily 'read' and understood by drivers and should not present them with any sudden surprises that should be addressed by appropriate signage or other measures.

7. Reducing conflicts: These cannot be avoided entirely but can be reduced by appropriate design, including staggering junctions or using guard rails to channel pedestrians to safer crossing points.

8. Roadside access: Use of lay-bys to allow villagers to sell produce or for buses or taxis to avoid carriageway restriction and improve visibility.

9. A Clear Zone (Safety Zone or Recovery Zone) on both sides of the road. Shoulders are usually classed as part of the Clear Zone.

10. Earthworks: Roadside embankment side slopes should if possible be a maximum of 1V:4H. Slopes as steep as 1V:3H may be acceptable, if there is a clear run-out area at the bottom of the embankment.

11. The 'Village Treatment' approach to traffic calming develops a driver-perception that the village is a low-speed environment and to encourage drivers to reduce speed because of this perception. To this end, the road through the village can be divided in three zones, namely, the approach zone, the transition zone and the core zone, with corresponding increases in safety measures.

12. Unsealed earth or gravel roads in dry seasons are likely to cause significant traffic safety issues related to dust (apart from any health safety issues). Sealing or other dust mitigation procedures are recommended at least through village or peri-urban areas.

13. Low-Water Crossings (LWCs). Traffic safety is a principal concern on low-water crossings, such as causeways, where water depth of 0.5–0.75 m has enough lateral force to push a vehicle off the ford. Despite warning signs and obviously unsafe road conditions

Table 3.5 Percentage Reduction in Crashes Due to Design Interventions

Road feature	Percentage reduction in crashes (%)
Appropriate road standard/classification	19–33
Improved horizontal alignment	20–80
Improved vertical alignment	10–56
Road structures	10–74
Improved visibility	2–75
Pedestrian facility	13–90
Cyclist facility	10–56
Traffic calming	10–80

Source: Data from PIARC (2019).

suggesting the crossing not be used in high water conditions, fatalities still occur at these sites. Prudent traffic safety design is required, including warning signs and water depth markers.

14. Low curbs or delineators for defining the edge of the structure, and object markers to define each corner of the structure. Where practical, use depth markers to indicate the depth of flow over the structure.

Table 3.5 summarises the beneficial impact that good, safe road engineering can have on reducing crashes. Many of these issues are closely associated with geometric design and are included within the recommendations outlined in Chapter 9.

REFERENCES

ADB. 2003. Road safety guidelines for the Asian and Pacific Region, Asian Development Bank, Manila, Philippines. https://hdl.handle.net/11540/257.

Cook, J. R., R.C. Petts and J.R. Rolt. 2013. Low volume rural road surfacing and pavements: A guide to good practice. Research report for AfCAP and DFID. https://www.research4cap.org/ral/Cook-etal-Global-2013-LVPGuideline-AFCAP-v130625.pdf.

Dingen, R. and J. R. Cook. 2018. Review of low volume rural road standards and specifications in Myanmar. AsCAP project report for DRRD. https://www.research4cap.org/ral/DingenCook-2018-ReviewLVRRStandardsSpecifications Myanmar-AsCAP-MY2118B-180425.pdf.

DoLI. 2012. Rural road standards. Dept of Local Infrastructure, Government of Nepal. https://doli.gov.np/doligov/download-documents/norms-and-specifications/.

Giummarra, G. 2001. Road classifications, geometric designs and maintenance standards for low volume roads. Research report AR 354, ARRB Transport Research, Vermont South, Victoria, Australia. https://trid.trb.org/view/712280.

Heydari, S., A. Hickford, R. McIlroy, J. Turner and M. Bachani. 2019. Road safety in low-income countries: State of knowledge and future directions. *Sustainability* 11:6249. www.mdpi.com/journal/sustainability.

Hills, B., C. Baguley and G. Jacobs. 1996. *Engineering Approaches to Accident Reduction and Prevention in Developing Countries*. UK: Overseas Centre, TRL. http://transport-links.com/wp-content/uploads/2019/11/1_566_PA3141_1996.pdf.

iRAP. 2019. The road safety toolkit. http://toolkit.irap.org.

PIARC. 2019. *Road Safety Manual*. https://roadsafety.piarc.org/en.

TRL. 1993. Overseas Road Note 31. In *A Guide to Structural Design of Bitumen Surfaced Roads in Tropical and Sub-tropical Countries*. 4th edition. UK: TRL Ltd for DFID. https://www.gov.uk/research-for-development-outputs/orn31-a-guide-to-the-structural-design-of-bitumen-surfaced-roads-in-tropical-and-sub-tropical-countries.

TRL. 1994. *Towards Safer Roads in Developing Countries: A Guide for Planners and Engineers*. TRRL for the UK Overseas Development Administration (DFID).

World Bank. 2019. *Guide for Road Safety Opportunities and Challenges: Low-and Middle-Income Countries. Country Profiles.* Washington, DC: World Bank. https://documents1.worldbank.org/curated/en/447031581489115544/pdf/Guide-for-Road-Safety-Opportunities-and-Challenges-Low-and-Middle-Income-Country-Profiles.pdf.

World Health Organization (WHO). 2018. *Global Status Report on Road Safety 2018: Summary.* Geneva: (WHO/NMH/NVI/18.20). Licence: CC BY-NC-SA 3.0 IGO). https://www.who.int/publications/i/item/9789241565684.

Chapter 4

Roads and the green environment

INTRODUCTION

Road improvements can not only bring substantial economic and social benefits, but they can also dramatically affect the landscape, destroying or impacting farmland, forests and other natural areas, as well as human and wildlife habitats (Kennedy 2004; USAID 2018). The effects on air and water quality, as well as noise, cultural and social impacts, are equally important. In most terrains, the alignment demands of a road will substantially modify the natural drainage regime and that impact must be carefully managed.

In terms of impact, it is important to consider the Whole Road Life-Cycle, from investigation through construction, operation, to maintenance and then upgrade. Each of these stages can have varying impacts on the whole range of environmental issues, as listed in Table 4.1.

ROAD ENVIRONMENT IMPACTS

Water resources

Water quality may be adversely impacted by soil erosion and the siltation of rivers, streams, lakes and wetlands adjacent to road construction and their maintenance. This may also be associated with poor management of fuel and lubricants at road construction camps or vehicle maintenance depots.

The natural flow of surface water and in stream channels may be disrupted by artificial changes in channel depth, width or shape. Road construction may also disrupt the rate of direction of ground water flow, much like a dam across a stream. This can, in turn, lead to erosion, deterioration of soil and vegetation, loss of water for drinking and agricultural or aquaculture use, and impacts on fish and wildlife.

Large quantities of water are needed to help prepare and compact the road component layers during road construction and maintenance. Although this demand for water is temporary, it may significantly affect local water supplies. In arid and semi-arid areas, drawing water for road improvements

DOI: 10.1201/9780429173271-4

Table 4.1 Likely Environmental Impacts

Stage	Pre-construction	Construction	Associated development	Operation	Maintenance
Environmental issues	Quarry and borrow pit development Ground investigation	Earthworks Drainage Bridges Site clearance Equipment/ site camps	Ribbon development Commercial development	Traffic Fuel spillage	Resurfacing Quarries Borrow pits
Water resources; quality and quantity	I	I	I	I	2
Soil erosion and landslip	I	I	I	2	2
Hydrology and flooding.	2	I	I	I	2
Health & safety	2	2	3	I	2
Damage to ecosystem & diversity	2	I	I	3	2
Deforestation	I	I	2	3	2
Resettlement	2	I	I	3	3
Socio-economic & impacts	2	2	I	I	3
Cultural heritage	I	I	I	3	3

Notes: I: potential major impact; 2: potential minor impact; 3: impact unlikely.

may decrease the amount of water available for community domestic use, aquatic species and farm production, especially if the water is taken during dry seasons. The concept of road water harvesting (Van Steenbergen et al. 2018), in which rainfall runoff from roads, is captured in the rainy season and used in the dry season, can bring many advantages:

- Reduced risk of road induced flooding and water logging.
- Reduce erosion and sedimentation.
- Reduced damage from uncontrolled runoff on unpaved roads.
- Create substantial opportunities for productive use of water.

Soil degradation

Soil is commonly used as the foundation for roads and structures, whether obtained from within the road alignment or imported from nearby borrow pits. As such, it both affects and can be affected by road works. The most

Figure 4.1 Multiple track widening, Myanmar.

direct impact on soils is erosion which results from interactions between soil, road structures, climatic conditions and water. The consequences can extend beyond the immediate vicinity of the road where, for example, erosion-induced reduction in slope stability can have an immediate impact. There can also be far-reaching effects on streams, rivers and irrigation structures that are some distance away.

Roads may also contribute to soil erosion through the development of multiple tracks, as road users try to avoid standing water, ruts and other surface defects. Multiple track development can occur wherever inadequate attention is paid to maintenance of the road surface, as shown in Figure 4.1.

Agricultural soil contamination can result from spills, runoff and the transportation of hazardous products during road construction and subsequent maintenance. The removal of agricultural soils, when roads are constructed in farming areas, represents a further impact.

Surface flow and flooding

Roads crossing areas subject to inundation flooding may act like dams that block surface water flows. This is especially true where roads have been constructed on raised embankments and insufficient attention has been made to provision of relief culverts. Figure 4.2 shows a long length of

Figure 4.2 Eroding embankment in flood area with no relief culverts.

eroding embankment with no relief culverts in the floodable Lower Mekong in Cambodia (SEACAP 2008). Under these circumstances, the embankment may act as a dam that holds back water such that land on one side of the road can become much wetter than it was before the improvement, while land on the opposite side may be drier. This may adversely impact crop production, the composition of species in the local ecosystem, as well as road stability.

Poorly installed or maintained culverts may concentrate water and form gullies upslope and/or downslope of the road. Subsequently, these gullies can contribute to erosion, landslip and loss of agricultural lands or damage to ecosystems.

Health and safety

Air pollution. The construction, operation and maintenance of a road may substantially change air quality by introducing significant amounts of pollutants into the air; these include:

- Dust: particularly from unsealed earth or gravel roads.
- Carbon particles.
- Lead compounds and motor oil droplets from exhaust systems.
- Oil from engine.
- Particulate rubber from tyres.
- Asbestos from brake and clutch linings.

One review of this pollution by the World Bank (Greening 2011) confirmed the potential adverse health, road safety, agricultural and environmental impacts from road dust, including dust from gravel roads. Particulates cause inflammation of the lungs and increase incidents of pneumonia. In addition, they may carry irritants and carcinogenic material into the respiratory system and encourage the development of diseases, such as bronchitis and lung cancer. Despite the potentially damaging impacts of the cumulative effects of dust on health (particularly children) and the potentially costly impacts to agricultural production, road safety and the environment, quantitative data on the impacts of dust from unpaved roads are extremely scarce. This is especially so in developing countries where dust impacts from unsealed roads are likely to be greatest.

Carbon monoxide (CO) is the most abundant pollutant in motor vehicle exhaust. Although inherently less noxious than other pollutants, it represents the greatest danger to health, since it is concentrated in areas of traffic through villages and peri-urban areas. The danger to humans arises from a strong affinity for haemoglobin – the oxygen-carrying red cells of the blood (Kennedy 2004). The known effects range from a headache to death, depending on the degree of concentration and duration of exposure. Other pollutants include nitric oxide (NO), sulphur dioxide (SO_2), benzene, carbon dioxide (CO_2), hydrocarbons (HC) and chlorofluorocarbons (CFCs).

Noise. Depending on local conditions and the vicinity of communities, noise may impact human health during ground investigation, construction and ongoing road use. The health of road construction and maintenance staff may also be adversely affected by noise produced from construction, road rehabilitation and maintenance.

Adverse impacts through greater contact with outside communities. Road improvements increase interactions amongst populations. New or upgraded rural roads serve as an entry point for new products and services and, in most cases, this is a positive development. However, the construction of a road does raise the spectre of unwanted influences and impacts, as communities become more open and accessible entities. This may include exposure to communicable diseases (AIDS or pandemics) and unwanted social patterns (ADB 2003; World Bank 2020).

Spread of water-borne diseases. Where sub-standard road design and maintenance result in poor drainage and areas of standing water, the risk of water-borne diseases such as cholera or malaria increases. The same is true for standing water found in open quarries and borrow pits.

Traffic hazards. As the number of motor vehicles on roads increases, road improvements, especially those that allow increased vehicular speed, can lead to increases in accident rates for both human and animal populations. Safe infrastructure design to accommodate pedestrians and cyclists is frequently lacking in LMICs at rural level.

Road works hazards. The operation of road works machinery often endangers both operators and labourers during construction and road maintenance. Poorly planned borrow pits and quarries for road work can also pose threats, ranging from falls from quarry faces to drowning in quarry pits that have become standing water reservoirs (Roughton International 2000).

Impacts on ecosystems and biodiversity

The effects of roads on plants and animals that make up the surrounding natural environment can be of both a direct and indirect nature. Direct impacts can result from both wildlife mortality as well as the loss of wildlife habitat. In addition, the severance of habitats, brought about through the construction of a road in a remote natural area, can often be as damaging as the actual amount of habitat loss. A new or upgraded road can isolate animal and plant populations living on either side of the right-of-way, leading in some cases to their extinction. Severance can also restrict the access of some animals to their usual areas of reproduction or feeding.

The indirect impacts to flora and fauna are related mainly to the increased human access to wildlife habitat and sensitive ecosystems that are facilitated by road provision. Other indirect impacts relate to the ecological disequilibrium that can, for example, come about as the result of wetland destruction, as described earlier (USAID 2016).

The construction of new roads may also introduce exotic or non-indigenous flora and fauna that may severely destabilise local plant and animal communities. Great care in this regard is required when selecting bioengineering stabilisation and protection options. Road access can also contribute to poaching and the trapping of endangered species or species with international trade value.

Deforestation

Opening up new roads or upgrading existing tracks for expanded agricultural development puts adjacent forests at risk, especially where no effective forest management systems are in place. Typically, the most significant impact on forests results from the clearing of land for farms. However, once a road is in place, it also provides access to people wanting to supply urban markets with wood products such as charcoal, fuelwood or construction materials, contributing further to deforestation, carbon emissions and a loss of carbon sinks.

Socio-economic impacts

The negative socio-economic impacts can sometimes be greater than the impacts on the natural environment. For example, upgraded roads and increased traffic can reduce accessibility to local activities adjacent to the right-of-way, and disrupt the traditional patterns of everyday life and business. These 'severance' effects are difficult to quantify and are a frequent cause of community concern with road projects in populated areas. Other negative socio-economic impacts can result from roads which may by-pass communities and bring about a reduction in business. Ironically, these same by-passes can result in positive environmental impacts to the same communities through a reduction in traffic noise, congestion and pollution.

Cultural heritage

The term 'cultural heritage' usually refers to cultural or historical monuments and archaeological sites. The construction of roads can have a direct impact on this heritage, and particularly on archaeological sites, even though their discovery can be a result of road construction. Physical damage to cultural sites can also be caused by quarry and borrow site works, as well as unregulated access to construction sites.

The development of new roads, or rehabilitation of existing ones, often improves personal livelihoods. Access to educational opportunities and to social services, including health care, is often a key rationale for road improvements. However, socio-cultural values may also be impacted, and the stability of communities adversely affected by exposure to social change.

Ecosystems can provide nonmaterial benefits, such as the aesthetic value of an area, to local residents and tourists. Construction of new roads or the realignment of existing roads may adversely affect scenic vistas. Under some circumstances, such damage can lower tourism revenues. The cumulative effects of poorly located and poorly managed quarries and borrow pits supplying building materials for road projects may also cause significant loss in scenic value.

ENVIRONMENTAL IMPACT ASSESSMENTS

Background

Environmental Impact Assessment (EIA) is a legal requirement in many countries and with most MDBs for most types of road development project. EIAs are commonly required for new roads and road improvements because of the range of potential environmental impacts associated with these projects. The main goal of an EIA is to positively influence development decision-making by providing sound information on environmental impacts and the means for preventing or reducing those impacts. Three major outputs of the EIA process are as follows:

1. An identification and analysis of the environmental effects of proposed activities.
2. An Environmental Management Plan that outlines the mitigation measures to be undertaken.
3. An environmental monitoring programme that outlines the environmental-related data that must be collected in conjunction with the project.

All three outputs are required for the EIA process to be effective. Under some regulations, the documentation for the EIA process requires that three separate documents be prepared; in others, all three may be presented as part of a single EIA document. Many governments and most bi-lateral

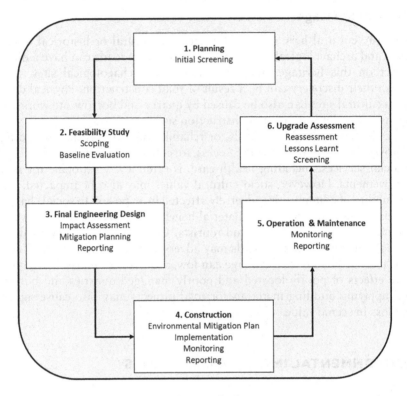

Figure 4.3 The EIA processes and the Project Cycle.

and multilateral funding or aid agencies have formal EIA guidelines that must be followed (ADB 2003; World Bank 2020).

Even when a full EIA is not required, as in the case of some smaller scale rural road improvement projects, an appropriate level of environmental analysis should still be undertaken based on the likely effects of the project. The EIA not only addresses more than just the natural or ecological environment, but it also considers wider community and socio-economic impacts. For this reason, the EIA needs to be managed as an integrated component of the overall project appraisal process.

The basic EIA process involves a series of stages of examining and acting upon relevant environmental information. Figure 4.3 shows how the EIA stages can link with the project-cycle stages of a typical road project. EIAs should be undertaken by an appropriately qualified team.

Initial Environmental Examination

In some EIA processes, an Initial Environmental Examination (IEE) may be required. After a project has been screened, an IEE can be undertaken to determine the probable environmental impacts associated with the project and ascertain whether a full-scale EIA is required. The IEE is usually conducted within a limited budget and is based on existing information from similar projects, allied to experienced professional judgement. The three primary objectives of the IEE are to:

1. Identify the nature and severity of specific, significant environmental issues associated with the project.
2. Identify easily implementable mitigation or offsetting measures for the significant environmental issues. If the IEE shows there are no significant environmental issues which need further study, then the IEE serves as the final EIA Report.
3. Develop the ToR for the full-scale EIA study, should more detailed assessment be needed, or any special topic reports which may be required instead of, or in addition to, the full-scale EIA.

Screening

The first stage of environmental assessment is screening to determine whether a full EIA is needed. If not, some form of environmental analysis will usually be necessary. This should be proportional to the scale of the road project and its likely environmental impacts. A full EIA must be undertaken in certain circumstances including:

- Where national/regional EIA regulations apply.
- Where a project funding agency requires EIA.

The environmental screening decision should be made in consultation with relevant parties such as the environmental regulatory body and any funding agencies. The screening decision may also be required to be made available to a range of stakeholders, including potentially impacted communities.

Based on the initial screening, the requirement from an EIA may be defined within a number of levels; for example, the ADB uses a threefold classifications as follows (ADB 2003):

Category A: Projects with potential for significant adverse environmental impacts. An Environmental Impact Assessment (EIA) is required to address significant impacts.

Category B: Projects judged to have some adverse environmental impacts, but of lesser degree and/or significance than those for category A projects. An Initial Environmental Examination (IEE) is required to determine whether or not significant environmental impacts warranting an EIA are likely. If an EIA is not needed, the IEE is regarded as the final environmental assessment report.

Category C: Projects unlikely to have adverse environmental impacts. No EIA or IEE is required, although environmental implications are still reviewed.

The World Bank and other groups and governments have similar classifications.

Scoping

Scoping is the process of determining the environmental issues to be addressed, the information to be collected, and the analysis required to assess the environmental impacts of a project. The primary output of scoping is the Terms of Reference (ToR) required to conduct an EIA and to prepare the EIA report.

Environmental scoping is carried out alongside or soon after screening and involves identifying the key environmental issues associated with the project in question. Good scoping will lead to a focused EIA and avoid time unnecessarily spent on minor or irrelevant issues. Scoping involves:

- Collaboration between environmental experts relevant to the project and its location.
- Consultation with relevant parties including relevant national regulatory bodies.
- Developing a work plan for the EIA.
- Proposing the assessment techniques for particular EIA topics (e.g. desk-study assessment, site surveys).
- Defining the assessment criteria for determining the significance of the impacts.
- Identifying the timing and arrangements for the EIA.

Baseline surveys

The evaluation of the existing environment potentially impacted by a project is commonly known as the baseline survey. It requires collection of information on the conditions making up the existing environmental model. This may be subject to change through natural events or human activities other than the project road. The future state of the environment model should also be assessed, assuming that the road project does not happen. This is particularly important where the construction is not due to be completed for a number of years.

Assessment methods and techniques

A number of guidelines and manuals have been developed for identifying and assessing the environmental impacts of road projects (TRL 2005; Montgomery et al. 2015; USAID 2018). Many of these are in the form of checklists or matrices and are quite detailed and extensive. Their main function is to serve as an aid for road planners to ensure that all the relevant impacts are considered in the preparation of an EIA. Table 4.2 provides a typical example of such a checklist.

Table 4.2 Typical EIA Checklist

Project activities	Impacts to be checked

Pre-construction impacts

Surveying and demarcation of centre-line.	Some minor loss of vegetation during demarcation, particularly bridge approaches.
Mobilisation of Contractor, presence of construction workers, associations with local people.	Spread of pandemics and HIV/AIDS.
Site clearance, digging, excavations.	Accidental disturbance of archaeological assets, sites or resources.
	Erosion of temporary access.
	Loss of riparian vegetation.
	Loss of vegetation and habitat through road widening, realignment of right-of-way.
	Impacts on wildlife through interruption of migratory routes and other habitat disturbances.
	Encroachment on irrigation structures from road widening and realignment adjusting.
	Encroachment on water supply systems.
	Damage of agricultural land through road widening and alignment adjustments.

Construction phase impacts

Restrictions on land use, land and resource acquisition for bridge approaches.	• Loss of land. • Removal of trees. • Relocation of houses.
Operation of construction plant and vehicles generating emissions.	• Emission of exhaust from vehicles and machinery. • Dust from aggregate crushing plant; stockpiles, generated by heavy vehicles transporting materials on roads; uncovered loads on trucks.
Site clearance, digging, excavations. Operation of construction plant and equipment creating noise.	• Noise in community. • Noise affecting wildlife. • Light attracting wildlife. • Impacts on construction workers.
Works in, or adjacent to, rivers and streams.	• Changes to river water flows, including levels and velocity. • Changes to channel depth, structure and location. • Changes to riverbank stability. • Damage to floodplain areas affecting flood cycles, temporary flood water storage and release, loss of soil fertility through silt deposition during floods. • Increased turbidity of river waters due to gravel extraction or bridge/causeway construction. • Increased silt deposition at culverts and bridges.

(Cotinued)

Table 4.2 (Continued) Typical EIA Checklist

Project activities	Impacts to be checked
Aggregate extraction.	• River channel degradation and changes in river morphology. • Quarries or borrow pits leaves unusable land, exposed water table, attracts rubbish dumping, reduces visual values.
Earthworks, cut and fill activities, construction of embankments.	• Soil erosion and silt deposition. • Increased surface water runoff. • Sediment contamination of rivers and water bodies. • Stockpile and staging areas lead to loss of land uses. • Erosion on natural hill slopes.
Runoff, discharges, generation of liquid wastes.	• Increased siltation at culverts and bridges. • Construction materials washed out into rivers. • Soil contamination from fuels, chemicals.
Emergency or accidental spills.	• Oil and other hazardous chemicals spillages resulting in pollution. • Hydrocarbon leakage/spills from construction camps/workshops. • Accidents place people at risk.
Encroachment into protected areas, disturbance of terrestrial habitats.	• Impacts on fisheries. • Terrestrial habitats become more fragmented. • Loss of primary forests. • Endemic, rare or endangered species affected. • Road workers hunting wildlife for consumption or sale. • Low-level crossings may interfere with migration patterns of freshwater species.
Encroachment into historical/cultural sites.	• Effects on cultural values. • Special areas affected.
Presence of vehicles and equipment in villages, use of people's land for access to construction site, traffic and safety issues.	• Traffic and access disrupted during construction. • Traffic safety affected. • Construction traffic safety issues. • Land erosion from temporary works.
Construction activities causing accidental damage to existing services.	• Water supplies contaminated or disrupted through breakage of pipelines or exposing water table during excavation for gravels.
Waste generated at construction camps.	• Contamination of local water supplies through waste. • Discharges of waste-waters/sewage from camps to rivers and smaller streams.
Presence of construction workers and construction camps.	• Social disruption. • Possibility of conflicts or antagonism between residents and Contractor. • Spread of communicable diseases. • Sexual exploitation in camps. • Impacts on general health and safety.

(Cotinued)

Table 4.2 (Continued) Typical EIA Checklist

Project activities	Impacts to be checked
Clearing of land and removal of crops and trees as part of works.	• People lose dwellings or other buildings. • People lose gardens or cash crops.
Operation of vehicles creating emissions.	• Hydrocarbons, carbon monoxide, nitrous compounds, sulphur dioxide and particulate matter increase through increased traffic.
Placement of bridges and crossings.	• Fords, bridges, causeways and other structures blocking water flow. • Restriction of natural meandering of streams. • Restriction of natural flood channels by filled approaches to bridges. • Reduced sediment deposition on floodplain or agricultural areas through restricted movement of flood flows on flood plain.
Runoff from road.	• Problems with runoff, loss of soils and other forms of erosion. • Water quality in rivers and other water bodies is affected by use of the new roads.
Improved access to previously inaccessible, areas.	• Hunting and poaching increase.
Spread of communicable diseases.	• Roads act as pathway for spread of communicable diseases.
Increased traffic.	• Increases in noise nuisance for residents. • Increased traffic volumes and higher speeds within village areas leads to accidents.
Routine and ongoing maintenance.	• Acquisition of new material source areas affecting properties.

One of the inherent difficulties in assessing the impacts of road projects is that some may be outside of the control of the road planner or designer. These include effects such as in-service noise and air pollution which may be more directly associated with the quality, maintenance and use of motorised vehicles than of the road itself.

EIAs can vary widely in terms of their length and level of detail. There is a tendency for EIAs, particularly those which have been prepared without scoping, to be presented in voluminous and unwieldy documents. A successful scoping process should eliminate this risk and ensure that the EIA concentrates on the significant impacts. At a minimum, a typical EIA report should have the following contents:

1. Introduction.
2. Description of the Project.
3. Description of the Environment.
4. Anticipated Environmental Impacts and Mitigation Measures.
5. Alternatives.
6. Environmental Monitoring.

7. Additional Studies.
8. Environmental Management Plan and Environmental Management Office.
9. Summary and Conclusions.

ENVIRONMENT IMPACT MANAGEMENT

The environmental plan

A central goal of the EIA process is to develop an implementable set of environmental protection measures set out in an Environmental Management Plan (EMP). The EMP outlines the mitigations and other measures that must be undertaken to ensure compliance with environmental laws and regulations, to reduce or eliminate adverse impacts, and to promote feasible environmental enhancement measures.

A well-structured EMP usually covers all phases of the Project Cycle and addresses all major environmental issues or impacts identified during the EIA process. The plan outlines environmental protection and other measures that will be undertaken to ensure compliance with environmental laws and regulations and to reduce or eliminate adverse impacts. The plan defines:

1. The technical work programme to mitigate environmental impacts, including details of the required tasks and reports, and the necessary staff and resources.
2. A detailed estimated cost for implementing the plan.
3. The proposed implementation of the EMP, including a staffing chart and proposed schedules of participation by the various members of the project team, and activities and inputs from relevant governmental agencies.

Environmental monitoring

Environmental monitoring involves the systematic collection of information and the management of the monitoring programme in order to determine:

1. The actual environmental effects of a project.
2. The compliance of the project with regulatory standards and the requirements of the EIA.
3. The degree of implementation of environmental protection measures and success of these measures.

The information generated by monitoring programmes provides the feedback necessary to ensure that environmental protection measures have been effective in helping achieve an environmentally sound project. Monitoring also provides valuable lessons to be learnt for future similar projects.

MITIGATION OF IMPACTS

Some general issues

Many of the physical environmental mitigation measures are related to good engineering practice in design, construction, and maintenance. Some key physical impact mitigation measures are outlined below.

Negative impacts on water resources can be mitigated in a number of ways. Often the most effective and basic of these is simply choosing or modifying the road alignment so that it has the least impact on these resources, has a minimum number of water crossings, and avoids potential ground water hazard areas.

Water flow speed reduction measures included in road design can substantially reduce erosion impacts. These include actions such as grass planting on slopes and the provision of check dams in drainage channels, to slow water velocity and reduce erosion. The construction of settling basins to intercept and remove silts, pollutants and debris from road runoff water, before it is discharged to adjacent streams or rivers, can also be effective. Interceptor ditches can also be used to reduce the flow of pollutants.

Measures for avoiding or minimising soil erosion include:

- Minimising cleared areas.
- Replanting cleared areas and slopes using bioengineering principles (ADB 2013).
- Construction of interception ditches, contoured slopes, vegetated riprap, vegetated gabion walls, etc.

To avoid damaging soil productivity, road planners should, wherever possible, avoid alignments through highly productive agricultural lands. Where this is not possible, excavated valuable topsoil should be stored and reused. With appropriate foresight, borrow pits can minimise their erosion and sedimentation impacts by constructing retention basins and settling ponds. When managed effectively, they may also be used to supplement water supplies during dry seasons or be managed to supplement dry-season water supplies. Quarries and borrow pits can be transformed into lakes, and roadside amenities provide appropriate safety measures that are in place (Gourley and Greening 1998). Frequently, however, borrow pits are just abandoned; for example, see Figure 4.4.

Mitigating the negative effects of pollutants is often beyond the competence or responsibility of the road engineer, since they are most directly connected with measures such as fuel technology and quality, as well as vehicle emissions standards, and vehicle inspection and maintenance requirements. Nonetheless, there are a number of actions that can be taken as part of the road planning and construction process to lessen the negative impacts on air quality. These include:

Figure 4.4 Abandoned borrow pit used for rural road materials.

1. Routing traffic away from sensitive community areas such as health centres and schools and sensitive ecosystems.
2. Use of vegetation to filter dust and particulate matter.
3. Ensuring that construction equipment is in efficient operating condition.
4. Look to Spot Improve, or seal, gravel or earth reads in sensitive areas (Chapter 16).
5. Locating processing facilities, such as asphalt batch plants, away from community areas.

The best way to mitigate against biodiversity impacts is to avoid them altogether by locating roads away from areas with sensitive flora and fauna populations. Where this is impossible, or too costly to do, there are a number of specific steps that can be taken in road design and operation to reduce impacts. These include:

1. Engineering road cross-section designs to maintain habitat and avoid vehicle – wildlife collisions through narrower clearing of right-of-way, lower vertical alignment and improved sight-lines for drivers.

2. Provision of wildlife or cattle underpasses or overpasses to facilitate animal movement, or by fencing to restrict movement and direct wildlife to designated crossing points.

For cultural mitigation measures, site management plans can be developed which can include requirements to:

1. Carry out mapping of archaeological sites.
2. Include clauses in construction contracts that cover archaeological investigation.
3. Excavate and relocate artefacts.
4. Stabilise embankments, soils and rock containing artefacts.
5. Control groundwater levels.
6. Control flora and fauna.
7. Establish monitoring and evaluation systems.

Planning issues

Many impacts can be avoided or minimised through careful attention in the initial planning and design stage. Relevant standards and specifications should be embedded into construction contracts or roadwork procedures for governments or communities. Environmental mitigation training and guidance should be provided prior to and during construction, operation and maintenance.

Hydrology studies should be done at an early stage in the Project Cycle, in association with climate impact assessment, to minimise possible effects on surface or sub-surface water resources, to ensure correct design of drainage structures and systems, and to reduce the potential for damage from unusually heavy rains, floods and major storms These studies should take into account anticipated future changes in the climate, as discussed in Chapter 17.

Construction

Once conditions or mitigation measures have been defined in the environmental review process, they should be included in technical specifications in all relevant contract documents related to the road construction or rehabilitation activities. Environmental clauses should be prescriptive and specify what needs to be done, where it needs to be done, and when and how the actions will take place. Clauses should also state who is responsible for environmental issues, what monitoring and reporting requirements there are, and what sanctions or legal recourses are available for work that does not meet the required specifications (USAID 2018).

Many negative impacts may be avoided by taking pre-emptive preventive measures when setting up a work site. Careful siting of borrow pits, stock-piling areas, depots, and work camps can preserve sensitive areas, reduce air and noise pollution, and minimise visual intrusion. Constraining the handling and use of hazardous materials can go a long way in reducing the risks of accidental spills.

Measures to prevent erosion are of major importance during the work phase and can include:

1. Prohibition of side-tipping of spoil materials, particularly on sloping ground, see Figure 4.5 as an example of poor environmental practice.
2. Ensure earthworks are not left exposed without drainage or protection during rain seasons.
3. Undertake bioengineering planting on cleared areas and slopes immediately after construction.
4. Temporarily cover soil with mulch or fast-growing vegetation, if necessary.
5. Intercept and slow water runoff.

Dust problems can be avoided by watering the site on a predetermined schedule and as required. Construction noise problems can be minimised by using well-maintained and 'silenced' equipment, operating within existing noise control regulations and limiting work hours near community areas.

Enforcement of environmental issues during construction is frequently a major challenge in situations where control of local contractors may be lax.

Figure 4.5 Side-tipping of rural road spoil materials in mountainous terrain.

The inclusion of environmental issues within the Contract Technical Audit is strongly recommended. The use of appropriately trained local community groups as environmental 'watch-dogs' can have a major beneficial effect.

Maintenance

Environmental damage from unpaved rural road construction is frequently the effect of insufficient or poor-quality maintenance. Good road maintenance practices that keep the road usable and durable, such as clearing drainage structures and restoring camber on unsealed roads, will minimise much of the environmental damage the road might cause. Periodic maintenance, which may involve significant equipment use, should be subject to environmental safeguards similar to those guiding construction. (See Chapters 21 and 22 for details on maintenance activities.)

A great deal of off-road driving near existing roads results from drivers' attempts to avoid deep ruts, potholes or flooding in the official roadway. Regular, correct maintenance of the road surface and drainage system will minimise this problem by preventing the flooding and the expansion of ruts. Wet-season traffic on roads designed only for dry-season use can severely damage the road surface and promote erosion. Closure and enforcement are possible management measures, but the best solution, if there is a significant demand during rainy seasons, is to upgrade the road for wet-season use.

SUSTAINABLE AND ENVIRONMENTALLY OPTIMISED DESIGN OPTIONS

Construction of rural roads will also have strategic environmental impacts, over and above the direct project impacts discussed previously in this chapter. The nature and scale of these impacts needs to be understood and taken into consideration when developing rural transport policies and choosing the materials and techniques to construct these rural roads.

The strategic issues centre around:

1. The use of non-renewable construction materials.
2. The recycling of construction materials or other waste products in rural road construction.
3. Minimising the carbon footprint of construction options.

The use of gravel as wearing course surfacing for rural roads has been the default options in most LMICs. Within the last two decades, however, there has been a growing realisation that continuing use of this non-renewable and wasting option for construction and ongoing maintenance is, in many physical environments, not sustainable in terms of a diminishing natural resource. Other much more environment friendly options are available, as discussed in more detail in Chapter 15.

Many developed countries now demand that new, upgraded or rehabilitated roads must contain at a significant percentage of recycled materials in their pavements. While this option is not currently realistic to such an extent in LMIC rural networks, it is nevertheless practical to pursue the policy in a limited way, for example the reuse of old gravel layers as a new sub-base, or re-employing old penetration macadam, or other degraded bituminous surfacing, within a base layer. In addition, there are now a limited number of other options available for the use of plastics in asphaltic surfacing that have potential application in rural Road Environments (Sasidharan et al. 2019).

Construction and maintenance of rural roads have variable carbon footprints depending on the materials involved and the construction plant used. Sturgis and Petts (2010) investigated the variable carbon footprint of three low volume rural road options. Of the three surfacing/paving options investigated using typical characteristics, concrete road surfacing was found to have the largest footprint at around 255,700 $kgCO_2e$ per km of road. The impact of a gravel-surfaced road, including periodic re-gravelling, was of a similar magnitude, at around 218,000 $kgCO_2e$ per km of road. The hand-packed stone road had a considerably lower footprint at 9,850 kg CO_2e per km of road, i.e. a factor of 20 less than the other two options.

REFERENCES

ADB. 2003. *Environment Assessment Guidelines*. Asian Development Bank, Manila. https://www.adb.org/sites/default/files/institutional-document/32635/files/environmental-assessment-guidelines.pdf.

Gourley, C. and P. A. K. Greening. 1998. Environmental damage from extraction of road building materials. DFID- funded Knowledge and Research programme. Report Ref R6021. https://assets.publishing.service.gov.uk/media/57a08d9940f0b652dd001a7a/R6021.pdf.

Greening, P. A. K. 2011. Quantifying the impacts of vehicle-generated dust. Transport Research Support Program as a joint World Bank/DFID initiative. https://documents1.worldbank.org/curated/en/325581468161082361/pdf/815700WP0Trans00Box379840B00PUBLIC0.pdf.

Kennedy, W. 2004. Chapter 5 Roads and the environment. In *Road Engineering for Development*. Second edition. Robinson R and Thagesen B, Eds.

Montgomery, R., H. Schirmer and A. Hirsch. 2015. Improving environmental sustainability in road projects. World Bank Group Report 93903-LAC. https://documents1.worldbank.org/curated/en/220111468272038921/pdf/939030REVISED0Env0Sust0Roads0web.pdf.

Roughton International. 2000. Guidelines on borrow pit management for low-cost roads. DFID KaR Project Report (Ref. R6852). https://assets.publishing.service.gov.uk/media/57a08d7e40f0b649740018ba/R68524.pdf.

Sasidharan, M., M. E. Torbaghan and M. P. N. Burrow. 2019. Using waste plastics in road construction. K4D Helpdesk Report. Brighton, UK: Institute of Development Studies. https://opendocs.ids.ac.uk/opendocs/bitstream/handle/20.500.12413/14596/595_Use_of_Waste_Plastics_in_Road_Construction.pdf?sequence=62&isAllowed=y.

SEACAP. 2008. Development of local resource-based standards: Study of road embankment protection. South East Asia Community Access Programme (SEACAP). UK-AID DFID for Royal Government of Cambodia, Tech Report 6. https://www.gov.uk/research-for-development-outputs/seacap-19-technical-paper-no-6-study-of-road-embankment-erosion-and-protection.

Sturgis, M. and R. C. Petts. 2010. Developing an approach for assessing the carbon impact of rural road infrastructure provision in developing countries, a proposed methodology and preliminary calculations, a discussion paper for gTKP.

TRL. 2005. Chapter 5 Environmental impact assessment, R8132. In *Overseas Road Note 5: A Guide to Road Project Appraisal*. UK: DFID. https://www.gov.uk/research-for-development-outputs/overseas-road-note-5-a-guide-to-road-project-appraisal.

USAID. 2016. USAID Mekong adaptation and resilience to climate change. Regional Development Mission for Asia. Final report contract number: AID-486-C-11-00004.

USAID. 2018. *Sector Environmental Guidelines: Rural Roads*. https://www.usaid.gov/sites/default/files/documents/1860/SectorEnvironmentalGuidelines_RuralRoads_2018.pdf.

Van Steenbergen, F., K. Woldearegay, M. Agujetas Perez, et al. 2018. Roads: Instruments for rainwater harvesting, food security and climate resilience in arid and semi-arid areas. In *Rainwater-Smart Agriculture in Arid and Semi-Arid Areas*. Kebede Manjur et al., Eds. Springer International Publishing. http://roadsforwater.org/wp-content/uploads/2019/01/Frank-et-al-_Roads_-Instruments-for-Rainwater_-2018.pdf.

World Bank. 2020. *Environmental and Social Framework (ESF) Implementation Update*. https://www.worldbank.org/en/projects-operations/environmental-and-social-framework.

Chapter 5

Appraisal of low volume rural road initiatives

INTRODUCTION

Road project appraisal is a key element of the Planning and Feasibility phases of the road cycle, and involves a gradual and sometimes iterative process of assessment and screening of potential options. In order to justify, or prioritise, investments in Low Volume Rural Road (LVRR) programmes, it is necessary to carry out appraisals, be it for initial construction, rehabilitation or upgrade project, or periodic maintenance programmes. This allows each option to be evaluated and prioritised against the limits on the available resources or against other options.

This chapter is primarily concerned with, and provides guidance on, the steps to be taken in the appraisal of LVRR initiatives. This guidance acknowledges that the financial and economic appraisal of any initiative needs to be supported by the complementary assessment of social, institutional, technical and environmental issues.

PRINCIPLES

Robinson (2004) made a distinction between the appraisal process for major (economic) roads and minor (social) roads. Major roads were defined as those for which the provision and management could be justified by economic cost–benefit analysis. Minor were defined as those that may not be justified on economic grounds alone. Although the latter may have an economic function, their purpose has a significant social function, and social benefits are needed to justify any investment in provision or management. LVRRs, in the main, fall into the 'minor' category.

Improvements in rural access typically give rise to socio-economic benefits that can arise from four principal sources:

1. Lower transport costs to due to smoother upgraded road surfaces or shorter routes.
2. Savings in time due to faster travel.

DOI: 10.1201/9780429173271-5

3. Economic development benefits and improved access to markets resulting in generated (new) traffic.
4. Social benefits due to improved access to schools, hospitals, skills acquisition, employment opportunities, etc.

The first two of these road-user cost factors are the most important in conventional high-volume road appraisal in developing countries. However, while lowering transport costs is still relevant to the appraisal of improvements to LVRRs, the other sources of benefit can be proportionately more important. Thus, whereas conventional appraisal of investments in highways can focus exclusively on the benefits due to lower transport costs, this is not appropriate for rural access appraisal (TRL 2005).

AN APPRAISAL FRAMEWORK FOR LVRRs

Key issues

A general investment appraisal of LVRRs can be facilitated by considering seven key issues identified as being the primary contributors to a road project's sustainable benefit, as shown in Figure 5.1 (Southern African Development Community, SADC 2003).

The evaluation of these issues can be used for assessment against a required level, or for comparative prioritisation, using a weighted point scoring basis derived for the particular programme or a national requirement at planning or feasibility stages. Similar weighted scoring systems have been developed appraising the need for sealed roads (Henning et al. 2006;

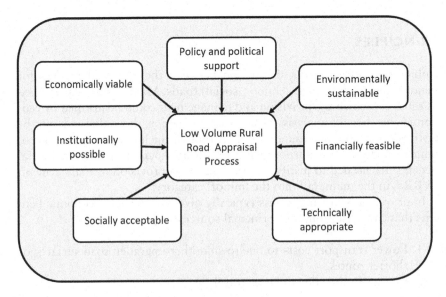

Figure 5.1 Key sustainability appraisal issues.

World Bank 2020). The World Bank (2020) Systematic Paving Decision (SPADE) model incorporates 41 decision factors grouped into five categories: country context, regional context, operational environment, road context, and engineering context. Points are assigned for factors with appropriate weighting. The model then combines the scores for a total weighted assessment. A system such as this can be simply adapted for a sustainability index based on the key issues in Figure 5.1, as described in the following sections.

Political and policy support

To be sustainable a road project should be compatible with identified national policy or development initiatives and have the full support of all relevant ministries or departments at central as well as at local levels. The central support acts as the top-down driver of the project and will also provide the necessary central guidance for financial and technical issues. Experience has shown that, to be fully sustainable, LVRR projects require buy-in and support from local district, communities and groups. A sense of community ownership of local roads can add significantly to bottom-up local focus on appropriate design and support to important sustainability issues such as routine and periodic maintenance.

Economic viability

At its simplest, the assessed benefits accruing from the project in terms of socio-economic development benefits (and their assessed value) must be greater than its initial and ongoing costs or the assessed consequences of a do-nothing option.

The purpose of carrying out the economic appraisal is primarily to demonstrate that there will be an adequate return in terms of the stream of benefits resulting from the investment. Often the appraisal will include an analysis of the net contribution that the investment will make to society as a whole.

Economic appraisal is a key element of the Planning and Feasibility phases of the Project Cycle and involves a sometimes iterative process of assessment and screening of potential options. At the end of each of these stages, potential projects may be either rejected or taken forward. It is important to appreciate that economic analysis depends on a significant range of assumptions regarding future events, interactions, conditions, value and costs.

More detail on the specific issue of LVRR economic appraisal is given later in this chapter. In addition, reference can be made to Lebo and Schelling (2001), TRL (2005) and World Bank (2005).

Institutionally possible

The road project should be within the technical and managerial capacity of the local road institutions to carry it forward throughout its life cycle from design to construction and maintenance. This view is linked to the concept of compliance with the Road Environments, as discussed in Chapter 6. Essentially, there needs to be a resource pool of road designers, contractors and road managers available with the necessary appropriate resources, knowledge, and experience to implement the proposed programme. This is an important issue when considering the implementation of innovative design or construction possibilities, such, for example, as those associated with climate resilience or new pavement surfacing options.

Identification of this issue will highlight the need for capacity development to be included within the overall project appraisal outcomes. This appraisal will also play a significant part in the selection of construction or maintenance contract models (see Chapter 18). Where local capacity deficiencies are apparent, the need for outside enhancement assistance or resources must be clearly identified and costed. Failure to recognise these issues and address them would almost certainly result in delay, cost escalation or reduced impact of the initiative.

Socially acceptable

The project should be capable of being positively accepted by local communities as being a required initiative and not seen as an unnecessary imposition. It should foster existing social safeguards on issues such as (1) measures that improve road safety; (2) follow-up training especially for vulnerable road users; (3) resettlement and land acquisition; (4) community acceptance and participation; (5) gender equality; and (6) protection of vulnerable groups.

In addition, what are termed 'Complementary Interventions' can significantly enhance the social acceptability aspects of an appraisal (see Chapter 6).

Technically appropriate

The road design concepts should follow national rural transport classifications, standards and specifications that reflect socio-economic and transport needs. Design options should be appropriate for local transport needs, be compatible with the Engineering Environment; constructible using readily available resources, and maintainable within the existing maintenance regime and resource base. Appraisals should be wary of investigation, design or construction methods that have been found to have been successful in other Road Environments but are, as yet, unproven in the environments in question.

The general level of technical appropriateness may be assessed using the Road Environment concept described in Chapter 6. Further details on the assessment of appropriate pavement technologies and on investigative techniques are included in Chapters 13 and 7, respectively.

Financially feasible

There should be adequate funding or funding mechanisms in place, or identified, for design, construction, management, and long-term maintenance of the road and its assets over its proposed design life. The financial assessment will determine whether the necessary finance resources can be made available to fund the initiative from local or external sources. The key question is whether or not the project 'business plan' is viable in cash and cash flow terms.

The costs of a road project must be assessed in terms of the total expenditure incurred over the project's design life. There are two main components of this:

1. Investment cost: development (construction and upgrading), renewal, rehabilitation.
2. Recurrent costs: operations and maintenance.

Once a capital investment has been made, this has an inevitable, ongoing consequence in terms of recurrent expenditure needs.

Road project cost estimating techniques range from the broad-brush category of 'global' estimation to the more detailed 'unit rate' technique. Global cost estimation is a simple technique that relies on historical data on costs of similar projects. These are related to the overall size or capacity of the asset provided. Examples are cost per kilometre of road or cost per square metre of bridge deck.

Using historical global data has dangers. It is not always clear what costs are included. For example:

- Are the costs of design and supervision included?
- Are tax and duties included?
- Are true overheads properly assessed?
- Are any project subsidies evaluated?

There is also the risk of not comparing equivalent projects. For example, are the levels of quality, pavement thickness the same; are terrain, earthworks, materials haulage, structure's characteristics and soil conditions comparable?

The unit rate estimation technique is based on the traditional bill of quantity approach to pricing construction work. Historical data on unit rates or prices are selected for each item in the bill, typically taking data from recent similar contracts or published information. The dangers are that the previous projects were not carried out in identical conditions, using identical construction methods and with the same scope or duration. The same factors affecting global data can apply. It must be appreciated that many road authority cost databases are not always inclusive of all components and are rarely kept up to date. However, experienced estimators with good intuitive judgement and the ability to assess the realistic conditions of the work can prepare reliable cost estimates (Robinson 2004). It is informative to compare previous similar project estimates and outturn costs for similar projects with the same authority. The complexities of works costing are discussed in Petts (2002).

Environmentally sustainable

Road construction, as well as its subsequent use and maintenance, should not cause significant environmental damage and should be compatible with current environmental applicable legislation. The relationships between the various stages of a road project and the green environment have been discussed in Chapter 4.

ECONOMIC APPRAISAL OF MAJOR ROADS

Although the focus of this chapter is on LVRRs, it is worth considering as background the main issues for the major roads, as defined by Robinson (2004). For these roads, the economic assessment of construction and rehabilitation projects is very well documented and reasonably straightforward. The economic appraisal is mainly concerned with savings in road-user costs due to the provision of an improved road facility. There may also be maintenance cost savings. Therefore, the analysis usually focusses on the three principal cost components of construction, maintenance and road-user costs, which constitute what is commonly referred to as the total (road) transport cost or the Whole Life Cycle (WLC) cost, for the selected analysis period, as shown in Figure 5.2.

The common measurements of economic viability are Net Present Value (NPV) and internal rate of return (IRR).

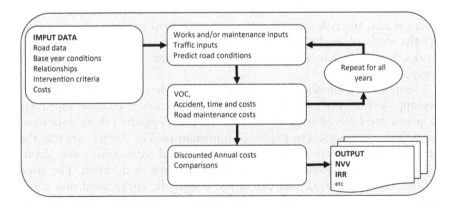

Figure 5.2 Framework for high-volume road economic evaluation.

Net Present Value (NPV) is defined as the difference between the discounted benefits and costs of a project or initiative using a selected discount rate. Thus, the benefits of paving a gravel road would be obtained by assessing the difference between the total transport costs for the sealed road improvement and those of the existing gravel surfaced road. Internal Rate of Return (IRR) of a project is defined as the discount rate at which the present value of costs equals the present value of benefits, that is, when NPV is zero. The solution is obtained by searching through a range of discount rates (from a large positive to a small value). A discount rate is applied to all future costs for each year of the analysis period to reflect the uncertainty of future conditions and the time value of money represented by the opportunity cost of the capital invested in a road project or initiative. Discounting is performed by multiplying the costs in a given year by the discount factor (*DF*) for that year, where:

$$DF = \left(1 + r/100\right)^N$$

where r is the discount rate (%) and N is the number of years from the base year.

In recent years, typical discount rates for transport projects in the major European countries have varied between 1.7% and 5.0% (OECD 2018). These rates are lower than those previously applied, probably due to the exceptionally low central bank interest rates following the 2008 global financial crisis. Discount rates may now change again as interest rates vary. For developing countries discount rates of 10%–20% are typically used to reflect the added risk and higher cost of capital, with the World Bank using between 10% and 14% worldwide for transport projects.

There is a substantial difference in the cost of money and perceived risk between developed and developing countries, where market credit rates can rise to more than 25% per annum in the latter. Furthermore, availability of commercial credit may be severely restricted in developing countries.

THE ECONOMIC EVALUATION OF LOW VOLUME RURAL ROADS

Background characteristics

Low traffic flows in rural roads are unlikely to generate the VOC benefits alone to justify the investments. The main characteristics of LVRRs that set them apart from High-Volume Roads are as follows (World Bank 2005):

1. The development of the road has a high potential to influence economic development in its catchment area through supply side effects. For example, improved accessibility may lead to changes in crop types – from subsistence crops to cash crops – or to improvements in health and education leading to more productive days' work per year and a

more skilled workforce, or commercial or residential development due to more reliable access.

2. The majority of the road users may travel using slow modes of transport; that is, the majority of the traffic may consist of pedestrians, motorcycles and non-motorised traffic (NMT).

3. Pedestrians, animals, NMTs and motorised vehicles tend to intermingle in the traffic stream. Consequentially, accidents tend to be dominated by single vehicle type accidents and accidents between motorised and non-motorised traffic. There may be periods during a year with disrupted passability due to weather.

Methodologies and tools

There are three commonly accepted general methods for economic assessment of LVRRs (World Bank 2005), which are:

- Consumer surplus methods.
- Producer surplus methods.
- Cost-effectiveness analysis.

There are two well-known, computer-based tools commonly used in road project economic appraisal:

- The Highway Development and Management Model, Version 4 (HDM4).
- The Roads Economic Decision Model (RED).

Highway development and management model, version 4 (HDM4)

The HDM4 model has been developed from research in East Africa, Brazil, and other locations. It is most appropriate for use in evaluating high (traffic)-volume road paved road investments (World Bank 2008; Kerali 2003) rather than LVRR interventions.

In a typical project economic evaluation, two project alternatives are evaluated: a 'without project scenario' and a 'with project scenario'. Annual road agency capital and recurrent costs and road-user costs are computed for both alternatives over a defined evaluation period, and total costs to society are compared for the two scenarios. The resulting stream of net benefits to society is used to compute the economic indicators that help determine the project's economic viability (NPV and IRR).

The main drawbacks for the use of this model on LVRR projects are that considerable effort and data are required to calibrate the model relationships to the local conditions and environment, and for evaluation inputs. For example, there can be considerable variation in maintenance regime assumptions; particularly for unpaved roads. The model is primarily VOC driven, which may be a subsidiary factor in LVRR considerations. VOC relationships vary considerably between countries with respect to vehicle type and availability, vehicle working life, mechanical maintenance and rehabilitation regimes. There are no condition deterioration relationships in the model for many of the possible LVRR road surface options.

The use of HDM4 therefore requires a long-term agency commitment to local research, data collection and analysis, and calibration to make the model truly up to date and representative of local conditions. HDM4 is not generally used in the assessment of LVRRs.

The Roads Economic Decision Model (RED)

The Roads Economic Decision Model (RED) not only simplifies the economic evaluation process but also at the same time addresses the following concerns related to low volume roads, and it:

1. Reduces the data input requirements.
2. Takes into account the higher uncertainty related to the inputs.
3. Allows for the incorporation of or modelling of induced traffic (based on a defined price elasticity of demand).
4. Quantifies the economic costs associated with the days-per-year when the passage of vehicles is disrupted by a highly deteriorated road condition.
5. Optionally, uses vehicle speeds as a surrogate parameter to road roughness to define the level of service of low volume roads.
6. Includes road safety benefits.
7. Includes in the analysis of other benefits (or costs) such as those related to non-motorised traffic, social service delivery and environmental impacts, if they are computed separately.
8. Presents the results with the capacity for sensitivity, switching values and stochastic risk analyses.

The RED model is implemented in a series of Excel workbooks that estimate vehicle operating costs and speeds, perform economic comparisons of investment and maintenance options, switching values and stochastic risk analysis (Archondo-Callao 2001, 2004).

Consumer surplus methods

Consumer surplus is a measure of the welfare that people gain from consuming goods and services. It is defined as the difference between the total amount that consumers are willing and able to pay for a good or service (indicated by the demand curve) and the total amount that they actually do pay (i.e. the market price).

Consumer surplus methods are well established and have been applied in road investment models, such as HDM4. The methods are reliable to apply to higher volume roads (greater than 200 VPD). However, its application to low volume roads encounters problems related to the small magnitude of user benefits and the stronger influence of the environment, rather than traffic, on infrastructure deterioration.

For traffic levels between 50 and 200 VPD, and particularly for unpaved roads, a modified and customised approach can be taken.

For traffic levels below 50 VPD, the consumer surplus approach needs to be augmented as the main benefits do not arise from savings in motor vehicle operating costs, but instead relate to the provision of access itself. The benefits of access are difficult to quantify. Traffic can in many LMICs comprise a significant percentage of non-motorised vehicles, such as haulage by mules, hand carts, or walking and head loading (porterage). In such cases the transport costs cannot be easily priced.

Producer surplus methods

Producer surplus models exist for forecasting generated traffic based on the anticipated response of farmers to road investment. However, the predictive accuracy of these models is poor, and a major limitation to their use is that they consider only agricultural freight (TRL 2005). The method requires assumptions concerning the impact of transport investments on local agricultural productivity and output that are difficult to assess, particularly in a situation where proposed projects are expected to open-up new areas, and adequate production data may be difficult to compile (Beenhakker and Lago 1983; World Bank 1998).

In cases where no road exists and a significant change in vehicle accessibility is planned, the producer surplus method can appear to be the most appropriate procedure. However, the following problems and limitations need to be taken into account:

1. The method only considers agricultural freight, which may account for less than 20% of rural road traffic. Passenger benefits and other non-agricultural benefits still need to be calculated separately.
2. The method requires a detailed knowledge of agriculture and its likely response to changes in input and output prices.
3. The data requirements for this method are usually very large.
4. The experience of the practical application of the method has been poor. The empirical justification for estimating changes in agricultural production can commonly be weak, and a failure to consider all the relevant costs of production has often led to benefits being grossly overvalued.

To the extent that low volume rural road investments are increasingly focused on social rather than economic objectives, the application and relevance of the producer surplus method has decreased. Generally, this method should only be used in special situations where the required knowledge and data are available or can be collected at reasonable cost.

Cost-effectiveness analysis

Cost-effectiveness techniques involve a comparison between the costs of a project and the achievement of stated objectives or outcomes. As such they have an intuitive appeal as they directly focus on delivering transport-related improvements to meet certain goals (e.g. maximising the number of people within 1 day's travel of a good road). They are also particularly strong in assessing the most effective measure for delivering a project whose benefits are not readily measurable in monetary terms – an area in which cost–benefit analysis is traditionally weak (Benmaamar 2003).

The principal difficulties associated with cost-effectiveness techniques are that if the project has multiple goals, weightings associated with each of those goals must be derived, often subjectively. There is, therefore, the potential that decisions based on cost-effectiveness techniques may be biased by the method in which the weightings are developed (e.g. consultation only occurs with sectors of the community who have a particular interest). Additionally, unlike cost–benefit analysis, with cost effectiveness there is no threshold that would indicate whether the opportunity cost of the investment is greater than the benefit that will be received. Therefore, while cost-effectiveness can inform the choice between alternatives, it is weak at informing the decision regarding whether to invest or not. Consumer surplus methods rely on the ability to be able to measure costs and benefits in monetary terms (World Bank 2005), which renders them problematic for projects where a significant component or the majority of benefits cannot be readily monetised (e.g. a Low Volume Rural Road).

Cost-effectiveness techniques are a useful tool for project screening or ranking. Such a screening process ensures that projects that are subjected to a more detailed analysis (including cost–benefit analysis) are those that best fit with the objectives of the investment (e.g. poverty alleviation).

ROAD MAINTENANCE

The appraisal of the maintenance elements of a road programme should be given a high priority. The pragmatic appraisal of the maintenance regime likely to be in place throughout a LVRR design life is crucial and often poorly assessed with undue credence given to anticipated, promised or proposed asset management improvements. Realistic and full appraisal of the existing road maintenance arrangements is an essential input to the project design process. This should include data collection and analysis on the target road(s) and comparable routes under the same authority. Existing resources, costs, outcomes and performances are a good indicator of possible future achievements, and their analysis may identify the need for capacity improvement support as part of the project initiatives.

Road maintenance programmes typically have higher economic returns than construction or rehabilitation projects. Rioja (2013) reviewed a selection of World Bank road maintenance projects and found economic rates of return of between 23% and 70%. Despite these high returns, the complexities of the institutional, management and resourcing challenges in developing countries require particularly careful project design to ensure that the intended benefits are achieved and sustained. The important issues around road maintenance and its management are discussed in Chapters 21 and 22.

REFERENCES

Archondo-Callao, R. S. 2001. Roads Economic Decision model (RED) economic evaluation of low volume roads. Africa region findings & good practice info briefs; no. 179. World Bank. https://openknowledge.worldbank.org/handle/10986/9820.

Archondo-Callao, R. 2004. The Roads Economic Decision Model (RED) for the economic evaluation of low volume roads. Software user guide & case studies, SSATP working paper no. 78, World Bank. https://documents1.worldbank.org/curated/en/764611468202186160/pdf/349290ENGLISH0ssatpwp78.pdf.

Beenhakker, H. L. and A. M. Lago. 1983. Economic appraisal of rural roads: simplified operational procedures., World Bank staff working paper no. 610. Washington, DC: The World Bank. https://trid.trb.org/view/1186001. http://transport-links.com/wp-content/uploads/2019/11/1_839_PA4163-04.pdf.

Benmaamar, M. 2003. A method for the appraisal of low volume roads in Tanzania. Paper for the XXII PIARC World Road Congress, Durban. http://transport-links.com/wp-content/uploads/2019/11/1_839_PA4163-04.pdf.

Kerali, H. 2003. Economic appraisal of road projects in countries with developing and transition economies. *Transport Reviews* 23(3):249–262. DOI: 10.1080/0144164032000068920.

Lebo, J. and D. Schelling. 2001. Design and appraisal of rural transport infrastructure, ensuring basic access for the rural communities, Technical paper no 496, World Bank, Washington, USA. https://documents1.worldbank.org/curated/en/227731468184131693/pdf/multi0page.pdf.

OECD. 2018. Cost-benefit analysis and the environment, further developments and policy use. https://www.oecd.org/governance/cost-benefit-analysis-and-the-environment-9789264085169-en.htm.

Petts, R. C. 2002. Low-Cost Road Surfacing (LCS) project. LCS working paper 3. Costing of roadworks. https://www.gtkp.com/assets/uploads/20100103-220522-2539-LCS%20Working%20Paper3-aa.pdf.

Rioja, F. 2013. *What Is the value of infrastructure maintenance? A survey.* Lincoln Institute of Land Policy. https://www.lincolninst.edu/sites/default/files/pubfiles/what-is-the-value-of-infrastructure-maintenance_0.pdf.

Robinson, R. 2004. Chapter 3 Economic appraisal. In *Road Engineering for Development*. Robinson R. and Thagesen B, Eds. Taylor and Francis. ISBN 0-203-30198-6.

SADC, 2003. Chapter 2 Regional setting. In *Low Volume Sealed Roads Guideline*. SADC-SATCC.

TRL. 2005. *Overseas Road Note 5, a Guide to Road Project Appraisal*. Wokingham, UK: TRL Ltd, 145 pp. ISBN: 0-9543339-6-9. https://www.gov. uk/research-for-development-outputs/overseas-road-note-5-a-guide-to-road-project-appraisal.

Henning, T., P. Kadar and C. R. Bennett. 2006. Surfacing alternatives for unsealed rural roads. World Bank transport note. TRN-33. Washington, DC: World Bank. https://documents1.worldbank.org/curated/en/473411468154454358/pdf/371920Surfacing0Alternatives01PUBLIC1.pdf.

World Bank. 2020. *To Pave or Not to Pave. Developing a Framework for Systematic Decision Making in the Choice of Paving Technologies for Rural Roads*. Mobility and transport connectivity series. https://openknowledge.worldbank. org/handle/10986/35163.

World Bank. 2005. *Notes on the Economic Evaluation of Transport Projects, a Framework for the Economic Evaluation of Transport Projects*, Transport note no. TRN-5. https://openknowledge.worldbank.org/bitstream/handle/10986/11 787/33945a10trn125120EENote2.pdf?sequence=1&isAllowed=y.

World Bank. 2008. *Applying the HDM-4 Model to Strategic Planning of Road Works, Rodrigo Archondo-Callao*, Transport note no. TRN-20. https://documents1.worldbank.org/curated/en/993961468338479540/pdf/463190NWP0 Box334086B01PUBLIC10tp120.pdf.

World Bank. 1998. *Handbook on Investment Operations*. Operational Core Services Network Learning and Leadership Centre. https://documents.worldbank.org/ en/publication/documents-reports/documentdetail/749061468740206498/ handbook-on-economic-analysis-of-investment-operations.

Chapter 6

The rural road environment

INTRODUCTION

It is now recognised that the life-time performance and service delivery of Low Volume Rural Roads (LVRRs) are influenced significantly by the impacts within what is termed the 'Road Environment', comprising engineering, operational and socio-economic factors (Cook et al. 2013). Together these factors should be crucial elements for under-pinning the rural road design process. It follows that a full understanding of the influence of the factors is essential if the roads to be constructed and maintained are to be Fit for Purpose, sustainable and provide value for money.

Figure 6.1 illustrates concept of the Road Environment schematically, by indicating the elements of the Engineering and Operational and Local Resource Environments that are detailed in the following sections.

This chapter describes these Road Environment factors and their relevance to the design process within the whole-life cycle.

THE ENGINEERING ENVIRONMENT FACTORS

Climate

The main features of climate that are of importance in the design of LVRRs are:

- Rainfall distribution.
- Rainfall intensity.
- Temperature.
- Occurrence of storm events.
- Evaporation.

Current and future climates will influence the supply and movement of surface water and groundwater, and impact upon the road in terms of direct erosion, flooding and the influences of the groundwater regime. Unpaved

DOI: 10.1201/9780429173271-6

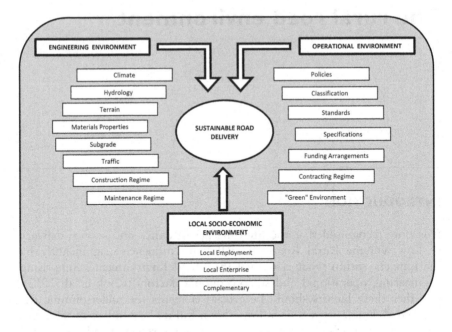

Figure 6.1 Road Environment factors.

road performance is particularly influenced by quantity and intensity of rainfall and runoff, and hence climatic factors can have a significant influence, for example, on the selection of pavement options (Cook and Petts 2005). For general assessment, rainfall, temperature and evaporation may be combined into a Climatic Index (Weinert 1974). The Weinert 'N' value has been adopted as part of the pavement design process and is calculated as follows:

$$N = 12Ej/Pa$$

where Ej=evaporation for the warmest month and Pa=total annual precipitation.

N-values less than 4 apply to a climate that is seasonally tropical and wet, whereas N-values greater than 4 apply to a climate that is arid, semi-arid or dry. The implication for pavement design is that drier climates ($N > 4$) could allow a reduction in pavement material strength specifications (SADC 2003). The assessment of future climate with respect to rural road design is discussed in Chapter 17.

Hydrology

Water is the single most important factor affecting performance of rural road pavements, earthworks, drainage and long-term maintenance costs. The interaction of water with elements of the road and its movement within and adjacent to the road structure has an over-arching impact on its performance. Changes in near-surface moisture condition can be the trigger of earthwork instability as well as initiating significant subgrade and earthwork volume changes in pavements underlain by 'expansive' clay materials.

There is a direct relationship between current and future climates and hydrology. The assessment of hydrological factors is a vital part of bridge and culvert design and, in particular, the assessment of future climate shock events through analysis of return periods and anticipated maximum daily rainfalls. Further detail on hydrology analysis and road design is contained in Chapter 11.

Terrain

Terrain reflects geological and geomorphological history, and has a direct influence on both horizontal and vertical road alignments and associated earthworks. It is a key issue in the application of geometric and safety standards and the amount by which they may need to be adjusted for specific locations. Severe terrain may necessitate some compromise on pavement and shoulder widths for LVRRs; otherwise, costs and environmental impacts may be excessive (see Chapter 9).

Terrain is a key factor in the investigation for natural materials resources and the associated environment compliance requirements. Appraisal of soil–rock mass geomorphology may be an important action in the initial assessment of geotechnical hazards along a road alignment. Terrain in conjunction with rainfall patterns and intensity can be a controlling influence on pavement option and the implementation of a Spot Improvement methodology (see Chapter 16).

Construction material properties

The nature, engineering character and location of construction materials are fundamental aspects of the rural Road Environment assessment and input into design. The pragmatic assessment of the usability of local marginal materials for use in rural roads is often a key issue where the expense of hauling-in international-specification materials may be a significant constraint. In LVRRs, where the use of local materials is a priority, the issue should be; 'what design options are compatible with the available materials?' rather than seeking to find material to meet standard specifications, as is the case with higher level roads and larger budgets. Further detail on construction materials is contained in Chapter 8.

Subgrade and foundations

The subgrade is essentially the foundation layer for the pavement, and the assessment of its condition is critical to the pavement design.

The combination of subgrade and traffic has long been an established relationship fundamental to the structural design of pavements. This still holds true for LVRRs, even although its importance in relation to other Road Environment factors may be less than for higher volume roads.

Weak, soft, or compressible foundations have a significant impact on the design and performance of embankments or culverts placed on them. In situ foundation strength is a key issue in the design of bridges.

Traffic

Although recent research indicates that the relative influence of traffic on the structural design of LVRRs is often less than that from other Road Environment parameters, consideration still needs to be given to its influence on road geometry. Identifying and mitigating the risk of axle overloading is important issues for light road pavements. The nature of the vehicles is also a major input into the geometric standards. Once selection of pavement type has been undertaken in line with the wider Road Environment, the detail of pavement structure still remains a direct function of traffic and subgrade strength, as indicated in Chapters 14 and 15. It is also a significant factor in assessing required road safety measures.

The construction regime

The construction regime can govern whether the road design is applied in an appropriate manner or not. Selection of specific design approaches should be guided by the assessed capacity and experience of the available contractors. Key elements for consideration include:

- Local construction practices.
- Experience of contractors or construction groups.
- Skills and training needs of labour force.
- Availability, use and condition of appropriate construction plant.
- Quality Assurance and compliance with specification.
- Experience and motivation of technical supervisors.

In many LMICs, some local contractors may have limited experience in road building procedures other than those associated with unsealed gravel wearing course construction. Issues surrounding the construction regime are presented in detail in Chapters 18–20.

Maintenance

All roads, however designed and constructed, require regular maintenance to ensure that their basic task is delivered throughout the design life.

Achieving this depends on the maintenance strategies adopted, the timeliness of the interventions, and the local capacity and available funding to carry out the necessary works. When selecting road design options, it is essential to pragmatically assess the actual maintenance regime, including current and likely future funding, that will be in place during its whole design life. Assessment of the existing maintenance regime and possible development of maintenance capacity should be an integral part of the road project design process. Selection of high-maintenance options, such as thin bituminous seals, should be avoided in low-maintenance capacity environments. Issues surrounding the maintenance regime are presented in detail in Chapters 20 and 21.

THE OPERATIONAL ENVIRONMENT FACTORS

Policies

National or sectorial policies will provide guidelines, requirements, and priorities for the road network decision-making processes. They provide the framework within which the road design must be undertaken. National policies, for example, on Climate Resilience or Environmental impact are important drivers of good road design practice.

Policy and associated road law are the necessary foundations for the effective and appropriate application of road network management. Policy frequently drives the strategic guidelines on which appropriate Fit for Purpose road design is based, by setting access targets within economic development programmes; for example, the Indian National Rural Roads Program (Pradhan Mantri Gram Sadak Yojana or PMGSY) or the National Strategy for Rural Roads and Access in Myanmar (PMGSY 2015; GoM 2017).

Classification

Road classifications based on task or function provide road planners and designers with a practical guidance framework to initially select and cost appropriate road options. Road classification essentially guides the road designer towards appropriate standards. Having a clear rural road classification linked to relevant standards facilitates design and construction within acceptable performance criteria. Issues around classification and design have been discussed in Chapter 3.

An appreciation of the governing road classification is an essential first step in developing the road design envelope within which to complete the design process. It may be that no appropriate classification exists, in which case amending a classification or developing one from basic principles would be an action for the planning or feasibility project stages.

Road design standards

Geometric design standards provide the link between the cost of building the road and the benefits to road users (Dingen and Cook 2018). From the road designer's point of view, the geometric standard governs the minimum level that must be achieved by the design. Geometric design covers road width, crossfall, horizontal and vertical alignments, and sight lines. Identifying, amending or developing appropriate safety standards is clearly linked to the issue of geometric standards, as indicated in the more detailed procedures in Chapter 9.

Additional standards can exist which will govern the whole range of investigation, design, construction and maintenance actions. The relevant standards need to be identified at an early project stage and either incorporated into the design process or modified to suit specific project parameters (justifiable Departures from Standards).

Technical specifications

Technical specifications define and provide guidance on the design and construction criteria for rural roads to meet their required level of service within the governing standards and classification. Road designs should be either in agreement with the local technical specifications or compliant with approved modifications. Specifications appropriate to the local engineering environment are an essential element of an effective enabling environment.

The automatic adoption of international or high-volume road technical specifications can cause problems for LVRR design and lead to poor construction and un-deliverable maintenance expectations. The review of existing technical specification and their assessment as to fitness for purpose is, therefore, a necessary early action in LVRR projects. Technical specifications for community access roads may have to be used by local small contractors or community groups in a labour-based environment. This may require the adoption of locally relevant specifications.

Funding arrangements

Available funding has an over-arching influence on the scale and nature of the roads that are feasible. The assessment of funding as a constraint on road design should include not only available budgets for design and construction but also those available for on-going road management and maintenance. In other words, the life-time costs or whole-life asset costs should be matched by whole-life funds.

Funding arrangements have a particular influence on the level of climate resilience that can be provided for rural roads. There has to be a balance between the level of service required from a rural road and the funding available to provide that service.

Some bi-lateral or MDB funders may require specific design issues to be considered: for example, sealing of pavements, NMT-friendly geometric designs, gender-focussed road safety or bridge options. Such requirements must be identified and included within the design concepts at an early stage.

The contracting regime

The nature of the general contracting regime for either construction or maintenance influences the design and construction of a road through the following issues:

- Local legislation and contract documentation.
- Contracting model; e.g., direct contracting (Force Account), traditional BoQ, and Performance Based.
- Governance and bureaucracy.
- State-owned or private contractors.
- National or international contractors.
- Arrangements for facilitating local SMEs or labour-based approaches.
- Use of local community groups for construction or maintenance.

The contracting regime should be defined and assessed as to its impact on the design, construction and maintenance. Contracting regime issues are discussed further in Chapters 18 and 19.

The green environment

Road construction and on-going road use and maintenance have an impact on the natural environment, including flora, fauna, hydrology, slope stability, health and safety. These impacts should be assessed, and adverse effects should be mitigated as much as possible by appropriate design and construction procedures, as laid out in Chapter 4.

THE SOCIO-ECONOMIC ENVIRONMENT

Principle

Low volume rural road design, much more than higher volume roads, is closely linked to local socio-economic issues. Primarily, this link is enshrined in the road task in terms of the people and purposes it should be designed to serve, as well as some additional issues such as agricultural development and diversification, gender equality, community strengthening, health and education.

Local employment

Local population groups can benefit significantly from temporary employment in road works projects and their follow-on maintenance. This provides an opportunity for the design engineer to consider local participation and find ways to help them access employment through selection of appropriate design options. Utilising local human resources should benefit both the road agency, or promoter, as well as the local population groups, provided design options are compatible with the available resources. For example, options such as hand-packed stone or concrete block pavements are much more local-resource friendly than heavy-machinery-based bitumen-sealed options.

Rural road projects should also be an opportunity for the design engineer to develop an understanding of any barriers to participation in employment by women and women's groups, and find ways to help them access employment without causing conflict or concern amongst the wider community. At a strategic level, utilising local resources is an important issue in terms of empowering local stakeholders to take ownership of their road network: a potentially important factor in meeting the challenges of routine maintenance and asset sustainability.

Development of local enterprise

Local economic development, as a whole, is a key high-level objective of most rural roads; in addition, there are opportunities to encourage the more formal participation of local groups (for example, Women's Groups, Youth Associations or Veteran Associations) in planning, construction and maintenance. The selection of specific pavement types or earthwork protection procedures (e.g., bioengineering) are important in this regard and, in the longer term, encourage ownership and support for routine maintenance.

Involvement in road projects will encourage development of Local Groups to becoming Local Enterprises, SMEs and local contractors.

Complementary interventions

Complementary Interventions take advantage of the presence of a road project to build-in aspects that will enhance the social, environmental and safety situation of communities affected by the road (Engineers Against Poverty 2016). These are additional to the statutory obligations of the road authority, designer or contractor. They could include, for example, issues such as road safety, road corridor enhancement, transport services, community health services or community school access (ERA 2016).

Complementary interventions should be demand driven, reflecting the needs expressed by local communities themselves, and are agreed through interaction and dialogue with local representative groups. They may enhance, but do not replace existing, safeguards required either by government regulations or by donor guidance documents (ADB 2013; World Bank 2016). The task of the design engineer is to materialise these desires and agreements into the LVRR design and its contract documentation.

ROAD INVENTORY DATA COLLECTION AND THE DESIGN PROCESS

Consideration of the Road Environment framework of factors provides the road designer and owner/promotor with a broad vision of the inputs required for sustainable road design and can form the basis for investigation and data collection throughout a road project.

The level of detail required will vary with the stage of a project as well as its complexity. Tables 6.1 and 6.2 summarise Road Engineering and Operational Environment data collection, and its application through the Project Cycle up to and including construction.

Table 6.1 Data Collection and Application for the Engineering Environment

Factor	Stage	Data collection and application
Climate	Planning	General climate issues and identification of data sources.
	Feasibility	Current and likely future climatic patterns, including severe climatic events. Initial likely level of climate resilience required.
	Design	Climatic patterns and the incidence of severe climatic events confirmed. Detailed vulnerability and risk assessment linked to required level of CR and adaptation options.
	Construction	Detailed climate data to avoid high-risk exposure of partially completed works to climate impacts. Monitoring of climate data on site.
Hydrology	Planning	Indications of any major flood problems or drought issues.
	Feasibility	Hydrological conditions prevailing over the proposed alignments by walkover survey and discussion with local groups. Access available records. Broad drainage options.
	Design	Ground water levels should be established, flooding defined. Current and future hydrological data for bridge, culvert design – rainfall intensity data.
	Construction	Monitoring a flow and flood data from site.
Terrain	Planning	Broad classification of project terrain and identification of mapping sources.
	Feasibility	Terrain groups and earthwork relevance along alignments; identify high-risk critical areas.
	Design	Gradients and earthworks based on detailed topographic survey. Detailed design data on critical risk areas.
	Construction	Monitoring of potential failures at identified risk areas.
Materials properties	Planning	Assess any major construction materials issues in terms of quality or quantity.
	Feasibility	Identify likely sources of material in terms of location, quality and quantity, and note any problem issues.

(Continued)

Table 6.1 (Continued) Data Collection and Application for the Engineering Environment

Factor	Stage	Data collection and application
	Design	Sources defined in terms of location, quality and quantity. Clearly establish that construction is possible with the available materials. Costs of extraction, haulage, processing and placement.
	Construction	Monitoring of materials quality and variation in properties from designated materials sources. Monitoring data from site laboratories.
Subgrade	Planning	Identification of issues with previous projects.
	Feasibility	Establish likely minimum strength values for subgrade along alignments. Identify problem areas likely to impact significantly on pavement design.
	Design	Design subgrade strengths should be selected based on updated or more detailed site work building on the feasibility data.
	Construction	Monitor actually encountered subgrade conditions against design assumptions.
Traffic	Planning	General traffic levels within existing road classification system.
	Feasibility	Likely traffic volumes in terms of PCU/esa/ADT and potential 'design vehicles.' Equivalent axle loads for prevailing vehicles. Initial traffic surveys.
	Design	Assumptions on traffic patterns should be cross-checked and, if required, additional traffic and axle surveys undertaken.
	Construction	Monitor construction traffic and impact on newly laid pavement layers.
Construction Regime	Planning	Assessment of general options: machine based, or Labour Based, or a combination.
	Feasibility	Level of experience of the potential contractors in terms of the likely pavement and surfacing options. Potential training needs for local contractors.
	Design	Contractors or contracting groups capable of undertaking the works identified. Training programmes for contractors or labour-based organisations defined.
	Construction	Construction regime data from Technical Audits. Monitor effectiveness of capacity building.
Maintenance regime	Planning	Identification of relevant authority, existing or future policies and funding capacity.
	Feasibility	Existing maintenance programmes, their funding and effectiveness. Identify agencies or groups responsible for specific elements of maintenance.
	Design	Maintenance commitments defined, and assess any shortcomings. Define any required training or capacity building programmes related to specific design options. Confirm whole-life costs.
	Construction	Collate as-built drawings for maintenance planning.

Table 6.2 Data Collection and Application for the Operational Environment

Factor	Stage	Data collection and application
Policies	Planning	Identify relevant policy documents, and determine key issues that will govern proposed project.
	Feasibility	Assess key policy issues against proposed project. Highlight potential problem areas. Ensure Feasibility Study (FS) designs are compliant with policy.
	Design	Cross-check that final road design is compliant with relevant policies.
	Construction	No specific actions.
Classification	Planning	Identify formal road classification or classification guidelines from the relevant road authority. Agree any required temporary guidelines. Identify likely classification for the proposed project.
	Feasibility	Define classification parameters for the project. Ensure Preliminary Engineering Design (PED) fits within required classification.
	Design	Cross-check that Final Engineering Design (FED) is compliant with relevant classification.
	Construction	No specific actions.
Standards (including road safety)	Planning	Identify the relevant standards that are legally in place and review whether or not they are appropriate for road(s) in question. Define problem issues.
	Feasibility	Assess in detail the relevant standards and how they will influence the design. Draft PED.
	Design	Agree any modifications to standards based on FS/PED reporting. Draft FED in line with agreed standards.
	Construction	Monitor any areas of concern in application of standards. Recover data from safety audit.
Specifications	Planning	Identify the relevant technical specifications that are legally in place, and review whether or not they are appropriate for road(s) in question. Define problem issues.
	Feasibility	Obtain and review reports/designs relevant to application of the existing technical specifications. Identify any problems. Report any potential need for specification adjustments.
	Design	Agree any modifications to specifications based on FS/PED reporting. Draft FED and contract documentation in line with agreed specifications.
	Construction	Monitor any potential issues with compliance with specifications.

(Continued)

Table 6.2 (Continued) Data Collection and Application for the Operational Environment

Factor	Stage	Data collection and application
Funding	Planning	Identify budget available and its constraints.
	Feasibility	Define details of funding constraints and how they apply to different aspects of the project. (For example, land acquisition, compensation, training, and climate adaptation.) Prioritise road assets in line with budget constraints. Draft FS/PED in line with funding requirements.
	Design	Draw up FED within budget constraints and any funding adjustments resulting from FS.
	Construction	Monitor expenditure against cost assumptions.
Contracting regime	Planning	Identify the relevant contractual and legal frameworks that will govern the works. Obtain copies of all relevant documentation. Obtain information on preferred contractual modes and limitations.
	Feasibility	Assess the existing documentation and identify any areas that will require discussion and possible modification. Review lessons to be learnt from previous application of the documentation in similar rural Road Environments. Draft outline contract documentation.
	Design	Agree any final amendment to contractual documentation and ensure engineering design is compatible.
	Construction	Monitor application of contract documentation through Technical Audits. Note lessons to be learnt for future projects.
Maintenance regime	Planning	Identify general asset management framework.
	Feasibility	Assess asset management/maintenance capacity and resources.
	Design	Ensure FED options are within maintenance capacity of road authority.
	Construction	Collate and define maintenance requirements fir rolling programme.
	Post construction	Assess maintenance effectiveness and collate lessons learnt. Feed back to upgrade phase.
The 'green' environment	Planning	Identification of general legal requirements and any likely major constraints and issues.
	Feasibility	Environmental impacts of the proposed design options within the framework of governing regulations.
	Design	Undertake any further required environmental impacts studies.
	Construction	Monitor impacts.

REFERENCES

ADB. 2013. *Social Protection Operational Plan 2014–2020*. Manila: Asian Development Bank. https://www.adb.org/sites/default/files/institutional-document/42704/files/social-protection-operational-plan.

Cook, J. R. and R. C. Petts. 2005. Rural road gravel assessment programme. SEACAP 4, module 4, final report. DFID-UKAID report for MoT, Vietnam. https://www.research4cap.org/index.php/resources/rural-access-library.

Cook, J. R., R. C. Petts and J. Rolt. 2013. Low volume rural road surfacing and pavements: A guide to good practice. Research report for AfCAP and UKAID-DFID. https://www.research4cap.org/index.php/resources/rural-access-library.

Dingen, R. and J. R. Cook. 2018. Review of low volume rural road standards and specifications in Myanmar. AsCAP project report for DRRD. https://www.research4cap.org/ral/DingenCook-2018-ReviewLVRRStandardsSpecifications Myanmar-AsCAP-MY2118B-180425.pdf.

Engineers Against Poverty. 2016. Maximising the social development outcomes of roads and transport projects. Guidance note for the Chartered Institution of Highways and Transportation. https://www.ciht.org.uk/import/pdf/maximising%20the%20social%20development%20outcomes%20of%20road%20 and%20transport%20projects.pdf.

Ethiopian Roads Authority (ERA). 2016. Part C, complementary interventions. In *Manual for low Volume Roads: Ethiopia*. www.gov.uk.

GOM (Government of the Union of Myanmar). 2017. *Nationals Strategy for Rural Roads and Access*.

PMGSY. 2015. National rural roads program (Pradhan Mantri Gram Sadak Yojana) programme guidelines. https://www.pm-yojana.in/en/pradhan-mantri-gram-sadak-yojana-pmgsy.

SADC. 2003. Chapter 5 Pavement design, materials & surfacing. In *Low Volume Sealed Roads Guideline*. SADC-SATCC. https://www.gtkp.com/assets/uploads/20100103-131139-9347-SADC%20Guideline-Part2.pdf.

Weinert, H. 1974. A climatic index of weathering and its application in road construction. *Geotechnique* 24:475–488.

World Bank. 2016. *Fact Sheet. The World Bank's New Environmental and Social Framework*. World Bank. http://pubdocs.worldbank.org/en/7483914 70327541124/SafeguardsFactSheetenglishAug42016.

Chapter 7

Site investigation

INTRODUCTION

Fit for Purpose rural road design depends on the collection, analysis and application of relevant information, and, consequently, site investigation is an essential part of the design process. Chapter 6 emphasised the importance of the Road Environment, and the collection of data should be based around its engineering and non-engineering components. This chapter, therefore, although focussing on road engineering issues, also includes some guidance on the investigation of non-engineering topics. Figure 7.1 outlines the key aspects of a site investigation.

In regard to engineering topics, this chapter has a focus on low-cost investigations appropriate to the needs of Low Volume Rural Roads (LVRRs) in Low and Middle Income Countries (LMICs), for example the use of desk-studies, trial pits, and walkover and mapping surveys.

The aim of this chapter is to provide guidance on the appropriate type and level of investigation that is required for the selection, design, construction and maintenance of LVRRs. The chapter also provides practitioners with the necessary tools to develop suitable ground investigation programmes and testing schedules, along with some guidance in managing the data obtained.

INVESTIGATION FRAMEWORK

Engineering investigations

The scale of investigations may vary from simple test pit and walkover surveys to complex phased programmes, involving expensive and sophisticated equipment, depending on the type, complexity and risk-assessments of the road involved. Important issues to be considered are the project stage within the overall road cycle and the data sets required for the design of the different road components.

DOI: 10.1201/9780429173271-7

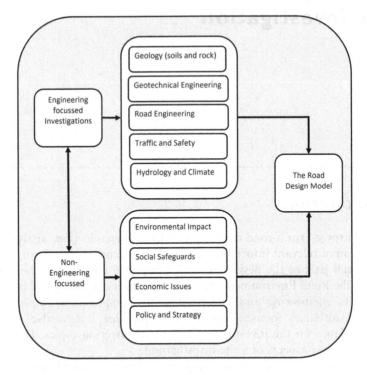

Figure 7.1 The site investigation framework.

In general terms, an engineering investigation will be seeking to collect, analyse and assess information in relation to:

- Selection of road alignment.
- Identification of geotechnical hazards.
- Climate vulnerabilities.
- Subgrade or embankment foundation conditions.
- Road-side slope stability for cuts and embankments.
- Construction materials.
- Hydrology input into hydraulic design.
- Selection and design of road pavement.
- Design of foundations for bridges and other structures.
- Traffic levels and vehicle types.
- Safety issues.

Non-engineering investigations

Non-engineering investigations should be run in parallel with engineering investigations and focus on key issues that are summarised in Table 7.1.

Table 7.1 Key Non-engineering Investigations Actions

	Non-engineering issue	Investigation aims	Reference
A	The economic framework	Identify and understand the project economic drivers and the influence they may have on the selection of design options.	Chapter 5
B	Financial constraints on design	Define confirmed budgets for design and construction as well as those available for ongoing road management and maintenance.	Chapters 5 and 13
C	Relevant policies and strategies	Identify and comment on all relevant rural road network strategy and policy documentation as they may impact on the road project. Assess and discuss impacts and consequences in formal meetings with road authorities.	Chapters 1 and 3
D	Green environment impact issues and constraints	Collect and assess sufficient data to be able to draft required Environmental Management Plans; through walkover surveys, structured local stakeholder meetings and discussions with environmental authorities.	Chapters 4 and 17
E	Impacts of the project on local communities	Assess any likely socio-economic impacts of the project on local communities – both negative and positive. Likely to require specialist household and local group surveys.	Chapters 4 and 5
F	Local stakeholder expectations	Identify the central expectations regarding the road project from local groups. A particularly key action when a Spot Improvement strategy is envisaged.	Chapters 5 and 16
G	Knowledge transfer requirements	Identify knowledge gaps that require addressing through training. These may be technical training issues for contractors/supervisors and/or information for local community groups.	Chapters 6 and 20
H	Liaison links with relevant stakeholders	Identify key local stakeholder groups and leaders. Establish working relationships such that feedback of genuine concerns is facilitated and important information on, for example, flood impacts and access is captured.	Chapters 16 and 17

The majority of non-engineering investigation are best initiated at planning and feasibility stages of a road project and best largely completed prior to final design.

Investigation aims

In the majority of cases, whatever the size of the project, the overall site investigation can be considered as a phased programme linked to the stages within the Project Cycle. Each stage will demand a different level of detail. Additional special investigations may be required for specific purposes during the Project Cycle; for example, for additional construction materials or to investigate slope stability issues during the construction phase.

Investigation budgets are likely to be very constrained for most rural road projects, and hence the procedures and techniques employed must be carefully selected and planned to obtain maximum information for the least cost.

An important aim of the investigation process is to collect and assess sufficient information such that Road Design Models (RDMs) can be assembled as an accessible knowledge base for making informed project decisions. The model should be considered a living document that is updated, as more information comes to hand through the Project Cycle (Figure 7.2).

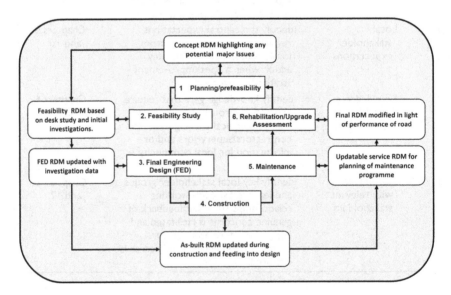

Figure 7.2   Development of the Road Design Model.

GROUND INVESTIGATIONS

Desk-study

Before any ground investigation can be planned and designed, it is vital to study all the relevant information that is available about the project area. This is done through a desk-study that entails the collection and analysis of information relevant to the engineering and non-engineering assessment of the site. This desk-study as well as highlighting key issues often eliminates unfavourable options from further consideration, thus saving a considerable amount of time and money. Table 7.2 summarises sources of desk-study information and their use.

Walkover surveys and geotechnical mapping

Walkover surveys are a low-cost means of collecting data and are particularly useful in the early stages of road investigations. Initial walkover surveys should focus on covering the entire length of the proposed alignment before concentrating on detailed locations. Key walkover survey types and objectives are summarised in Table 7.3.

The walkover survey should, in conjunction with the desk-study, establish the key physical, geotechnical, and engineering aspects of the alignment, either pre-existing or proposed. On some projects, some preliminary geotechnical or engineering geological mapping may be required to determine the extent of potential hazards (Geological Society 1972). The location of existing or potential borrow areas or rock quarries for construction materials may also be an element of the walkover process (Roughton International 2000). Standard forms and procedures, including the use of GPS equipment, can be utilised for collecting walkover information in a logical manner; examples are included in Appendix A.

Traffic surveys

Traffic surveys are used for estimating the types and amounts of vehicles (motorised and non-motorised) and pedestrians likely to use the road. Data from these surveys will be principally analysed in terms of:

- Passenger Car Units (PCU); mainly for geometric design and transport service issues.
- Average Annual Daily Traffic (AADT) adjusted for vehicle type, mainly for pavement surface and structural design.
- Equivalent Standard Axles (esa); for pavement structural design.
- Axle loads; for pavement and bridge design.

Table 7.2 Desk-Study Information

Information	Functional use
Official Government, Road Authority or Funding Agency policies and statutory requirements	These are the frameworks within which the road project has to be designed. They include policies/requirements on road safety, classification, environment protection and climate resilience.
Statistics and future development plans	Population data and demographics; socio-economic and household survey information input into defining road task. Identification of future activities within, or impacting upon, the planned road corridor.
Road design guides or manuals; relevant standards and technical specifications	Information to guide aspects of geometry, drainage, pavement, materials and structural design of roads and bridges. Detailed descriptions of procedures and associated standards.
Previous design or investigation reports	Valuable information sources that will provide a range of information such as: soil and rock type, strength parameters, hydrological issues, construction materials, information on local road performance and issues. These reports may help reduce the scope or better target the nature of the site investigation.
Remote sensing; photographs, satellite images and drone information	Provide geologic and hydrological information which can be used as a basis for site reconnaissance. Tracking site changes over time. Can save time during construction material or geo-hazard surveys.
Topographic maps	Provide essential base maps. Key initial data on site topography. Identify physical features; identification of access areas and restrictions and can provide initial ideas on cut and fill balance before visiting the site.
Geologic reports and maps	Identifies rock types, fracture, orientation and groundwater flow patterns. Provides information on corridor soil and rock type and characteristics, potential sources of materials, hydrological issues and environmental concerns.
Soil maps, reports	Provide information on near-surface soils to facilitate preliminary borrow source evaluation, existence of duricrusts (laterite/calcretes) and potential problem soils (e.g. Black cotton soils).
Current and future climatic data	Mean annual/monthly; rainfall and temperature. Maximum daily rainfalls. Major storm events. Max and min temperatures; evaporation. All required to assess current and future climatic impacts on Road Environment.
Land use maps/reports	Identify the physical and biological cover over corridor, including water, vegetation, bare soil, and artificial structures. Useful for assessing any changes with time – e.g. deforestation – that could impact the erodibility of the alignment earthworks.
Local knowledge	Traffic classification, traffic variation, road user demand, hazards and ground instability, local road performance including flooding, and maintenance history, accident black spots, water sources, local weather conditions and drainage characteristics. Identification of specific problems and hazards along proposed alignment; local sources of materials and previous performance.

Table 7.3 Key Walkover Surveys

Survey	Description
1. Initial site visit	A broadly-based walkover or 'drive and stop' survey to identify general Road Environment, key features and potential hazards or engineering challenges.
2. Alignment inventory	Walkover survey to systematically log, on a strip map, all the key features of the alignment or existing road. Include adjacent current or planned land use and irrigation that may impact road performance.
3. Road condition	Systematic visual survey of existing road condition. An essential step if a Spot Improvement strategy is being considered.
4. Climate impact	Required to identify areas vulnerable to current and future climate impact. Gathering of local knowledge is key part of this survey. May be combined with surveys 2 and/or 3.
5. Hazard	Detailed additional walkover of hazards identified a part of surveys 2 or 4. May include some simple geotechnical mapping.
6. Hydrology	Walkover survey to log all stream/river crossing and their size, and other potential locations of water crossing. Note issues such as gradients, soil types, existing condition of structures, and evidence of flood levels.

Standard traffic data collection forms are in include in Appendix A. Detailed advice on the design and implementation of traffic surveys are laid in the following UK Overseas Road Notes (ORNs):

- ORN 20. Management of rural road networks (TRL 2003).
- ORN 31. A guide to the structural design of bitumen-surfaced roads in tropical and sub-tropical climates (TRL 1993, updated as TRL 2023).
- ORN 40. A guide to axle load surveys and traffic counts for determining traffic loading on pavements (TRL 2004).

Pitting, boring and drilling

Sub-surface investigations involve the physical sampling, examination and in situ testing of the soils and rocks underlying and adjacent to the route corridor, in order to determine geotechnical and engineering properties relevant to appropriate design (FHWA 2017).

These investigations should provide a description of ground conditions relevant to the proposed works and establish a basis for the assessment of the geotechnical and road engineering parameters relevant to the stages of the Project Cycle. They may also be required to provide relevant information on groundwater and hydrology needed for geotechnical design and construction. Ground investigations for construction materials determine the nature and extent of proposed construction materials sources as well as their relevant geotechnical parameters. Specialist investigations may be required to collect information about identified geo-hazards and large structures. Table 7.4 outlines the most common sampling and testing techniques used in LVRR investigations, and Figure 7.3 illustrates some typical investigation equipment

Table 7.4 Standard Ground Investigation Techniques

Technique	Purpose	Application
DCP survey	In situ test for strength characteristics. Depth normally 0.8 m, or up to 1.5 m with extension rods and occasionally up to 2.5 m in some weak or soft conditions.	Light and portable, gives information on state of near surface ground or existing pavement layers (Figure 7.4). Testing is quick and simple. Used for pavement design, Quality Control and light foundation investigations (TRL 2006).
Vane shear test	In situ shear strength in soft clays.	Especially good for assessing soft clays for embankment foundations. Equipment is easily portable. Can be used in conjunction with boreholes.
Cone Penetration Test (CPT) or piezocone	In situ strength and compressibility of soils.	Light portable machines as well as heavier machines are available. Good reliable information in soft to stiff clays and loose to dense sands. Used in areas under moderate to high embankment and for structural foundation investigations.
Test pits	Visual examination of in situ soil profiles; normally 1–2 m deep, although up to 5 m possible with adequate support and safety precautions.	Provides a ground profile and allows good undisturbed sampling as well as large bulk samples for testing subgrade and potential fill material. DCP profiles can be undertaken from pit bottoms.
Auguring	Provides in situ information on material present. Wide range of options from hand-held to machine driven.	Can extend to 15–20 m depth if machine driven. Hand-held very useful for during walkover surveys. Used for establishing soil profiles and depth to bedrock. Highly disturbed samples. In situ testing possible.
Boring and drilling	A number of options available from soil wash boring to high-quality rock coring to recover samples for examination and testing.	Boring and drilling in most LVRR projects limited to structure sites, deep earthworks and special purposes (e.g. landslide or deep weak soil investigations). Frequently used in conjunction with in situ testing.
Standard Penetration Testing (SPT)	In conjunction with boreholes can provide in situ test strength results in most materials and can be used in weak rocks.	Used in conjunction with auguring or boring holes. Used for gauging ground strength for structure foundation investigation and high earthworks.

Source: Sabatini et al. (2002), SANRAL (2013) and FHWA (2017) provide details of the above techniques.

Figure 7.3 Typical investigation equipment. (a) DCP being used to investigate an existing alignment for upgrade. (b) A light 'shell and auger' rig. (c) Typical small drilling/boring rig; breaks down into portable elements for difficult terrain. (d) Small mobile drill rig.

The use of a technique or a combination of techniques for a specific road will be a function of the scale, nature and geotechnical environment of that road. Ground investigations need to be carefully planned and must take into account the nature of the ground, the nature and phase of the project, and the project design requirements. Results from the desk-study and walkover surveys must be fed into the planning of cost-effective ground investigations.

Allied to all site investigation techniques should be a logical and well implemented procedures for soil and rock description and classification (Norbury 2010).

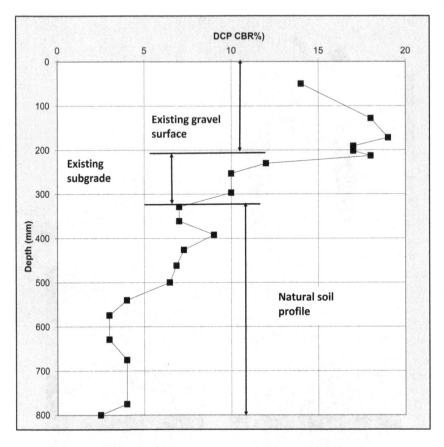

Figure 7.4   DCP profiles.

Geophysical methods

Geophysical surveys, as outlined in Table 7.5, can very usefully be employed as a cost-effective means of defining soil–rock boundaries and either interpolating between boreholes or used as the basis for making more effective selection of borehole locations (Anderson et al. 2008).

Geophysics, when appropriately applied, can enhance the reliability, speed and cost-benefit of geotechnical ground investigations. Geophysical surveys are commonly employed for the following reasons:

1. Sub-surface characterisation of soil–rock masses, including identification of bedrock depth, soil–rock internal boundaries and water tables.
2. Estimating engineering properties of soil–rock profile materials through correlation of geophysical properties such as seismic velocity and resistivity with soil–rock strength and stiffness.

Table 7.5 Commonly Used Geophysical Investigation Techniques

Method	Application	Limitations
Seismic refraction. Impact applied to ground surface (e.g. from heavy hammer) and seismic energy refracting off sub-surface interfaces is recorded on the ground surface using geophones positioned along a line.	• Depth to bedrock or thickness of overburden. • Thickness of layers. • Relative strength or stiffness of soil/rock layers. • Depth to water table.	• Does not work well if strength decreases with depth. • Requires distinct layer discontinuities. • Will not recognise weaker layers underlying stronger.
Cross-hole seismic. Energy sources and geophones are placed in boreholes and/or on ground surface; interval travel times are converted into seismic wave velocity as a function of depth in the borehole.	• Relative strength or stiffness of soil/rock layers. • Depth to water table. • Correlation of lithologic units between boreholes.	• Requires drilling of boreholes. • Significant support field equipment.
Resistivity. DC current is applied to the ground using electrodes. Voltages are measured at different points on the ground surface with other electrodes positioned along a line.	• Depth to water table. • Inorganic groundwater contamination. • Groundwater salinity. • Overburden/soil layer thickness. • Delineation of certain some features (e.g. sinkholes).	• Difficult resolution between groundwater level and soil-rock layers. • Resolution decreases significantly with increasing depth. • Resolution is difficult in highly heterogeneous deposits.
Ground Penetrating Radar (GPR). Electromagnetic energy is pulsed into the ground, which reflects off boundaries between different soil layers and is measured at the ground surface.	• Depth to water table. • Thickness of pavement layers. • Near surface void detection.	• Not effective below the water table or in clay. • Depth of penetration is limited to about 5–10 m.

3. Identification for sub-surface geotechnical hazards such as karstic sinkholes or old mining activity.

It is important that the geophysical methods are applied in the appropriate circumstances and that the limitations of each geophysical method are recognised (Sabatini et al. 2002).

LABORATORY TESTING

Aims

Laboratory testing is concerned with characterising the likely condition and performance of the soil–rock materials and masses along a road alignment or with the suitability of various soils and rocks as construction materials. Within an overall aim of assuring that selected materials and designs are capable of carrying out their function, testing is undertaken for a number of reasons.

- Characterisation of soil–rock masses and materials underlying the alignment.
- Assessment of geotechnical properties influencing earthwork cuts and fills.
- Assessment of road bridge and culvert foundation conditions.
- Assessment of geotechnical properties associated with natural hazards.
- Identification of potential material resources.
- Proving quality and quantity of material reserves or processed materials.
- Construction Quality Assurance.
- In service monitoring.

To be effective, laboratory testing programmes should take into account not only the selection of appropriate tests but also the capacity of the laboratory and staff to undertake the tests and manage the quality of the data produced. It is essential for a project-selected laboratory to translate numerical test output data into believable and useful project information. Particular emphasis should be placed on the selection of appropriate tests and the need for effective Quality Management throughout the whole testing and reporting process. Head in his three volumes (1986, 1992, 1996) provides a comprehensive review of geotechnical laboratory testing.

Testing programmes

Laboratory testing programmes vary greatly in size and scope depending on the type of the road project and associated works. Testing should not be commissioned on an arbitrary or ad hoc basis but should be part of a rationally designed programme. Clear objectives should be identified and test programmes designed with these in mind.

In the majority of cases, no single test procedure will satisfy requirements, and a battery of test procedures will be needed. An appropriate test programme will include a logical selection and sequence of tests that is function of the geotechnical environment, the nature of the investigation and the road design requirements. The relationships between in situ conditions and those experienced by the sampled and tested material need to be taken into account when developing test programmes, particularly within fabric-sensitive Tropically Weathered In Situ Materials (TWIMs).

Test identification and selection

Materials test procedures may be considered as falling into categories that reflect the nature of the test (Cook et al. 2001). These, in general, are:

1. Physical or index tests: associated with defining or indicating inherent physical properties or conditions.
2. Simulation/behaviour tests: associated with portraying some form of geotechnical or engineering character either directly or by implication.
3. Chemical tests: aimed at identifying the occurrence of key chemical compounds.
4. Petrographic tests: assessments associated with analysing or describing fabric or mineralogy.
5. Concrete tests: associated with assessing the properties of concrete mixes and their constituent parts.
6. Bituminous tests: associated with assessing the properties of bitumen to be used pavements surfacing.

Low-cost index tests have an important role to play in LVRR investigations. However, care needs to be taken in tropical environments when applying empirical correlations between index tests and behaviour tests that have been derived for sedimentary soils.

The majority of laboratory tests in developing countries are governed by strict procedures that, in the main, have been originally derived from British (BSI 2022), American (ASSHTO, ASTM), European (Eurocode 7 2004 or French (AFNOR) Standards. In most cases, they have been incorporated into national standards, sometimes, however, with local amendments. Commonly used laboratory procedures in LVRR investigations and construction are listed in Table 7.6.

As-built road quality is dependent on the processes of selection, winning, hauling, spreading and compaction, and attempts need to be made to replicate and to predict their impacts through pre-treatment programmes prior to testing, for example, by subjecting samples due for particle size analysis, to a compaction cycle prior to sieving. Aggregate impact testing (e.g. AIV), abrasion, soaking, drying or slake durability pre-treatments could also be used in appropriate circumstance on samples prior to a main simulation test.

Materials falling within the hard soil to weak rock category are likely to cause particular difficulty with respect to sample disturbance and/or choice of test procedure. The climatic environment may influence the selection of

Table 7.6 Commonly Used Test Procedures in LVRR Investigations

Test	Test Procedure	
Laboratory/site procedure	ASSHTO	ASTM
Soil classification	M145	D3282
Sampling and sample description	M145	D2488
Sample preparation		D421
Moisture content		D2216
Water absorption and SG		C127 & C128
Particle size distribution by sieve	T88	D422/D6913
Particle size distribution <200 mesh	T11	D1140
Atterberg limit test method	T89, 89	D4318
Compaction test (modified)	T180	D1557
Compaction test (standard)	T99	D698
California bearing ratio test method	T193	D1883
Bulk density – aggregate	T19	C29
Soil particle density (SG)	T100	D854
Specific gravity of coarse aggregate	T85	C128
Specific gravity of fine aggregate	T84	C127
Making & curing concrete test specimens – lab		C192
Concrete compressive strength (cylinders)		C39
Organic impurities in fine aggregate		C40
Concrete slump test		C143
Los Angeles abrasion test – fine		C131
Los Angeles abrasion test – coarse		C535
Sodium or magnesium soundness		C88
Flakiness index		D4791
Fine aggregate particle size (washing)		C117
Aggregate sieving fine and coarse		C136
Schmidt re-bound hammer		D5873
Bitumen content		D4
Penetration of bituminous materials		D5
Softening point of bitumen		D36
Density of bituminous materials		D70
Ductility of bituminous materials		D113
Marshall flow		D5581
Preparation of specimens for Marshall apparatus		D6926
Marshall stability and flow		D6927

procedures, or their modification, in terms of moisture condition during test, i.e. soaked, unsoaked, or some intermediate moisture condition.

The laboratory environment

The quality of testing programme depends upon the procedures in place to assure that tests are conducted correctly in an appropriate and controlled environment and utilise suitable equipment that is mechanically sound and calibrated correctly. The condition of test equipment and the competence levels of the laboratory staff are crucial in this regard. Procedures that assure effective Quality Management of the test data are a necessary requirement.

Testing equipment should be as specified in standard or modified standard procedures and must be suitably calibrated. Appropriate manuals and procedure standards must be readily available to staff, and these should be up-to-date originals and not photocopies.

Calibration is the process applied to check the measurements provided by an apparatus or device in order to verify its accuracy. This may be undertaken by checking against either a similar instrument with a traceable calibration, or directly utilising standard test materials or a calibration apparatus (Head 1992).

The climatic environment within a testing laboratory needs to be monitored, and regular measurements should be made. It may be required for some test procedures, particularly within the tropics, to have specific areas set aside for tight climatic control by air-conditioning. Lack of laboratory climatic control can lead, for example, to significant changes in moisture condition to samples both prior to and during test procedures (Head 1994).

Testing engineers and technicians are required to be adequately trained and certified competent in the test procedures for which they are responsible. In larger laboratories, there is likely to be a wide range tests undertaken, and an ongoing record of staff competence and training is recommended (ReCAP 2017).

INVESTIGATION IMPLEMENTATION

Level of investigations

The size and cost of investigations vary depending on the type of road and the complexity of the geotechnical and other environments. The following figures are based on Institution of Civil Engineers (2012) as a percentage of capital cost expenditure:

- Bridges: 0.2%–0.5%.
- Road: 0.2%–2.0%.
- Embankments: 0.1%–0.2%.

Table 7.7 Relative Importance of Investigation Activities during the Project Life-Cycle

Project phase	Investigation activity			
	Desk-study	Walkover mapping	Ground investigation	Laboratory testing
Planning	A	B	C	C
Feasibility study	A	A	B	B
Final engineering design	B	B	A	A
Construction	C	B	B	A
Maintenance	C	B	C	C
Rehabilitation	B	A	B	B

Note: A: primary activity; B: support activity; C: supplementary activity.

Difficult LVRR investigations, for example in mountain roads or roads on soft clays in coastal areas, will be at the higher end of the range, or in some cases even higher. Ideally, sufficient budget should be allocated for a thorough investigation to facilitate cost-effective road design and construction, and to reduce the possibility of unexpected conditions being encountered during construction of the works. These can frequently lead to costly delays amounting to much more than would have funded a properly programmed and conducted investigation. Table 7.7 summarises the relative importance of investigation types in relation to investigation stages.

Table 7.8 summarises some guidance on the extent of typical investigations in anticipated normal ground conditions. Additional or deeper investigation would be required in areas of special ground conditions.

Main investigation information

Within the overall framework of the main LVRR investigations, there are a number of design areas requiring specific information to be covered. Table 7.9 provides a checklist of the main data set requirements.

Climate resilience investigations

The recovery of information on climate threats and the adaptations to counter them should be an integral part of the investigation process and based on the approaches to climate resilience that are detailed in Chapter 17. A checklist for data collection for climate resilience assessment is summarised below (CSIR 2019; PIARC 2015):

- Identify the general climatic environment of the project and its principal characteristics.
- Define the location of all project roads within the environment.
- Identify specific climate threats.

Table 7.8 Some Minimum Requirements for Ground Investigations

Asset	Location and extent
Pavement	Feasibility studies; combination of pits and DCP at 0.5–1 km spacing and change of soil types. Detailed design studies at 200 m spacing with DCP, closer in difficult ground conditions. Sampling for soil index tests and CBR/MDD – number depending on soil, type and what DCP design approach is adopted. Boreholes/auger holes in areas of suspect ground.
Deep cut	At least 1 borehole per 200 m to Depth $(D) = 2$ m (or $D = 0.4$ maximum cut height) below foundation level. Test pits at 200 m. DCP at 50–100 m spacing can be used for interpolation of soil depths.
Embankment	At least 1 borehole per 200 m to $D = 6$ m (or $D = 0.8$–1.2 times maximum embankment height) below foundation level. Test pits at 200 m. DCP at 50–100 m spacing can be used for interpolation of soil depths. Undisturbed sampling for settlement characteristics. Additional boreholes/auger holes in areas of suspect ground.
Low Water Crossing (LWC)	No sub-surface ground investigation normally required unless poor ground identified during walkover.
Culvert	One trial pit at outlet. 1.5 m. Possible DCP in weak ground.
Vented ford	Two pits (only one required if ford is shorter than 15 m). At each end of the vented section, preferably one on the upstream and one on the downstream side $D = 1.5$ m. Possible DCP in weak ground.
Large bore culvert	Pits at each entry and exit location to 1.5m (deeper in poor ground conditions). Possible DCP in weak ground.
Bridge	One borehole (or pit if bedrock close to surface) at each abutment and each pier. To firm strata (minimum of 3 m into foundation strata). Geophysics may be used to interpolate bedrock if required.
Borrow pit	An exploration phase of 2–3 pits through depth of usable material to identify possible sources. A more detailed proving phase with pits at 50 m spacing is commonly recommended. Large bulk sampling will be required for the whole suite of properties required for earthworks or pavement usage and the development of haulage diagrams (see Chapter 8).

- Define the relevant current future climate parameters.
- Assess the overall condition of the road.
- Identify vulnerabilities of the roads or road assets to the climate threats.
- Identify and prioritise climate impact risks.
- Collect data on climate impact reduction options from similar projects or roads.
- Collect and assess cost data on likely options.

Table 7.9 Summary of Information Required for Main LVRR Investigations

Item	Information required
• Alignment	• Are there alternative alignments to consider? • Information on alternatives or minor adjustments to principal alignment. • Potential hazards; landslips, areas of soft ground, swelling clays, very difficult terrain. • Rock and soil type and their boundaries. • Vertical alignment gradients. • Flood areas (climate vulnerability). • Cut-fill balance.
• Pavement	• Traffic including vehicle types. • Risk of axle overload. • Subgrade types and strengths. • Likely water table and variation. • Condition of any existing pavement.
• Earthworks	• Cut and fill depths. • Likely soil-rock profiles in cutting cross sections. • Foundation conditions for embankments. • Embankments on sloping ground. • Potential instability locations. • Minimum embankment heights for flooding (climate vulnerability). • Ground water conditions. • Condition and geometry of existing cuts and fill. • Geotechnical analysis of major earthworks. • Identification of any special investigation requirements. • Location of spoil areas, if required. • Percentages of reusable material from cut excavation.
• Drainage	• Rainfall patterns – current and future (climate resilience input). • Side drainage; types required and dispersal constraints. • Erosion occurrences/risks. • Culvert foundations. • Culvert location and spacing (link to terrain and climate resilience). • Earthwork drainage requirements. • Hydrology analysis. • Bridge conditions, adequate geometry.
• Construction materials	• Definition of requirements – types and quantities. • Assessment of current sources; quality and quantity – for construction and maintenance. • Identify possible new sources; quality and quantity. • Shortfalls of specific material types. • Performance of materials from existing sources. • Location of defined sources – haulage plan. • Requirements for stabilisation. • Define stabilisation/modification types. • Sources of bitumen/cement. • Is bitumen emulsion a valid option.

Structures investigations

The engineering investigation for significant structures may be a distinct operation within the overall road investigations, particularly so for any major bridges. The objective of a structure's site investigation is to provide sufficient relevant information, such as to facilitate the design process detailed in Chapter 12. Bearing capacities are particularly important in this regard where large, localised loads are expected (e.g. bridge abutments and piers), where foundations must be selected and designed to support these loads, as well as being able to resist any watercourse surface erosion. Assessments undertaken at sites of proposed structure locations should be sufficiently detailed to ensure:

- Identification of the best location for any new structure.
- The appropriate type of structure is chosen.
- The appropriate type of foundation is selected.
- The structure is adequate for the purpose (traffic, water flows and size).
- The design should not need to be significantly changed during construction.

Assessments may be required for either a new structure or the upgrading of an existing structure. Table 7.10 summarises data requirements for LVRR small structures' design.

TROPICALLY WEATHERED SOIL–ROCK MATERIALS AND MASSES

The nature of tropically weathered soil–rock profiles

Many Low and Middle Income Countries (LMICs) are located within tropical or sub-tropical regions, and sometimes distinct engineering behaviour of Tropically Weathered In Situ Materials (TWIMs) can be important issue that needs to be addressed in LVRR investigations (Geological Society 1997).

Weathering can be considered as a combination of two processes: physical degradation and chemical decomposition. Tropical weathering is dominated by the decomposition process and may be thought of as a chemical reaction whereby the parent material is seeking to reach equilibrium with the governing environment. Weathering in most developed countries is dominated by physical processes of disintegration, transport and sediment deposition, although chemical decomposition does play a part. Consequently, investigation and testing procedures have largely been developed for a combination of sedimentary soils and physically weathered to fresh rocks rather than for tropical or chemically weathered materials.

Table 7.10 Structures Design Data Requirement

Item	Information required
Materials and equipment	• Availability and quality of local materials (e.g. masonry stone, aggregate, timber, locally manufactured brick and blockwork). • The imported and delivery costs of steel to site, steel-fixing skills available. • The strengths achievable, delivery/import delays, types of concrete and experience, Quality Control and possible testing arrangements. • Specialist equipment is available/would be required for construction?
Traffic	• What is the largest type of vehicle that uses the road? • Is a one lane, alternate traffic flow option feasible? • Traffic density, does it vary e.g. seasonally or on market days in the local area? • Will the vehicle size or loading increase if the road or structure is improved? • Any logging, quarries, mining or other industries in the area – existing or planned?
Ground conditions	• The depth to firm strata or rock. • Type of foundation material available to build on. • Level of the water table. • Compressibility or strength of subsoil? • Water/soil chemistry; is it aggressive to building materials?
Watercourse details	• Is the stream perennial or seasonal? • Type of watercourse? • Stability of watercourse and bed stable, e.g. in rock? • Low water level? • Maximum current flood levels (frequency of occurrence and duration). • Maximum future climate change flood levels (frequency of occurrence and duration). • Watercourse cross sections at potential site? • Gradient of watercourse upstream and downstream of the crossing point? • Evidence of course/bank or level changes, erosion/deposition at, or near, the site. • Is there sometimes floating debris in the water? • What is the water velocity during floods? • What is the longitudinal section or profile along the watercourse? • Is the watercourse used for private or commercial traffic with headroom requirements? • Size and amount of sediment supplied from catchment area.
Catchment details	• Area of catchment. • Shape of catchment. • Gradient of terrain. • Permeability of soil? • Vegetation coverage and type (land use changes?) • Rainfall intensity. • Is the vegetation coverage changing rapidly, e.g. Deforestation?

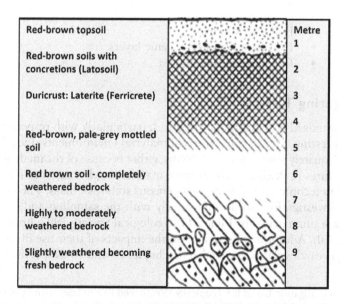

Red-brown topsoil		Metre
		1
Red-brown soils with concretions (Latosoil)		2
Duricrust: Laterite (Ferricrete)		3
		4
Red-brown, pale-grey mottled soil		5
Red brown soil - completely weathered bedrock		6
		7
Highly to moderately weathered bedrock		8
Slightly weathered becoming fresh bedrock		9

Figure 7.5 Typical horizons within a mature tropically weathered soil-rock profile.

Tropical weathering commonly results in the formation of a chemical, and consequently a geotechnical, continuum of gradual transitions. For convenience, this continuum is commonly described in terms of a number of horizons within a vertical profile (Geological Society 1997). Such profiles may contain duricrusts (e.g. laterites), mature soils, immature soils and highly weathered rock (saprolites), in addition to fresh or slightly altered parent material (Figure 7.5).

Behaviour

It is recognised that there are several TWIMs whose behaviour is noticeably different from normal sedimentary soils from which the standard principles of soil mechanics were derived. Their behaviour may not, as in traditional soils, be based on factors such as stress history, grain size, relative density and mineralogy, but more on factors such as relict fabric, mineral bonding and soil suction (Vaughan et al. 1988).

The material and mass characteristics of TWIM profiles may have behaviour patterns that have both detrimental and beneficial effects on their civil engineering performance (Cook and Younger 1991). Non-standard TWIM characteristics may be summarised as follows:

Material:
- Complex/unusual clay mineralogy.
- Leached or voided fabric.
- Weakly bonded fabric.

Mass:
- Relict structure.
- Development of pedogenic layers.
- Gradational boundaries.

Investigating TWIMs

Existing standard investigation procedures, particularly with respect to sampling and testing, are, in some tropical material environments, incapable of coping adequately with TWIM problems, either because of the inadequacy of the procedures in dealing with sensitive materials or because of an inability to represent a complex and non-homogeneous soil–rock mass. The problems raised in investigations to deal effectively with the sampling and testing of TWIMs are summarised in Table 7.11 (Geological Society 1997; Brand 1985; Wesley 2010). Additional discussion on the impacts of their use in construction and as earthworks are included in Chapters 8 and 10, respectively.

Table 7.11 Investigation Issues for Tropically Weathered In Situ Soil-Rock Profiles

TWIMs characteristic	Potential problem	Mitigation
1. Unusual clay mineralogy	Some minerals, such as halloysite, affect moisture content determination because of the irreversible loss of bound water from the crystal structure when the soil is dried at the 'standard' temperature of 105 degrees centigrade. As these conditions are not reproduced in the field, an incorrect interpretation of moisture contents may result. The change from halloysite to meta-halloysite under standard drying temperatures also physically change the apparent grain size from clay to silt/sand. Impacts on some standard correlations between index testing and engineering behaviour.	Use an air-dried procedure for determination of moisture contents and also for the preparation of other tests such sieving, compaction and CBR. Cross-check use of standard correlations.
2. Leached or voided fabric	Differential chemical weathering between different minerals can result in a voided sensitive physical structure that is very difficult to obtain and test undisturbed samples and test; for example the early decomposition of feldspar minerals in a coarse granite may leave the resistant quartz grains in such a sensitive fabric.	Use only the highest quality sampling classes. Figure 7.6 shows high-quality sampling technique in highly weathered granite.

(Continued)

Table 7.11 (Continued) Investigation Issues for Tropically Weathered In Situ Soil-Rock Profiles

TWIMs characteristic	Potential problem	Mitigation
3. Weakly bonded fabric	In parallel with (2) above, the chemical weathering may leave fine grain soil fabric dependent on weak chemical bonds for an apparent initial strength under compression which may rapidly deteriorate beyond a certain point (collapse fabric).	Determine behaviour using an adaption of the oedometer test involved flooding of the sample at constant pressures (Jennings and Knight 1957).
4. Relict mass structure	In situ tropical weathering of rock masses can leave the original geometry of discontinuities in place as relict structure that can have a significant impact on the soil-rock mass behaviour, for example in cut-slope faces.	Field mapping of available relevant exposures and/or an extrapolation of slightly weathered or fresh rock structure in highly weathered masses.
5. Pedogenic layers	The leaching and precipitation processes that form material such laterite or calcrete commonly lead to the occurrence of strong layers overlying weaker soils. Misinterpretation of such layers from borehole or geophysical data can lead to issues with foundations, excavatability and materials quantities being over-estimated.	Mapping of the site in question along with a good desk-studies to indicate the likelihood or not of pedogenic layers. Note that seismic refraction surveys cannot define weaker layers underlying stronger.
6. Gradational-variable boundaries	Tropical weathered in situ weathered of most rock types is a continuum from fresh to completely weathered. The thickness and lateral variations can be extensive depending on a number of factors including rock type, historical climate, and terrain. A full profile may be very compressed; less than 1–2 m for a mudstone for example to 100 m plus, for some granites. Interpretation of sub-surface investigation data is not straightforward.	A good understanding of the geology of the area is essential, combined if possible with mapping of relevant exposures.

(a)   (b)

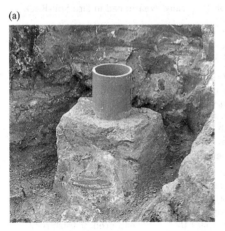 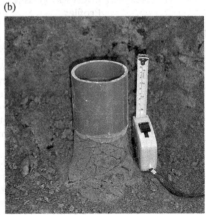

Figure 7.6   High quality sampling in tropically weathered Grade V granite. (a) Initial block cut. (b) Hand trimming into tube.

DATA MANAGEMENT

Managing uncertainty for the RDM

Data management includes the process of acquiring, validating, storing, organising, retrieving and delivering data. Efficient data management is a pre-requisite to establishing an efficient Road Design Model (RDM) to drive sustainable design.

Site investigations do not produce absolute certainty, where the actual volumes of soil-rock that are examined or tested are miniscule in relation to those involved in the project as a whole. The management of investigation information and its Quality Assurance is the key to minimising uncertainty as much as possible and maximising the value of the RDM based on recovered data.

Some important issues with respect to data collection, analysis and the RDM:

- Start data collection as early as possible.
- Start with essential background data and work down to particular detail.
- Don't leaving out data that doesn't fit initial concept.
- Don't work back from a preconceived solution.
- Avoid inclusion of superfluous data.
- Don't include data without a confidence level.
- Convey uncertainty in a manner that is understandable to the client/users.
- Within a model all data do not have uniform uncertainty.
- Assess project vulnerability to data changes.

- The purpose of the investigation must be clearly stated.
- Sources of data must be defined/identified.
- Check the accuracy and quality of data before using, importing or otherwise processing.
- The relationship of data to the site must be clear.
- The reasons for omitting some data must be justified and reasonable.
- Data quality must be satisfactory and defined if possible.
- Clearly identify areas of deep uncertainty (e.g. future climate changes).
- Analysis is appropriate to data reliability and accuracy.
- Uncertainties and related risks must be defined.

Laboratory data management

Test results are a function of controllable influences (test procedures) and non-controllable influences (the nature of the materials), and it is absolutely essential that the latter be clearly identifiable and not confused with the former. This is of particular importance when dealing with materials that may have unpredictable behaviour patterns.

With regard to laboratory testing, management should form part of an overall approach in which information recovered from the desk and field studies is closely integrated with the selection and interpretation of laboratory tests. The observed nature of a material can be of particular use in making decisions on the applicability of particular tests, and the use of descriptive information in laboratory testing procedures is strongly recommended.

The use of standard test procedures and reporting forms is normal laboratory practice, and their use should be strictly adhered to. Quality Assurance procedures should be identified and adhered to. Quality Control issues that cause difficulty for the production of meaningful geotechnical information include:

1. Sample misplacement: Responsibility for samples, from when they were taken in the field until they are finally disposed of, needs to be clearly identified.
2. Repeatability and reproducibility: Tests on duplicate samples should be used as control checks, particularly if several laboratories are being used.
3. 'Black-box' computer programmes: Some of the more sophisticated soils procedures, e.g. triaxial or oedometer tests may have computer control for testing by software that analyses and reports results. The validity of these programmes requires checking, particularly with respect to assumptions that are made regarding sample characteristics.
4. Non-standard procedures: If any non-standard, or modified, procedures are being used, this needs to be clearly stated on any test procedure and data record sheets.

REFERENCES

Anderson, N., N. Croxton, R. Hoover and P. Sirles. 2008. Geophysical methods commonly employed for geotechnical site characterization. TRB Circular number E-C130. http://www.trb.org/Publications/Blurbs/160352.aspx.

Brand, E. W. 1985. Geotechnical engineering in tropical residual soils. In *Proceedings 1st International Conference on Geomechanics in Tropical Lateritic and Saprolitic Soils, Brasilia*, vol. 3.

BSI. 2022. BS 1377-2:2022 Methods of test for soils for civil engineering purposes. Classification tests and determination of geotechnical properties.

Cook, J. R., C. S. Gourley and N. E. Elsworth. 2001. Guidelines on the selection and use of road construction materials in developing countries. TRL research report R6898 for DFID. https://assets.publishing.service.gov.uk/media/57a08d94e5274a27b2001917/R6898.pdf.

Cook, J. R. and J. S. Younger. 1991. Highway construction aspects of the Quaternary engineering geology of West Java. *Geological Society, London, Engineering Geology Special Publications* 7:475–483. https://www.lyellcollection.org/doi/10.1144/gsl.eng.1991.007.01.44.

CSIR. 2019. Climate adaptation: risk management and resilience optimisation for vulnerable road access in Africa. Climate threats and vulnerability Assessment. ReCAP Project GEN2014C for UKAID-DFID. https://www.research4cap.org/index.php/resources/rural-access-library

EUROCODE 7. 2004. *Geotechnical Design. Part 1 General Rules, Part 2 Design Assisted by Laboratory Testing and Part 3 Design Assisted by Field Testing*. European Committee for Standardization. CEN.

FHWA. 2017. Geotechnical engineering circular No. 5: Geotechnical site characterization. Publication no. FHWA NHI-16-072. U.S. Department of Transportation. https://www.fhwa.dot.gov/engineering/geotech/pubs/nhi16072.pdf.

Geological Society. 1972. The preparation of maps and plans in terms of engineering geology. *Quarterly Journal of Engineering Geology* 5:293–381. London, UK. https://www.lyellcollection.org/doi/10.1144/gsl.qjeg.1972.005.04.01.

Geological Society. 1997. *Tropical Residual Soils*. Fookes PG, Ed. A Geological Society Engineering Group. Working party revised report. Geological Society Professional Handbooks. UK: Geological Society.

Head, K. H. 1986. *Manual of Soil Laboratory Testing. Vol 3. Effective Stress Tests*. Pentech Press.

Head, K. H. 1992. *Manual of Soil Laboratory Testing. Vol 1 Soil Classification and Compaction Tests*. 2nd edition. Pentech Press.

Head, K. H. 1996. *Manual of Soil Laboratory Testing. Vol 2 Permeability, Shear Strength and Compressibility Tests*. 2nd edition. John Wiley & Sons.

Institution of Civil Engineers. 2012. Chapter 46 Ground exploration. In *Manual of Geotechnical Engineering Vol 1: Geotechnical Engineering Principles*. https://www.icevirtuallibrary.com/doi/book/10.1680/moge.57074.

Jennings, J.E.B. and K. Knight, 1957. The prediction of total heave from the double oedometer test. *Trans. South African Institution of Civil Engineers*, V(7): 285–291

Norbury, D. 2010. *Soil and Rock Description in Engineering Practice*, 288 pp. Whittles Publishing. ISBN 978-1904445-65-4.

PIARC. 2015. International climate change adaptation framework for road infrastructure report 2015RO3EN, Paris. https://www.piarc.org/en/order-library/23517-en-International%20climate%20change%20adaptation%20framework%20for%20road%20infrastructure.htm.

ReCAP. 2017. Capacity building and skills development programme for the laboratories of the Local Government Infrastructure and Transportation Research Centre (LoGITReC) in Tanzania. Research for community access partnership, Ref AfCAP/TAN/2095A for UKAID-DFID. https://www.gov.uk/research-for-development-outputs/capacity-building-and-skills-development-programme-for-the-laboratories-of-the-local-government-infrastructure-and-transportation-research-centre-logitrec-in-tanzania-final-report.

Roughton International. 2000. Guidelines on materials and borrow pit management for low cost roads. DFID KaR Report R6852. https://www.gov.uk/research-for-development-outputs/guidelines-on-materials-and-borrow-pit-management-for-low-cost-roads.

Sabatini, P. J., R. C. Bachus, P. W. Mayne, J. A. Schneider and T. E. Zettler. 2002. Geotechnical engineering circular no. 5. Evaluation of soil and rock properties. Report N. FHWA-IF-02-034. U.S. Department of Transportation. Federal Highway Administration.

SANRAL. 2013. Chapter 7 Geotechnical investigations and design considerations. In *South African Pavement Engineering Manual*. South African National Roads Agency Ltd.

TRL. 1993. Overseas Road Note 31. *A Guide to Structural Design of Bitumen Surfaced Roads in Tropical and Sub-tropical countries*. 4th edition. UK: TRL Ltd for DFID. https://www.gov.uk/research-for-development-outputs/orn31-a-guide-to-the-structural-design-of-bitumen-surfaced-roads-in-tropical-and-sub-tropical-countries.

TRL. 2003. Overseas Road Note 20. *Management of Rural Road Networks*. TRL for DFID. https://www.gov.uk/research-for-development-outputs/management-of-rural-road-networks-overseas-road-note-20.

TRL. 2004. Overseas Road Note 40. *A Guide to Axle Load Surveys and Traffic Counts for Determining Traffic Loading on Pavements*. Transport Research Laboratory Ltd. https://www.gov.uk/research-for-development-outputs/orn40-a-guide-to-axle-load-surveys-and-traffic-counts-for-determining-traffic-loading-on-pavements.

TRL. 2006. *UK DCP 3.1 User Manual. Measuring Road Pavement Strength and Designing Low Volume Sealed Roads Using the Dynamic Cone Penetrometer*. UK: TRL Ltd, for DFID. https://www.gov.uk/research-for-development-outputs/uk-dynamic-cone-penetrometer-dcp-software-version-3-1.

Vaughan, P. R., M. Maccarini and G. Mokhtar. 1988. Indexing the engineering properties of residual soil. *Quarterly Journal of Engineering Geology* 21:69–84.

Wesley, L. D. 2010. *Geotechnical Engineering in Residual Soils*. New York: John Wiley & Sons Ltd.

Chapter 8

Natural construction materials

INTRODUCTION

Construction materials are an important element in the Low Volume Rural Road LVRR environment, and their identification and characterisation are vital factors in the development of the cost-effective design of sustainable rural roads. A key objective in sustainable rural road design and construction is to best match the available construction material to the road task and the local environment. When reserves are limited, their cost-effective usage is a priority. Hence the necessity of applying locally relevant specifications and either adapting designs or improving materials to suit (Cook et al. 2001). The frequently distinct engineering behaviour of naturally occurring construction materials within sub-tropical and tropical regions has been identified as a significant factor in determining the long-term engineering success or failure of road projects in developing countries.

This chapter gives guidance on the identification, selection, and management of appropriate road construction materials with respect to key characteristics that govern their performance. Examples are given of typical material requirements and specifications. These are for general guidance and need to be assessed and adapted for the specific Road Environments. The comments and recommendations in this chapter are equally relevant to materials selected for maintenance as well as construction.

THE GEOLOGICAL CONTEXT

Geological background

The geological background of natural road building materials used in road construction has a profound influence on the engineering performance of these materials. It is, therefore, important to have a knowledge of the geological character of rocks and soils, particularly when considering marginal or problematic materials. There are three principal rock groups (Keary 2005):

1. Igneous rocks: formed from the solidification of molten magma and volatiles (ash) originating from within the earth's crust or the underlying mantle.
2. Sedimentary rocks: formed from the consolidation, compaction and induration of the eroded and weathered products of existing rocks.
3. Metamorphic rocks: formed by the influences of heat and/or pressure on pre-existing rocks.

Within this geological framework, general soil and rock behaviour can be considered a function of:

- Mineralogy of the constituent particles.
- Morphology (shape) of the constituent particles (texture).
- Physical relationship of the constituent particles to one another (fabric).
- Nature of any relevant discontinuities.

Typical examples of how the geological framework at a range of scales can impact performance as a construction material are illustrated in Figure 8.1 as follows:

1. Figure 8.1a: A very widely to widely jointed fresh, strong granite mass will provide a wide range of processed rock from fine aggregate to very coarse riprap.
2. Figure 8.1b: A moderately to closely joint unweathered, fresh, strong dolerite mass would not be so suitable for large riprap but could provide sound processed aggregate.
3. Figure 8.1c: A sequence of laminated to thinly bedded mudstone, siltstone, sandstone has a possible use only as a poor rock fill material.
4. Figure 8.1d: Fresh dolerite material scale sample with a tight isotropic texture and fabric indicates a potential to produce a well-shaped aggregate.
5. Figure 8.1e: Magnified thin section (under crossed polar light) of a quartzite; shows a tight crystalline fabric likely to produce a strong aggregate.
6. Figure 8.1f: Magnified thin section (under crossed polar light) of quartz–mica schist, with laminar fabric likely to produce very poorly shaped platy aggregate, becoming fissile and degradable if weathered.

Tables 8.1 and 8.2 summarise principal rock and soil types, respectively, and their potential use as road construction materials.

Figure 8.1 Scale of rock structure fabric and texture. (a)Very widely jointed granite, mass scale. (b) Close to moderately jointed dolerite, mass scale. (c) Thinly bedded to laminated shale and sandstone, material scale. (d) Tight crystalline fabric; dolerite, material scale. (e) Tight crystalline fabric, quartzite in thin section, microscopic scale.(f) Laminated quartz mica schist, thin section, microscopic scale.

Table 8.1 Common Rock Types and Their Potential Uses

Rock type	General description	General material uses
Granite	Medium-coarse-grained, light-coloured igneous rock. Susceptible to deep and variable weathering in tropical zones.	Strong durable rock used for block stone, concrete and surfacing aggregate, good shape. Poor bitumen adhesion when rock contains large feldspar crystals.
Dolerite	Medium-grained tightly crystalline dark basic minor intrusive rock.	Very strong durable rock used for block stone, concrete and surfacing aggregate, good shape and good bitumen adhesion. Potential problems with in-service deterioration if weathered.
Andesite	Fine-grained intermediate lava composed essentially of plagioclase feldspar and mafic minerals (hornblende, biotite, augite).	Strong durable rock used for block stone, concrete and surfacing aggregate. Possible poor aggregate shape.
Basalt	Fine-grained dark basic lava may contain deleterious inclusions (amygdaloidal basalt).	Strong durable rock used for block stone, concrete and surfacing aggregate. Possible anisotropic character and poor aggregate shape. Aggregates can be susceptible to disintegration problems in service, depending on occurrence of clay minerals and weathering.
Sandstone	Fine to coarse-grained detrital sedimentary rock with clasts may be composed of quartz, feldspar or rock particles, fabric may be cemented by silica, iron oxides or carbonates.	Variable strength and durability a function of fabric and matrix. May be interbedded with weaker materials. Strong quartz sandstone has possible use as an aggregate and some types of rock fill.
Siltstone	Similar to sandstone but with predominantly silt-sized particles. Tends to be interbedded with other sedimentary materials, including sandstone and mudstone.	Variable but generally poor strength and durability. A function of fabric and matrix. May be interbedded with weaker materials. Unlikely to be of use as an aggregate through poor shape and strength. Poor quality rock fill.
Limestone and dolomite	Consist essentially of crystalline calcium carbonate. If magnesium carbonate, then the term dolomite is appropriate.	Can be very strong to strong and durable. Used as block stone and aggregate. May contain minor amounts of non-carbonate detritus. Tendency to polish as a road surface aggregate.

(*Continued*)

Table 8.1 (Continued) Common Rock Types and Their Potential Uses

Rock type	General description	General material uses
Schist	A medium- to high-grade metamorphic rock characterised by the parallel alignment of moderately coarse grains usually visible to the naked eye. The preferred orientation described as schistosity.	Strong to very strong but anisotropic. Moderate durability poor particle shape. Possibility of free mica being produced during processing. Possible use as rock fill.
Gneiss	Medium to coarse mineral grains with a variably developed layered or banded structure, minerals tended to be segregated, e.g. quartz/feldspar/mafic mineral banding.	Strong to very strong but tends to be anisotropic. Moderate durability poor particle shape. Use as rock fill. Possible use as marginal aggregate if well processed. Potential shape problems.
Quartzite	A metamorphic rock formed from a quartz rich sandstone or siltstone. Contains more than 80% quartz.	Strong to very strong. Good durability, possible use as aggregate. Abrasive to construction plant.
Mudstone	Very fine grain sedimentary rock comprising predominantly clay-size particles.	Usually very low to low strength. Unsuitable as an aggregate. Potential for slaking and swell/shrink in wet climates. Possible use as embankment fills and possible selected fill/capping layer material.
Shale	A low-grade metamorphic rock derived from alteration of mudstone/siltstone in which cleavage planes are pervasively developed throughout the rock.	Variable weak to strong anisotropic strength. Unlikely to be used as an aggregate. Poor durability and shape. Marked tendency to split along cleavage planes (fissile). Requires care in compaction for embankment fill to prevent a voided rockfill. Potential for slaking and swell/shrink in wet climates. Possible use as sub-base material in dry climates.
Weak limestones	Poorly indurated or compacted limestone. Consists essentially of crystalline calcium carbonate as limestone.	Low particle strength and possible in-service deterioration. Use as embankment fill and selected fill/ capping layer material. Possible selected use as sub-base or base material for low volume roads.
Weak sandstones	Poorly indurated fine- to coarse-grained detrital sedimentary rock with clasts may be composed of feldspar or weak rock particles, likely to be weakly cemented.	Low particle strength and potential for in-service deterioration. Use as embankment fill and selected fill/ capping layer material. Possible use as sub-base or base material for LVRRs in dry climates.

Table 8.2 Principal Soil Types in Road Construction

Rock type	General description	Potential use
Soil from highly weathered rocks (saprolite)	Soil-like material within the weathering profile that has retained original relict structure fabric of the parent rock.	Generally used for common fill. Problems resulting from over-compaction and breakdown of material fabric. High mica content in some weathered rocks.
Residual soil	True residual soil has a new-formed fabric to replace the original rock fabric.	Used for common fill. Generally, less problems than with saprolite soil, although those derived from some volcanic rocks can have distinct behaviour patterns derived from distinct mineralogy.
Residual gravel	Concentrations of weathering resistant quartz within residual soil profiles.	Usability as-dug is a function of the ratio of fines to gravel. Commonly used as sub-base or gravel wearing course (GWC) and, if processed or stabilised, as base material.
Transported fine soils (clays-silt)	Colluvium, alluvium, with character a function of parent materials and transportation–deposition environment.	Commonly used as common fill and lower pavement materials depending on their grading, plasticity.
Transported coarse soils (sand-gravel-cobble)	Colluvium, alluvium, coastal deposits with character a function of parent materials and transportation–deposition environment.	Can be used as pavement materials depending on their grading, particle strength, plasticity and shape. After processing may be used as concrete aggregates.
Laterite	Formed within some residual soil profiles by the movement and the re-precipitation of iron-rich groundwater into sand-gravel size nodules or an indurated duricrust cap; also known more correctly as ferricrete.	'As dug' materials highly variable in strength, grading and durability. Commonly used as sub-base, base and GWC. Can be modified/stabilised with lime or cement. Higher plasticity materials will be subject to significant loss of strength on saturation (CIRIA 1988).
Calcrete	Formed within some residual soil profiles by the movement and the re-precipitation of calcium rich groundwater into sand-gravel size nodules or an indurated duricrust cap.	'As dug' materials highly variable in strength, grading and durability. Commonly used as sub-base, base and GWC. Can be modified/stabilised with lime or cement. Higher plasticity materials will be subject to significant loss of strength on saturation.

MATERIAL PROPERTIES

Material requirements

It is important to use materials appropriate to their role in the road, that is, to ensure that they are neither sub-standard nor wastefully above the standards demanded by their engineering task, and that appropriate construction material sources are utilised to maximum long-term advantage (Roughton International 2000; Cook et al. 2002). The requirement tasks for various types of natural road construction materials are generally summarised in Table 8.3, and some typical uses are illustrated in Figure 8.2;

a. Granular base, sub-base and imported subgrade materials.
b. Laterite gravel for use as a gravel surfacing.
c. 10–12 mm sealing aggregate.
d. Stone for gabion basket fill.
e. Stone aggregate in bituminous pre-mix.
f. Stone riprap preventing road embankment toe erosion.

Common fill

The location and selection of fill material for low volume roads does not generally pose significant problems, provided the following are avoided:

• Organic and peaty soils.
• Clays with high liquid limit and plasticity (swelling clays).

Table 8.3 Key Material Characteristics

Road use	Key specification criteria
Common fill	Plasticity, swell potential, compacted strength, max particle size.
Imported subgrade	Plasticity, swell potential, compacted strength, max particle size.
Sub-base/base	Plasticity, swell potential, compacted strength, grading.
Gravel wearing course (GWC)	Plasticity, swell potential, durability, compacted strength, grading.
Pavement surfacing aggregate	Fines, grading, shape, particle strength and durability, bitumen adhesion, polishing, abrasion.
Stone surfacing	Strength, durability, shape.
Filter material	Fines, grading.
Concrete aggregate	Fine: fines, grading, mineralogy, deleterious inclusions. Coarse: strength, shape, durability, deleterious inclusions, silica reaction (mineralogy).
Gabion fill	Strength, durability, size, shape.
Rock fill	Strength, durability, size, shape.
Riprap	Strength, durability, size, shape.
Stone masonry	Strength, durability, size, shape.

Figure 8.2   Some typical requirements for construction materials. (a) Granular base, sub-base and imported subgrade materials. (b) Laterite gravel for use as a gravel surfacing. (c) 10-12mm sealing aggregate. (d) Stone for gabion basket fill. (e) Stone aggregate in bituminous pre-mix. (f) Stone riprap preventing road embankment erosion Insert descriptions (JRC Notes).

- Fabric sensitive soils; some tropically weathered in situ soils (e.g. volcanic ash derived soils).
- Highly micaceous soil, e.g. soils derived from weathered mica schists or gneisses (Fookes and Marsh 1985).

Where possible, fill should be taken from within the road alignment (balanced cut-fill operations) or by excavation of the side drains (except in areas of expansive soils). Opening new borrow pits for fills should be limited as much as possible on environmental impact grounds and then appropriately reinstated after use (Gourley and Greening 1998; Roughton 2000).

Typical target specification:

CBR: 3%–8% @ 93%–95% AASHTO T180 compaction

Swell in CBR mould: <1.5% @ 93%–95% AASHTO T180 compaction

Particle size: <2/3 of layer thickness

Imported subgrade or improved fill

The main aim of the subgrade is to provide a uniform adequately strong foundation on which to place the main structural layers of the road pavement. Where in situ and alignment soils are weak, or where fill material is below required quality, import of improved subgrade as a capping layer may be necessary. The use of a capping layer can also provide cost economies by allowing some reduction overlying pavement thickness design. Typical target specification:

CBR: 8%–15% @ 93%–95% AASHTO T180 compaction

Swell in CBR Mould: <1.5% @ 93%–95% AASHTO T180 compaction

Particle size: <2/3 of layer thickness

Gravel wearing course

Ideally, an unsealed gravel wearing course (GWC) should be durable and of consistent quality to ensure it wears evenly. The desirable characteristics of such a material are:

- Good skid resistance.
- Smooth riding characteristics.
- Cohesive properties (suitable amount of plasticity).
- Resistance to ravelling and scouring.
- Wet and dry stability.
- Low permeability.
- Load spreading ability.

For ease of construction and maintenance, a wearing course material should also be easy to grade and compact. There is a requirement for there

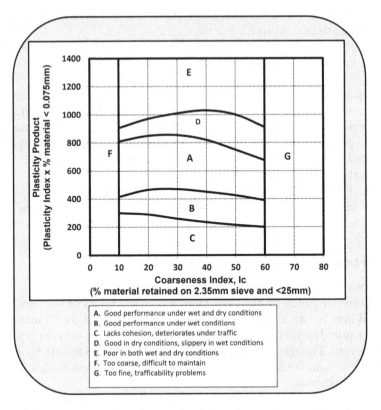

Figure 8.3   Index properties and gravel surfacing performance.

to be sufficient plasticity to hold the GWC material cohesively together following compaction. Figure 8.3 illustrates empirical relationships between index properties and GWC performance.

Typical target specification:

Liquid Limit (WL): <40%

Plasticity (Ip): >5 to <15

CBR: 25%–30% @ 98% ASSHTO T-180 compaction

Swell in CBR mould: <1% @ 98% ASSHTO T-180 compaction

Particle Size: Grading Envelope (Table 8.4 shows typical grading)

Because of the frequent limited extent of natural gravel deposits, it is important to assess potential source quantities for both the initial construction and essential ongoing periodic maintenance re-gravelling requirements.

Sub-base and base aggregate

A wide range of materials including crushed rock, lateritic, calcrete and quartzitic gravels, river gravels and other transported and residual granular

Table 8.4 Typical Grading Envelopes for Granular Base, Sub-base and Gravel Wearing Course

Sieve mm	Base % passing (by weight)	Sub-base % passing (by weight)	GWC % passing (by weight)
50	100	100	100
28	75–95	75–100	
20	60–90	60–100	75–100
14			65–90
10	40–75		55–80
4.75	25–55	30–100	40–60
2.00	20–45	17–75	25–45
0.425	10–25	10–50	15–39
0.075	5–15	5–25	7–20

materials, resulting from weathering of rocks, have been used successfully as base and sub-base material. Sub-base and base materials are expected to meet requirements related to maximum particle size, grading, plasticity and CBR.

Typical target specifications for sub-base:

Plasticity (Ip): <12%

CBR: 25%–30% @ 95% ASSHTO T-180 compaction

Swell in CBR mould: <1%

Grading: See Table 8.4

Maximum size: <60 mm or <50% layer thickness

Typical target specifications for base:

Los Angeles Abrasion (LAA): <30%

Plasticity (Ip): <12%

Soaked CBR: 55%–80@ 98% ASSHTO T-180 (a function of traffic/climate)

Swell in CBR mould: <0.2%

Particle size: Grading Envelope, Table 8.4

Bituminous surfacing aggregate

The requirements for aggregate to be used in a bituminous surfacing layer are that they must be durable, strong and should also show good adhesion with bituminous binders. It should also be resistant to the polishing and abrasion action of traffic. Adhesion failure implies a breakdown of the bonding forces between a stone aggregate and its coating of bituminous binder, leading to physical separation. Mechanical failure by fretting and subsequent ravelling of the surface is a consequence of adhesion failure. Poor adhesion can be due to excessive fines or mismatch between the chemistry of the aggregate and the binder.

Table 8.5 Common Grading for Bitumen Seal Aggregates

	Chippings nominal size Percentage passing	
Sieve (mm)	14–10mm	10–6mm
20	100	
14	85–100	100
10	0–30	85–100
6.3	0–7	0–30
2	0–2	0–2

Typical target specification:
Particle size: Commonly, 14–10 mm and 10–6 mm; see Table 8.5
LAA: <25%
Flakiness Index: <25%

Surfacing block or paving stone

Surfacing block or paving stones should be a strong, homogenous, isotropic rock, free from significant discontinuities such as cavities, joints, faults and bedding planes. Rocks such as fresh granite, basalt, andesite and crystalline limestone have proven to be suitable materials. Quartzite rock is generally not suitable, nor is any rock that polishes or develops a slippery surface or erodes under traffic.

Typical target specifications:
Uniaxial compressive strength >75 MPa;
LAA: <25%;
Sodium Sulphate Soundness: <10% loss.

Aggregates for structural concrete

Fine concrete aggregate is normally naturally occurring sand, although finely crushed stone can also be used, with particles up to about 2 mm in size. Coarse concrete aggregate is normally stone with a range of sizes from about 5 to 25 mm (sometimes larger); it may be naturally occurring processed gravel or more commonly crushed or hand-broken quarry stone (see Table 8.6). Uncrushed rounded natural gravel can compromise the aggregate interlock required for strong concrete.

Aggregates must be entirely free from soil or organic materials as well as fine particles such as silt and clay, otherwise the resulting concrete will be of poor quality. Some aggregates, particularly those from salt-rich environments, may need to be washed to make them suitable for use.

Aggregates can be crushed and screened by hand or by machine. Both reinforced and mass concrete has a range of potential uses within rural

Table 8.6 Typical Gradings for Concrete Aggregate

ASSHTO sieve (mm)	Coarse aggregate				Fine aggregate
	Nominal grading: Size down (mm), % passing by weight				Grading % passing by weight
	37.5 mm	*25.0 mm*	*19.0 mm*	*12.5 mm*	
50					
37.5					
25	100				
19	95–100	100			
12.5		95–100	100		
9.5	35–70		90–100	100	
4.75		25–60		90–100	
2.36	0–5		20–55	40–70	100
1.18		0–10	0–10	0–15	95–100
0.600		0–5	0–5	0–5	80–100
0.300					50–85
0.150					25–60
0.075					10–30

road projects, including pavement slabs, bridges, culverts, retaining walls and drainage linings. They can also be used in paving blocks, slabs or geocell concrete for surface paving (Chapter 15). Each of these may have specific aggregate specifications.

Filter/drainage material

Filter materials have crucial roles in assisting in controlling the ingress and flow of water and in the reduction of pore water pressures within earthworks, retaining structures and the pavement. Filter materials can account for a significant proportion of the construction material costs, particularly in wetter regions where road designs need to cater for the dispersion of large volumes of water, both as external drains and as internal layers within wet-fill embankments. No-fines concrete can be used as a porous medium, for example, as a backing to retaining walls and abutments.

Filter materials vary in grain size depending on design tasks, but typical requirements are:

1. Single (or near-single) sized grading, ideally equidimensional particles, to ensure required permeability.
2. Particles resistant to loads imposed by the road design.

3. Particles need to be resistant to breakdown due to wetting and drying and weathering during design life.
4. As-placed material must be resistant to internal and external erosion.
5. Particles should generally be inert and resistant to alteration by groundwater.
6. Surface coatings such as clay, iron oxide, calcium carbonate, gypsum are undesirable.

Gabion fill

Rock infill for gabion baskets or mattresses is generally defined as comprising fragments ranging between 100 and 250 mm in size with a variation of 5% oversize and/or 5% undersize rock. The finer material may be used, provided it is not placed at the exposed surface. Rocks should be strong, angular to round, and durable so that they will not disintegrate under wetting–drying cycles during the life of the structure. Minimum stone size must be compatible with the wire or steel mesh size so that stone losses do not occur.

Embankment rock fill

Strong angular rock is preferably cubic and not elongated or platy in shape. It is recommended that, in general, fragments should not exceed a maximum dimension of two-thirds of the placed layer thickness. Fragments should be intact and preferably homogeneous without planes of weakness (bedding or schistosity). Fragments should be durable so that they will not disintegrate under wetting–drying cycles during the life of the structure (U.S. Dept of Interior 1998).

Riprap

Riprap for use as slope protection should comprise strong, durable rock fragments. They should be angular and not subject to breaking down when exposed to water or weathering. It is recommended that the specific gravity should be at least 2.5. The sizes of stones used for riprap protection are determined by purpose and specific site conditions (Smith 1999; U.S. Dept of Interior 1998).

Masonry stone

Masonry stonework should comprise strong angular rock and be durable so that it will not disintegrate under wetting–drying cycles during the life of the structure.

MATERIALS MANAGEMENT

Extraction and processing of materials

The efficient implementation of extraction and processing techniques is an essential pre-requisite to maximising the benefit from efficient use of natural non-renewable resources. Removing naturally occurring material from the ground and transforming it into a satisfactory product useful for road construction involves processes of extraction and, possibly, of modification through processing operations (Figure 8.4). The extraction and processing of road construction materials can be large components of cost in the overall project budget.

There are five main types of material extractive operation:

- Quarrying: extraction of drilled and blasted material, e.g. hard rock.
- Borrow pitting: extraction of unconsolidated material, e.g. gravels and weak rocks.

Figure 8.4   Typical rural road material resources. (a) Well-established rock quarry being worked in benches. (b) Unsafe locally developed rock source. (c) Borrow pit for common fill. (d) Alluvial source of cobble and boulders for crushing and sorting Insert descriptions - JRC notes.

- Cut to fill operations along a road alignment.
- Mining: underground material extraction, either by shaft or by adit.
- Dredging: extraction of unconsolidated material from under water

Processing of the as-excavated resource is undertaken to produce specification-compliant construction materials by means of mechanical alteration (crushing), physical selection (sorting) or blending. In general terms, materials utilised for common fill would normally require no processing in contrast to high-quality hard-rock aggregates which could be subjected to several phases of crushing and sorting. The amount of processing required is a function of the relationship between the as-extracted character and the required mechanical, chemical and physical properties. Processing is used to:

- Improve the overall grading characteristics of the material.
- Remove or break down oversize particles.
- Vary the fines content.
- Improve particle shape.
- Break down and remove weak particles.

Processing plants can be fixed or mobile. Fixed plant is more common in large, established quarries, while semi-mobile plant is more appropriate for major construction projects where the life of the quarry is directly related to the duration of the project. Small portable or tractor attached stone crushing equipment is also available and is suitable for small scale and maintenance operations. Table 8.7 describes typical crusher options, Figure 8.5 illustrates examples of typical processing plant, and Figure 8.6 presents a flow chart for a typical small processing operation.

USE OF LOCALLY AVAILABLE MATERIALS

Marginal materials

PIARC (WRA) has defined non-standard and non-traditional marginal materials as follows:

> ...any material not wholly in accordance with the specification in use in a country or region for normal road materials but which can be used successfully either in special conditions, made possible because of climatic characteristics or recent progress in road techniques or after having been subject to a particular treatment.

(PIARC-WRA 1989).

Table 8.7 Rock Crusher Types (McNally 1998; Smith and Collis 2020)

Crusher type and use	Description	Use and limitations
Jaw crusher. Primary crushers, small versions may be used as secondary crushers.	Rock is broken by compression-release cycles between plates, one fixed and one moving on opposite sides of a wedge-shaped chamber. This narrows downwards so that after blocks are split on the compression stroke the resultant pieces slip further down to be split on further strokes until released through the base.	Double toggle machines are able to exert larger compressive forces and able to handle most blocky materials (up to 3 m). In most quarrying operations the single toggle machines operate satisfactorily – well suited to small- and medium-sized operations, including mobile plant.
Gyratory crusher; usually as primary crushers; small versions may be used as secondary crushers.	A gyratory crusher essentially two cones one inverted within the other. The inner solid cone moves eccentrically and around the fixed outer bowl resembling a pestle in a mortar.	Less tolerant of oversize than jaws. Work best when they are choke fed. Outputs more fines than jaw, particle size range is narrower than from impactor but wider than a rolls crusher.
Rolls crusher; primary crusher.	Machines can be fitted with single, double or multiple rollers, although double rolls are most common; one fixed and the other spring-loaded.	Use of the rolls crusher is limited to weaker rock (UCS < 100 Mpa) and non-abrasive rock such as limestone and shale. Cheap in relation to capacity and easily transportable. Well suited to processing weak or fissile rock, which might block jaw or gyratory crushers.
Cone crusher; usually used as secondary or tertiary crushers.	Similar in operation to small gyratory crushers in having an oscillating inner and a static outer cone.	They have large capacities in relation to small size. Compared to impact crushers they produce a narrower range of sizes, less fines and more flaky particles.
Impact crusher; usually used as secondary or tertiary crushers.	Rocks broken by the action of rapidly rotating or beaters attached to a central shaft that may be horizontally or vertically mounted. The feed particles cascade into the crushing chamber and shatter on impact with the beaters or are deflected by them to strike hardened breaker plates lining the chamber.	Relatively light and cheap for their capacity and do not require elaborate foundations. Their main disadvantage is the cost of wear on the breakers and plates and the consequent downtime for replacement. Generally limited to rocks with UCS < 150 MPa and free quartz content <5%–7%. Good option for improving aggregate shape.

Figure 8.5   Materials processing plant. (a) Large established rock crushing and sorting operation. (b) Primary jaw crusher. (c) Small mobile crusher and screen. (d) Semi-mobile single crusher and vibrating screens for processing indurated laterite.

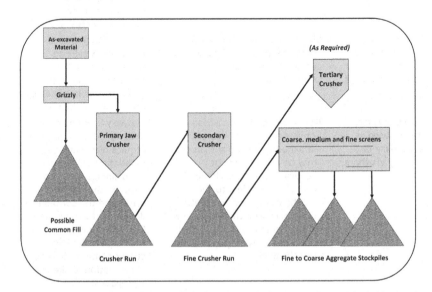

Figure 8.6   Aggregate processing summary flow chart for a rock quarry.

In LVRR design and construction, consideration of marginal materials is largely concerned with naturally occurring granular materials that do not comply with accepted specifications, but which can perform adequately in service within identifiable limits. The use of locally available non-standard natural materials is an important aspect of the appropriate Fit-for-Purpose road design concept.

Appropriate use

In many cases, the problem is not so much as how to identify the best material for a certain usage, but to decide what is the boundary between acceptable and non-acceptable, and how best to use what is acceptable. The boundaries of acceptability are not fixed but shift in response to aspects of the Road Environment, road design and technological innovation. Current international specifications tend to exclude the use of many naturally occurring, unprocessed materials (natural soils, gravel–soil mixtures and gravels) in pavement layers in favour of more expensive crushed rock, because they often do not comply with established international requirements. However, empirical research work has shown quite clearly that so-called 'non-standard' or marginal materials can often be used successfully and cost-effectively in LVRR pavements provided appropriate precautions are observed (Cook et al. 2001; ReCAP 2020). For high-volume roads, the specifications aim to minimise the risk of any failure by insisting on achievable high materials requirements, with consequent high construction costs. For LVRRs, the traffic impacts far less, and consequential economic, transport and engineering risks are lower, and this can justify the adoption of lower cost non-standard, but appropriate, solutions.

The adoption of a local-solution approach provides the scope to consider a pragmatic reduction in specification standards when considering particular material types within defined environments. Recognising the material's 'fitness for purpose' is central to assessing the appropriate use of non-standard materials. It is necessary, however, to engender confidence in the use of materials that would normally have been classified unacceptable or, at best, borderline, and this requires a sound knowledge of their properties and behaviour within the prevailing Road Environment.

If the project is in an area where acceptable construction materials are scarce or unavailable, consideration should be given to:

- Modifying the design requirements.
- Modifying the material properties (mechanical or chemical stabilisation).
- Increased material processing (crushing, screening, blending).

Assessing marginal materials

Cook and Gourley (2002) laid out a framework for the assessment of marginal materials (Table 8.8). At the start of the assessment approach, there may be a requirement to justify the investigation of the non-standard material and its use. This initial justification can involve economic, technical and social issues. This is then followed by environmental and technical assessments. The following are some basic assessment criteria:

Table 8.8 A Marginal Material Assessment Sequence (Cook and Gourley 2002)

	Module	*Purpose*
A	Requirement for use of non-standard material.	Identify clear technical, economic, or environmental reasons for considering the use of non-standard materials. Leading to justifiable reasons for considering further the use of the non-standard material.
B	Definition of non-standard properties.	Identification and quantification of the engineering properties causing a material to fail compliance criteria and to be defined as non-standard. *Leading* to a definition of the non-standard characteristics of the material.
C	Evaluation of existing information.	Evaluation of all technical, economic, and environmental data relevant to the possible use of the material in question as pavement aggregate. *Leading to* identification of the options for continuing the investigation into use or non-use of the material.
D	Evaluation of options.	Evaluation of the option, or options, identified on the basis of existing information. Leading to a decision to pursue one of the following: • An evaluation of engineering uncertainty. • A re-evaluation with a modified environment. • A recommendation for non-use.
E	Evaluation of engineering uncertainty.	Define the uncertainties associated with the use of the non-standard material and identify any consequent engineering and socio-economic risk. Leading to a decision on whether the engineering uncertainties are an acceptable project risk or not, namely: • Acceptable risk provided materials that comply with a modified design or construction specification. • Do not approve for immediate use but recommend that full scale trials be undertaken may be approved for downgraded use in the interim. • Risk unlikely to be acceptable in any event.
F	Approval or non-approval for use.	Decide, on the basis of the information now available, whether to use or not use the non-standard material within the defined Road Environment. Leading to a recommendation for use or non-use.

Economic

- Available standard sources involve long hauls.
- Higher quality materials may be required for other, higher standard, developments.
- Standard sources carry development, cost and time implications.

Environmental

- There are environmental impact issues associated with standard material sources (health, safety, pollution, erosion, natural beauty).
- Pollution and carbon deficits from longer aggregate haulage distances.

Technical

- There is already some evidence that the non-standard material may perform adequately.
- There are no better alternative sources of sufficient quantity (Table 8.8).

MATERIAL IMPROVEMENT

General

Natural materials can be modified to make them suitable for use in road pavements. The common practice is to refer to materials whose properties have been artificially improved as having been 'stabilised' rather than 'modified'. Some guidance documents refer to stabilisation where the character, strength and durability have been significantly altered while referring to modification as a process where smaller percentages of additive are added to 'modify' properties rather than radically alter them. To avoid any confusion, this chapter uses the term 'stabilisation' for all materials improvement.

Mechanical stabilisation

The simplest method of increasing the strength is to stabilise it mechanically. In areas where good-quality materials are not readily available, it may be possible to blend two different materials to produce an acceptable product. Blending of materials is normally carried out for two main reasons namely to:

1. Improve the acceptability of cohesive soils of low strength by adding coarse material, or
2. Improve the grading of fines-deficient granular materials by adding a fine material.

Mechanical stabilisation is usually found to be the most cost-effective process for improving poorly graded materials; however, this cannot always be achieved. It is important to consider the practical limits of this type of processing. For example, production of a uniform mixture by the addition of granular material to a clay-rich one may produce a uniformly graded material, but one in which the clay fraction may play a too-dominant role in determining the properties of the material.

For LVRRs, mechanical stabilisation is normally carried out at the road site by mixing the component materials together with grading or rotovating equipment. Both large-scale and intermediate equipment options are available commensurate with the scale of the works. Alternatively, mixing may be successfully undertaken by stockpile mixing and turn-over off-site.

Chemical stabilisation principles

Chemical stabilisation normally involves the incorporation of relatively small percentages of lime or cement (Sherwood 1993). These stabilisers are called hydraulic binders, which 'set' in the presence of water. They can dramatically increase the strength of unbound materials, making them more suitable for use in the main load bearing layer of a road pavement, or they can be mixed with soils in small amounts, which merely alter the physical characteristics of the soil, such as plasticity or moisture or moisture condition, rather than to significantly strengthen it.

The choice of stabiliser is largely dependent on the properties of the unstabilised material. Materials with a low plasticity, and therefore low clay content, are more suitable for cement stabilisation. Materials with higher plasticity and a more cohesive nature are better stabilised with lime. Important issues in the stabilisation process are as follows:

1. Ensure complete mixing of the component materials; this is especially vital for fine-grained host materials.
2. The curing time required for most stabilisation methods will have implications for subsequent construction or non-diverted public traffic on the completed work.
3. It is also important to test the efficacy of the stabilisation process before the next pavement layer is placed.

It is not unusual for designers to achieve stabilisation by using both cement and lime for a particular project, with the aim of achieving the desired strengths and other engineering properties at the lowest cost. Table 8.9 summarises the ranges of suitability for the different stabilisation methods.

Several other chemical methods, principally aimed at improving the durability of unsealed roads or the 'compactability' of pavement layers, are marketed by various commercial organisations, often promising 'universal'

Table 8.9 Guide to Selecting a Method of Stabilisation (After Austroads 2018)

	Plasticity index (Ip)					
	Applicability[b]					
Stabilisation type	Ip < 10	10 < Ip < 20	Ip > 20	Ip < 6 PP < 60[a]	Ip < 10	Ip > 10
Cement	A	A	B	A	A	A
Lime	B	A	A	C	B	A
Bitumen	B	B	C	A	A	C
Bitumen-Cement	A	B	C	A	A	B
Mechanical	A	C	C	A	A	B

[a] PP; plasticity product = Ip × % passing 75 micron
[b] A: usually suitable. B: doubtful. C: usually not suitable.

application. Due to the highly variable characteristics of the host materials, unscientifically supported claims should be treated with extreme caution and piloted on the actual target materials before any large-scale adoption. Even if effective, the whole-life costs of these treatments are usually not justifiable for LVRR applications. The US Federal Highways Administration has undertaken independent review of a range of these alternative options (FHWA 2014). This document includes a useful summary and assessment of range of chemical 'stabilisation' options under the following main categories:

- Water and water with surfactant.
- Water absorbing.
- Organic non-petroleum.
- Organic petroleum.
- Synthetic polymer emulsion.
- Concentrated liquid stabiliser.
- Clay additive (used for mechanical stabilisation).

This guide also includes a process for selecting an appropriate chemical treatment for a specific set of unpaved road conditions using ranked potential performance.

Cement stabilisation

Addition of cement to base materials results in a reduction in plasticity and swell, and an increase in strength and bearing capacity. CBR values well in excess of the minimum requirement for unstabilised materials normally result. The amount of cement added is usually less than 5% by weight. The initial chemical reactions occur quite quickly, and hence the processing of the materials has to be completed quickly, and construction must be finished within 2 hours (the onset of initial set). The cement then must be allowed to cure for a period, usually, 7 days. Although not essential, the use of a batching plant to blend the cement with the host material and water rather than mixing on the road gives a more consistent mix and a better result, although cost-wise this may not be possible for most LVRR projects.

Cement can be used to stabilise most soils. The exceptions are those with a high organic content, which retards the hydration process, and those with clay content outside the normal road materials specification range and where it is difficult to mix the soil–cement mixture evenly.

Lime stabilisation

Typically, 3%–5% by weight of hydrated lime stabiliser is commonly necessary to gain a significant increase in the compressive and tensile strengths. The Initial Consumption of Lime (ICL) test involves the addition of lime to relevant material sample until a maximum pH of 12 is achieved; that percentage of lime can then be taken as an appropriate amount to lime to add in practice (Sherwood 1993). This is an indicative and possible conserve test to be used to give a rapid indication of the minimum amount of lime that needs to be added to a material to achieve a significant change in its properties. Care is needed when using different sources of hydrated lime, particularly when using locally produced lime in rural areas, as percentage of calcium carbonate may be variable between sources.

The gain in strength with lime stabilisation is slower than that for cement, and a much longer time is therefore available for mixing and compaction. Lime has a much lower specific gravity than cement so, for a given percentage mass, a higher volume is available, and it is therefore generally easier to achieve uniform mixing.

Quicklime has a much higher bulk density and is less dusty than hydrated lime, but is generally not used in road projects due to its caustic nature and consequent health and safety issues.

Bitumen stabilisation

It is also possible to stabilise marginal materials with the addition of small percentages of slow setting bitumen emulsion. This process is only effective with sand materials with little or no fines and is likely to be more costly than the other stabilisation options. Therefore, it should only be considered in, for example, coast or desert areas where sand is the only available material for construction purposes within reasonable haul distances. Trials research, however, has shown it to be an effective procedure in the appropriate circumstances (Intech-TRL 2007).

Bitumen emulsion is also successfully used in the recycling of old bituminous pavements. The old pavement is broken up, processed in situ to the basic aggregate sizes and the new emulsion added to renew binding properties before layer shaping, compaction and sealing with some form of surface dressing. This approach has significant environmental advantages over traditional 'dispose and replace' approaches and can be achieved using heavy or intermediate equipment methods.

REFERENCES

Austroads. 2018. *Appropriate Use of Marginal and Non-Standard Materials in road Construction and Maintenance*. Technical report AP-T335-18. https://austroads.com.au/publications/asset-management/ap-t335–18.

CIRIA. 1988. *Laterite in Road Pavements*. Charman J Ed. Special publication 47. UK: Construction Industry Research and Information Association.

Cook, J. R., E. C. Bishop, C. S. Gourley and N. E. Elsworth. 2002. Promoting the use of marginal materials. TRL Ltd DFID KaR Project PR/INT/205/2001 R6887. https://www.gov.uk/research-for-development-outputs/promoting-the-use-of-marginal-materials.

Cook, J. R. and C. S. Gourley. 2002. A framework for the appropriate use of marginal Materials. Mongolia: World Road Association (PIARC)-Technical Committee C12. http://transport-links.com/wp-content/uploads/2019/11/1_796_PA3890.pdf.

Cook, J. R., C. S. Gourley and N. E. Elsworth. 2001. Guidelines on the selection and use of road construction materials in developing countries TRL Ltd DFID KaR Project R6898. Microsoft Word – postqatextGuidelines3-8-00.doc (publishing. service.gov.uk).

FHWA. 2014. Unpaved road dust control and stabilization treatment selection guide. Report reference: FHWA-FLH-14-002, US Federal Highways Administration.

Fookes, P. G. and A. H. Marsh. 1985. Some characteristics of construction materials in the low to moderate metamorphic grade rocks of the Lower Himalayas of East Nepal. 2: Engineering characteristics. *Proceedings of the Institution of Civil Engineers* 70:139–162. https://www.icevirtuallibrary.com/doi/epdf/10.1680/iicep.1981.1961.

Gourley, C. and P.A.K Greening. 1998. Environmental damage from extraction of road building materials. DFID- funded Knowledge and Research programme. Report Ref R6021. https://assets.publishing.service.gov.uk/media/57a08d9940f0b652dd001a7a/R6021.pdf.

Intech-TRL. 2007. Rural road surfacing research. RRST construction guidelines. Report for UKAID-DFID and Ministry of Transport, Vietnam. https://www.research4cap.org/ral/IntechTRL-Vietnam-2007-RRST+Construction+Guidelines-SEACAP1-v070910.pdf.

Keary, P. 2005. *The Penguin Dictionary of Geology*. ISBN 9780140514940, p 336.

McNally, G. H. 1998. *Soil and Rock Construction Materials*. London: E & F N Spoon.

PIARC. 1989. *Marginal Materials. State of the Art.* Brunschwig G, Ed. France: World Road Association-PIARC Paris. https://www.piarc.org/en/order-library/4064-en-Marginal%20Materials%20-%20State%20of%20 the%20art.

RECAP. 2020. Low volume road design manual: Section B design. UKAID-DFID for Department of Rural Road Development, Ministry of Construction, Myanmar. https://www.research4cap.org/ral/DRRD-2020-LVRRDesignManual-SectionB-Ch5t011-AsCAP-MYA2118A-200623-compressed.pdf

Roughton International. 2000. Guidelines on borrow pit management for low-cost roads. DFID KaR project report (Ref. R6852). https://www.gov.uk/research-for-development-outputs/guidelines-on-materials-and-borrow-pit-management-for-low-cost-roads.

Sherwood, P. T. 1993. *Soil Stabilisation with Cement and Lime. State of the Art Review.* HMSO, London: TRL Limited.

Smith, R. 1999. *Stone: Building Stone, Rock Fill and Armour Stone in Construction.* UK: Geological Society Engineering Geology, Spec Pub 16.

Smith M. R. and L. Collis. 2020. *Aggregates: Sand, Gravel and Crushed Rock Aggregates for Construction.* 3rd edition. UK: Geological Society Engineering Geology, Spec Pub 17.

U.S. Dept of Interior (UDSI). 1998. *Earth Manual.* 3rd edition. Bureau of Reclamation.

Chapter 9

Geometric design

INTRODUCTION

Geometric design is the process whereby the layout of the road in the terrain is designed to meet the needs of the road user. Geometric design covers road width, cross-fall, horizontal and vertical alignments, and sight lines as well as related elements of road safety (Giummarra 2001).

The geometric design of a rural road normally must comply with the geometric standards that are related to the road classification, as discussed in Chapter 2. The geometric design standards provide the link between the cost of building the road and the benefits to road users. Usually, but not always, the higher the geometric standard, the higher the construction cost and the greater the road-user benefits. A national 'geometric standard' is not a specification, although it could be, and often is, incorporated into specifications and contract documents. A standard should relate to a minimum level of service required for a particular category of road to be fit-for-purpose (Falck-Jensen 2004a).

Geometric design is one of the first design actions within the LVRR design process and should aim to provide the following:

- Minimum levels of safety and comfort for road users.
- A framework for economic design.
- Consistency of alignment.

This chapter provides summary guidance on the basic elements of geometric design, including influencing factors and aspects of traffic characteristics, cross section, alignment, and safety. Most national road design manuals include guidance on geometric standards to be adopted for different classes of road. The following is a selection of recent LVRR manuals or reports that contain geometric design recommendations:

- Afghanistan: MRRD (2020).
- Cambodia: SEACAP (2009).
- Liberia: MPW Liberia (2020).

DOI: 10.1201/9780429173271-9

- Malawi: MoTPW (2020).
- Myanmar: RECAP (2020).
- South Sudan: AfCAP (2012)

FACTORS INFLUENCING GEOMETRIC DESIGN

Key issues

Geometric standards provide the framework for economic design and ensure a consistency of alignment. The principal factors that affect the optimum geometric design of a rural road are:

- Cost.
- Terrain.
- Pavement type.
- Traffic (volume and composition).
- Safety.

Cost

Road width (running surface and shoulders) is one of the most important geometric properties since it is directly related to the cost of construction and maintenance. A DFID-funded review of the standards adopted by a range of countries or organisations indicated a logical relationship between road width and traffic, as summarised in Figure 9.1 (SEACAP 2009).

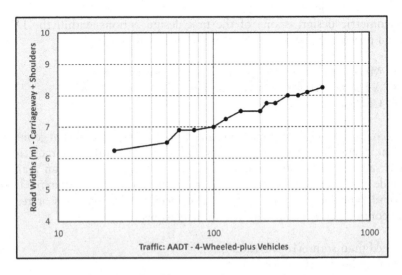

Figure 9.1 Daily traffic and road width.

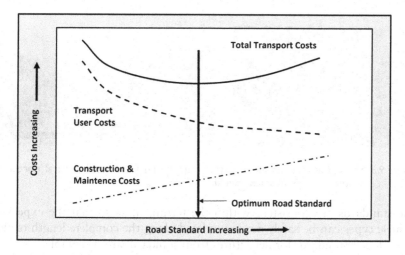

Figure 9.2 Transport costs and geometric standards.

There is a necessary trade-off between road standards and road-user costs (TRL 1988). As road standards increase, agency costs (road construction and maintenance costs) increase, but road-user costs normally decrease. As a result, the total costs calculated as a function of 'road standard' first decrease, pass through a minimum, and then increase again. From an economic point of view the standards adopted should be those applicable to the minimum total costs, as schematically illustrated in Figure 9.2. As to be expected, the position of this minimum is strongly dependent on traffic level. Hence, from an economic point of view, standards should increase as traffic increases.

While economic assessments linked to road geometry are now common in road appraisals carrying high volumes of traffic, they are less common for LVRRs (see Chapter 5). This is because such roads are not justified solely on standard traffic-based economic assessments. LVRRs commonly carry a traffic mix that can include are large percentage of bicycles, motor-cycles, motorcycle taxis, tractor-trailers, animal-drawn carts and pedes-trian. They serve social functions that are vital, but which are difficult to fit into traditional economic theory. For LVRRs the connection between traf-fic and justifiable geometric standard is more difficult to formally establish. Their standards are largely based on historic consensus, strongly influenced by factors based on safety and fitness for purpose.

Terrain

Terrain has a major effect on road costs to such an extent that the same stan-dards cannot be applied in across different geomorphological units. Severe ter-rain may necessitate some compromise on carriageway and shoulder widths, otherwise costs and environmental impact may be excessive. Provided that

(a)

(b)

Figure 9.3 Stacked alignments in very steep terrain. (a) Road stack in rural northern Vietnam. (b) Road stack, Nepal.

the standards are consistent within the terrain units so that the expected vehicle types can be safely accommodated along the complete length of the road, lower standards (or departures from standard) are acceptable.

There is a consensus on the definitions of broad terrain classes, based on the number of 5-m contours crossed per kilometre in a straight line linking the two ends of a road section. Common definitions are:

1. Flat: Level or gently rolling terrain; 0–10 number 5-m contours per km. The natural ground slopes perpendicular to the ground contours are generally below 3%.
2. Rolling: Rolling to hilly terrain where the slopes generally rise and fall gently but where occasional steep slopes may be encountered; 11–25 number 5-m contours per km. The natural ground slopes perpendicular to the ground contours are generally between 3% and 25%.
3. Mountainous: Rugged, hilly, and mountainous terrain which can definite restrictions on the standard of alignment obtainable for LVRRs and often involves long steep grades and limited sight distances. More than 26 number 5-m contours per km. The natural ground slopes perpendicular to the ground contours are generally above 25%.
4. Escarpment: Escarpment situations are those very steep slopes where it is required to consider stacked (or hairpin) road alignments or side-long ground where earthwork quantities may be very large (transverse terrain slope >70%), as illustrated in Figure 9.3.

Pavement type

For a similar 'quality' of travel, there is a difference between the geometric design standards required for an unsealed road (gravel or earth) and for a sealed road. This is because of the very different traction and friction properties of the two types of surfaces and the highly variable nature of natural materials. Thus, higher geometric standards are generally required for unsealed roads. A road that is to be sealed at later date should be designed to the higher unsealed geometric road standards. Pavement options are considered in detail in Chapter 13.

Traffic volume and composition

For rural roads the volume of traffic is usually low, and congestion issues largely arise not from traffic volume but from the disparity in speed between the variety of vehicles and other road users which the road serves. Nevertheless, it is still the size of the largest vehicles that use the road that dictates many aspects of geometric design. These are commonly referred to as the 'design vehicles' for different road standards. Such vehicles must be able to pass each other safely and to negotiate all aspects of the horizontal and vertical alignment. In some countries historical precedent has meant that the truck population in rural areas is predominantly one or two types and sizes of vehicle. This makes it relatively easy to select a typical vehicle for setting geometric standards. Conversely some countries have a wide variety of truck sizes, and selecting a suitable truck size for geometric design is more difficult.

An appropriate assessment of traffic is a key aspect of geometric design, and it is worth noting the difference between the nature of traffic data required for road width design and the nature of the data required for pavement structural design. The numbers and type of vehicles, including non-motorised traffic (NMT), carried daily are required for the former, while a design-life assessment of traffic load (axle loads) is normally required for the latter.

The road space occupied by different types of vehicle is an essential element in designing for capacity, and the number of vehicles that the road can carry in a unit of time (vehicles per hour or per day) is a crucial factor. A small truck requires more physical space to operate than a motorcycle, independent of whether the truck is empty or fully loaded. In contrast, an animal-drawn trailer occupies less physical space than a pick-up but is likely to occupy more time-related space due to its low travel speed. Hence, for LVRRs geometric design it is the speed at which types of vehicles can travel as well as their physical dimensions that are important. Vehicles that are slow-moving cause congestion problems because of their speed rather than because of their size. Therefore, in order to quantify traffic for normal capacity design the concept of equivalent Passenger Car Units (PCUs) is often used (see Chapter 3, Table 3.3).

Another important aspect of geometric design concerns the ability of vehicles to ascend steep hills. Roads that needed to be designed for very heavy vehicles or for animal-drawn carts often require specific geometric standard amendments to address this.

Safety

Experience has shown that simply adopting 'international' design standards from developed countries will not necessarily result in acceptable levels of safety on rural roads (SEACAP 2009). The main reasons include the completely different mix of traffic, including relatively old, slow-moving, and overloaded vehicles, a large number of motorcycles and bicycles, poor driver training, and poor enforcement of regulations. In such an environment, traffic safety assumes significant importance in relation to geometric design. LVRR traffic safety issues are discussed in Chapter 3.

KEY GEOMETRIC DESIGN ELEMENTS

Design speed

Design speed is the expected maximum safe speed which will be adopted by the majority of vehicle drivers, and which can be easily and comfortably maintained over the section of road being designed under conditions where the features of the road govern the speed, rather than traffic. In the case of many LMIC LVRRs, a maximum design speed of 50 km/h may be adopted on mixed traffic safety grounds.

The concept of design speed allows the key elements of geometric design to be selected for each class of road in a consistent and logical way. Design speed is relatively low in mountainous terrain to reflect the necessary reductions in standards required to keep road costs to manageable proportions. It is higher in rolling terrain and normally highest of all in flat terrain. The question that must be answered by the designer is simply the selection of design speed for each sector of road. It is generally agreed within most design guidelines for LVRRs that design speed should be related to traffic level and terrain, Table 9.1.

Table 9.1 Recommended Design Speeds

Author	Traffic (AADT of four-wheeled vehicles)				
	<20	20–100	100–400	400–1,000	>1,000
ORN 6	–	60/50/40	70/60/50	85/70/60	100/85/70
ARRB unsealed	Minimum 70/70/40; Maximum 120/100/70				
SFRDP		70/70/50	70/70/50		
Thailand		Minimum 60/50/30 Maximum 80/60/50			
Cambodia based on Australian recommendations[1]		40/30/20	60/50/40	60/50/40	70/60/50
Lao	50/40/30	50/40/30			
Other	25				

Source: Data extracted from SEACAP (2009b).

Design speed affects the following geometric standards:

- Horizontal alignment.
- Vertical alignment.
- Curvature.
- Sight distance.
- Superelevation (SE) on horizontal curves.

Road width

For the levels of traffic and vehicle types associated with the majority of LMIC LVRRs, single-lane operation is considered adequate, as there will be only a moderate probability of vehicles meeting, and the few passing manoeuvres can be undertaken at very reduced speeds. Provided sight distances are adequate for safe stopping, these manoeuvres can be performed safely, and the overall loss in efficiency brought about by the reduced speeds will be small.

In some cases, single-lane roads may not allow the larger commercial vehicles to easily pass in opposite directions or to overtake, hence passing places may need to be provided. The increased width at passing places should allow two vehicles to pass at slow speed and hence depends on the design vehicle. For example, trucks or buses of 2.5 m width require a safe minimum passing width of 6.0 m. When required, passing places would normally be provided every 300–500 m, depending on the terrain and geometric conditions. Passing places should be built at the most practical places rather than at precise intervals, provided that the distance between them does not exceed the recommended maximum.

For LVRRs, consideration has also been given to the movement of pedestrians, cyclists, and animal-drawn vehicles either along or across the road. Conflicts between slow- and fast-moving traffic need to be assessed, and increased widths of both shoulders may be necessary. The increase in width will vary with the relative amounts of traffic, their characteristics, the terrain and location (e.g. in village areas).

For single-lane roads, while there is good international agreement about the normal minimum carriageway width, namely 3.0 m, there is less agreement about the width of shoulders. This is mainly due to large variations in non-motorised traffic (NMT) between countries and regions, which is the key factor in assessing recommended shoulder widths.

Where NMT is primarily comprised of pedestrians, the width of the shoulder can be relatively low, but where there is a significant number of wheeled NMT, shoulder widths of 1.5 m are recommended. In a study in Lao PDR, shoulders of 1.5 m were recommended if the number of non-four-wheeled vehicles exceeded 150 per day. Below this level, 1.0 m shoulders were recommended (SEACAP 2008).

Cross-fall

Adequate cross-fall (camber) is essential to provide adequate surface drainage, while not being so great as to be hazardous to vehicles. The ability of a pavement surface to shed water varies with its smoothness and integrity, and hence there are different values recommended for paved and unpaved (earth gravel) roads. On unpaved roads, the minimum acceptable value of cross-fall is governed by the need to carry surface water away from the pavement structure effectively, with a maximum value above which either erosion of material or wet surface slipperiness starts to become a problem for traffic.

The optimum value of camber for earth or gravel surfaces varies considerably but normally lies between 4% and 7%, while there is general agreement that camber should be 2%–4% on sealed or concrete roads. There may, however, have to be some trade-off with road safety regarding shoulder camber (see Chapter 3). This is also discussed in more detail in Chapters 14 and 15. Shoulders having the same surface as the running surface should have the same camber. Unpaved shoulders on a sealed road should have shoulders that are 2% steeper, in other words 5% if the running surface is 3%.

Adverse cross-fall and superelevation

Adverse cross-fall arises on curves when the cross-fall or camber causes vehicles to lean outwards when negotiating a curve. This affects the cornering stability of vehicles and impacts safety. The severity of its effect depends on vehicle speed, the horizontal radius of curvature of the road, and the side friction between tyres and road surface. For reasons of safety, it is normally recommended that adverse cross-fall is removed where necessary on all roads regardless of traffic, as indicated in Table 9.2.

Some cross-fall is necessary for drainage, and hence flat sections without are not acceptable. Instead, a single value of cross-fall is designed in the proper direction such that the cross sectional shape of the road is straight across the running surface with the cross slope being the same as that of the inner side of the cambered road. For unpaved roads the recommended cross-fall should also be the same as the normal camber or cross-fall values.

Table 9.2 Adverse Cross-Fall to Be Removed If Radii Are Less Than Shown

Design speed (km/h)	Minimum radii (m)	
	Paved	Unpaved
<50	500	700
60	700	1,000
70	1,000	1,300
85	1,400	
100	2,000	

To remove adverse cross-fall the basic cambered shape of the road is gradually changed as the road enters the curve until it becomes simply cross-fall in one direction at the centre of the curve.

For sealed roads the removal of adverse camber may not be sufficient to ensure good vehicle control when the radius of the horizontal curve becomes too small. In such a situation additional cross-fall may be required. This is properly referred to as superelevation (SE).

For LVRRs, design speeds of below 30–40 km/h are below those for which superelevation is generally recommended on unsealed roads. However, the removal of adverse cross-fall on horizontal curves below 500 m radius is recommended.

Gradient

Gradient is a major aspect of vertical alignment and is related to vehicle performance, safety and level of service. For the low levels of traffic flow with only a few four-wheel-drive vehicles, the maximum traversable gradient is commonly noted as 20%, and two-wheel drive trucks are similarly recorded as successfully tackling gradients of 15%, except when heavily laden (TRL 1988). Bearing in mind the likelihood of heavily laden small trucks and animal-drawn carts, the LVRR standards generally have a proposed general recommended limit of 10%, but with an increase to 15% for short sections in areas of difficult terrain.

High gradients on unpaved surfaces lead to high rates of material erosion and surface gullies in high rainfall areas. Limiting gradient to 7-10% or specification of durable surfaces should be considered for these road sections.

Stopping sight distance

The distance a vehicle requires to stop safely is called the stopping sight distance. In order to ensure that the design speed is safe, the geometric properties of the road must meet certain minimum or maximum values to ensure that vehicle drivers can see far enough ahead to carry out normal manoeuvres safely, such as overtaking another vehicle or stopping if there is an object in the road. This mainly affects the shape of the road on the brow of a hill (vertical alignment), but if there are objects near the edge of the road that restrict a driver's vision on approaching a bend, then it also affects the horizontal curvature. Of lesser importance is passing sight distance – the distance needed to see ahead for safe overtaking.

A driver needs time to react to an upcoming situation and then the brakes need time to slow the vehicle down, hence stopping sight distance is extremely dependant on the speed of the vehicle. Finally, the surface characteristics of the road affect the braking time, so the values for unpaved roads differ from those of paved roads, although the differences are small for design speeds below 60 km/h. In order to calculate the stopping sight distance, assumptions must be made about all of these factors; Table 9.3 shows a range of values that are commonly assumed.

Table 9.3 Assumptions Used for Calculating Stopping Distances

Parameter	Values used
Driver reaction time	2.0–2.5 seconds
Driver eye height	1.0–1.15 m
Object height for stopping	0.1–0.2 m
Object height for passing	1.0–1.3 m
Longitudinal friction factor	0.43–0.60

Table 9.4 Stopping Sight Distances (m)

Stopping distances	Design speed (km/h)					
	30	40	50	60	70	80
Stopping distance range (m)	25–35	35–55	50–75	65–100	85–130	115–160
Recommendations: unsealed (m)	35	50	70	93	120	150
Recommendations: sealed (m)	30	40	55	72	95	120

As a result of these assumptions, the ranges of stopping sight distances are obtained. Values towards the higher and the lower end of these ranges are recommended for unsealed and sealed roads, respectively, as shown in Table 9.4.

Road curvature

Horizontal curvature

Horizontal curves are designed to ensure that vehicles can negotiate them safely. The main factor is the minimum radius of curvature. This is determined by two main considerations, namely, the design speed and the cross-fall or superelevation. The friction between the road surface and the vehicle wheels also has an effect, and hence the minimum values of curvature are higher for unsealed roads than for sealed roads. For each design speed, the minimum horizontal radius is determined by the cross-fall or superelevation – the higher the cross-fall, the smaller the radius of curvature that can be negotiated safely by the vehicles.

The values in Table 9.5, based on international practice, show a range because of the slightly different assumptions that were made in their derivation. As indicated in the tables, the use of a higher value of superelevation makes it possible to introduce a smaller horizontal curve based

Table 9.5 Range of minimum values of horizontal radii of curvature

	Minimum Horizontal Radius (m) Design Speed (km/h)				
	30	*40*	*50*	*60*	*70*
Paved Roads					
Super-elevation 4%	30m	55m	95m	150m	210m
Super-elevation 7%	25m	35m	70m	105m	180m
Super-elevation 10%	20m	35m	60m	95m	145m
Unpaved roads					
Super-elevation 4%	35m	65m	95m	175m	250m

Table 9.6 Minimum Values of L/G for Crest Curves

	Minimum values of L/G for crest curves Design speed (km/h)					
	30	*40*	*50*	*60*	*70*	*80*
Sealed roads	2	4	7	12	21	34
Unsealed roads	3	6	11	20	34	53

on the same design speed. This can be used for sealed roads but not for unsealed roads.

Road widening may be considered necessary on LVRRs where the horizontal curve is tight (<50 m) and where long vehicles are anticipated. This widening would normally be in the range 0.5–1.0 m, depending on curve radius.

Vertical curvature

The vertical alignment is more complicated than horizontal alignment, since an extra variable is involved, namely the difference in gradient ($G\%$) between the uphill and downhill sides. The minimum length of the curve (L metres) over the crest of a hill between the points of maximum gradient on either side is related to G and to the stopping sight distance and therefore to the design speed. The minimum value of the L/G ratio can be tabulated against the stopping sight distance, and therefore the design speed, to provide the designer with a value of L for any specific value of G. International comparisons give the values shown in Table 9.6.

Table 9.7 Minimum Values of L/G for Sag Curves

| | Minimum values of L/G for sag curves Design speed (km/h) | | | | | |
	30	40	50	60	70	80
Minimum L/G	0.7	1.3	2.2	3.5	4.8	7.5

Sag curves

Sag curves are the opposite of crest curves, with vehicles travelling downhill and then uphill. In daylight the sight distance is normally adequate for safety and the design criterion is based on minimising the forces that act upon the driver and passengers when the direction of travel changes from downhill to uphill, although on rural roads this is less important than road safety issues. However, at night time the key issue on sag curves for motorised vehicles is the illumination provided by headlights to see far enough ahead. To provide road curvature that allows the driver to see sufficiently far ahead using headlights while driving at the design speed at night is usually too expensive for rural roads. In any case, the driving speed should be much lower at night on such roads, particularly those carrying significant NMT. As a result of these considerations, it is normally recommended that the minimum length of curve is determined by the driver discomfort criterion. The results are shown in Table 9.7. In practice, a minimum length of curve of 75 m will cope with almost all situations.

SUMMARY

Departures from standards

It is anticipated that there may be LVRR design situations where the alignment must deviate from the official standards. These are likely to be in areas of difficult terrain, for example, where compliance with gradient or road curvature standards would incur unjustifiable costs. Usually, the designer would be required to justify the departures to the road authority and possibly propose additional warning signage to be applied in the interests of safety (Falck-Jensen 2004b).

Examples of geometric standards

Geometric standards are variable and are a function of specific road, road network, and national requirements and policies. Tables 9.8–9.12 present examples from a typical LVRR road design manual to illustrate how road geometry varies depending on key factors (SEACAP 2009).

Table 9.8 Geometric Design Standards for Rural Road (AADT 150-300)

Design element	Unit	Flat	Rolling	Mountain	Escarpment	Populated areas
Paved road						
Design speed	km/h	70	60	50	25	50
Width of running surface	m	6.5[a]	6.5[a]	6.5	6.5	6.5
Width of shoulders	m	1.25[a]	1.25[a]	0.5	0.5	1.25[b]
Total width	m	9.0	9.0	7.5	7.5	9.0
Minimum stopping distance	m	110	90	70	25	65
Minimum horizontal radius for SE=4%	m	195	135	85	15[c]	85
Minimum horizontal radius for SE=7%	m	170	120	75	17[c]	NA
Minimum horizontal radius for SE=10%	m	150	105	70	22[c]	NA
Maximum desirable gradient	%	4	7	10	12	4
Maximum gradient	%	7	10	12[d]	12[d]	6
Minimum crest vertical curve	K	21	12	7	4	7
Minimum sag vertical curve	K	4.8	3.5	2.2	1.3	2.2
Normal cross-fall	%	3	3	3	3	3
Shoulder cross-fall	%	6	6	3	3	6
Unsealed road						
Road width	m	7.0[c]	7.0[c]	7.0	7.0	7.0[a]
Minimum stopping distance	m	125	105	75	28	70
Minimum horizontal radius	m	245	175	110	23[c]	110
Maximum desirable gradient	%	4	6	6	6	4
Maximum gradient	%	6	9	9	9	6
Maximum superelevation	%	6	6	6	6	6
Minimum crest vertical curve	K	34	19	11	6	11
Minimum sag vertical curve	K	4.8	3.5	2.2	1.3	2.2
Normal cross-fall[e]	%	6	6	6	6	6

[a] If the number of large vehicles is >40, then running surface width should be increased to 7.0 m.
[b] Parking lanes and footpaths may be required.
[c] On hairpin stacks the minimum radius may be reduced to 15 m.
[d] Length not to exceed 200 m.
[e] Cross-fall can be reduced to 4% where warranted (e.g. for safety reasons).

Table 9.9 Geometric Design Standards for Rural Road (AADT 75-150)

Design element	Unit	Flat	Rolling	Mountain	Escarpment	Populated areas
Paved road						
Design speed	km/h	70	60	50	25	50
Width of running surface	m	6.0	6.0	6.0	6.0	6.0
Width of shoulders	m	1.0	1.0	0.5	0.5	1.0[b]
Total width	m	8.0	8.0	7.0	7.0	8.0
Minimum stopping sight distance	m	110	90	70	25	65
Minimum horizontal radius for SE=4%	m	195	135	85	20[c]	85
Minimum horizontal radius for SE=7%	m	170	120	75	18[c]	NA
Minimum horizontal radius for SE=10%	m	150	105	70	16[c]	NA
Maximum desirable gradient	%	4	7	10	12	4
Maximum gradient	%	7	10	12[d,e]	12[d,e]	6
Minimum crest vertical curve	K	21	12	7	2	7
Minimum sag vertical curve	K	4.8	3.5	2.2	1.3	2.2
Normal cross-fall	%	3	3	3	3	3
Shoulder cross-fall	%	6	6	3	3	6
Unsealed road						
Design speed	km/h	70	60	50	25	50
Road width	m	7.0	7.0	6.5	6.5	7.0[b]
Minimum stopping sight distance	m	125	105	75	28	70
Minimum horizontal radius	m	245	175	110	23[c]	110
Maximum desirable gradient	%	4	6	6	6	4
Maximum gradient	%	6	9	9	9	6
Maximum superelevation	%	6	6	6	6	6
Minimum crest vertical curve	K	34	19	11	3	11
Minimum sag vertical curve	K	4.8	3.5	2.2	1.3	2.2
Normal cross-fall[f]	%	6	6	6	6	6

[a] If there are more than 30 large vehicles, then Table 9.8 should be used.
[b] Parking lanes and footpaths may be required.
[c] On hairpin stacks the minimum radius may be reduced to 15m.
[d] Length not to exceed 200m and relief gradients required (<6% for minimum of 200m).
[e] If the number of large vehicles AADT is <20, this can be increased to 15%.
[f] Cross-fall can be reduced to 4% where warranted (e.g. for safety reasons)..

Table 9.10 Geometric Design Standards for Rural Road (AADT 25-75)

Design element	Unit	Flat	Rolling	Mountain	Escarpment	Populated areas
Sealed road						
Design speed	km/h	60	50	40	20	50
Width of running surface	m	3.3	3.3	3.3	3.3	3.3
Width of shoulders	m	1.5	1.5	1.0	1.0	1.5[b]
Total width	m	6.3	6.3	5.3	5.3	6.3
Minimum stopping sight distance	m	85	70	50	17	65
Minimum horizontal radius for SE=4%	m	135	85	50	15[c]	85
Minimum horizontal radius for SE=7%	m	120	75	45	15[c]	NA
Minimum horizontal radius for SE=10%	m	105	70	40	15[c]	NA
Maximum desirable gradient	%	4	7	10	12	4
Maximum gradient	%	7	10	12[d]	15[d]	6
Minimum crest vertical curve	K	12	7	4	2	7
Minimum sag vertical curve	K	3.5	2.2	1.3	0.7	2.2
Normal cross-fall	%	3	3	3	3	3
Shoulder cross-fall	%	6	6	3	3	6
Unsealed road						
Design speed	km/h	60	50	40	20	50
Road width[b,f]	m	6.0	6.0	6.0	6.0	6.0[e]
Minimum stopping sight distance	m	95	75	55	20	70
Minimum horizontal radius	m	175	110	70	15[c]	110
Maximum desirable gradient	%	4	6	6	6	4
Maximum gradient	%	6	9	9	9	6
Maximum superelevation	%	6	6	6	6	6
Minimum crest vertical curve	K	19	11	6	3	11
Minimum sag vertical curve	K	3.5	2.2	1.3	0.7	2.2
Normal cross-fall	%	6	6	6	6	6

[a] If there are more than 20 large vehicles, then Table 9.9 should be used.
[b] Parking lanes and footpaths may be required.
[c] On hairpin stacks the minimum radius may be reduced to 13 m.
[d] Length not to exceed 200 m and relief gradients required (<6% for minimum of 200 m).
[e] If there are less than ten large vehicles AADT, then DC1 may be used.
[f] Road widths may be reduced to address specific local conditions, especially in mountainous areas.

Table 9.11 Geometric Design Standards for DC1 (AADT < 25)

Design element	Unit	Flat	Rolling	Mountain	Escarpment	Populated areas
Design speed	km/h	50	40	30	20	40
Road width	m	4.5	4.5	4.5	4.5	4.5
Minimum stopping sight distance	m	70	55	35	18	50
Minimum horizontal radius	m	110	70	35	15[a]	70
Maximum desirable gradient	%	4	6	6	6	4
Maximum gradient	%	12[b]	12[b]	12[b]	12[b]	6
Minimum crest vertical curve	K	11	6	3	2	6
Minimum sag vertical curve	K	2.2	1.3	0.7	0.5	1.3

[a]On hairpin stacks the minimum radius may be reduced to 13 m.
[b]Length not to exceed 200 m.

Table 9.12 Minimum Standards for Very Low Volume Basic Access

Characteristic	Minimum requirements	
Radius of horizontal curvature	12 m absolute but up to 20 m depending on expected vehicles	
Vertical curvature K value for crests K value for sags	2.5 0.6	
Maximum gradients Open to all vehicles Restricted to cars and pick-ups	14% 16%	
Minimum stopping sight distance	Flat and rolling terrain Mountainous Escarpments	50 m 35 m 20 m

REFERENCES

AfCAP. 2012. South Sudan low volume roads design manual. UKAID-DFID for Ministry of Roads and Bridges, S Sudan. https://www.gov.uk/research-for-development-outputs/south-sudan-low-volume-roads-design-manual-volume-1-2-and-3.

Falck-Jensen, K. 2004a. Chapter 11 Geometric design controls. In *Road Engineering for Development*. Robinson R. and Thagesen B., Eds. Taylor and Francis. ISBN 0-203-30198-6.

Falck-Jensen, K. 2004b. Chapter 12 Geometric alignment design. In *Road Engineering for Development*. Robinson R. and Thagesen B., Eds. Taylor and Francis. ISBN 0-203-30198-6.

Giummarra, G. 2001. Road classifications, geometric designs and maintenance standards for low volume roads. Research report AR 354, ARRB transport research, Vermont South, Victoria, Australia. https://trid.trb.org/view/712280.

Ministry of Transport and Public Works (MoTPW) Malawi. 2020. Low Volume Roads Manual Vol 2.Geometric Design and Road Safety. pdf.https://www.research-4cap.org/ral/RoadsAuthorityMalawi-2020-LowVolumeRoadsManual-Volume2-200820-compressed.pdf.

MPW (Ministry of Public Works, Republic of Liberia). 2020. Manual for Low Volume Roads Part A – Policy, Geometric Design and Road Safety. G https://www.research4cap.org/ral/MinPublicWorksLiberia-2019-ManualforLVR-PartA-200125--compressed.pdf.

MRRD (Afghanistan). 2020. Low volume rural roads guideline and standards. Volume 2 geometric design and road safety. Islamic Republic of Afghanistan: UKAID-DFID for Ministry of Rural Rehabilitation and Development. https://www.research4cap.org/ral/MRRD-2020-LVRRGuidelineandStandards-Vol2GeometricDesign-AsCAP-AFG2155A-200831-compressed.pdf.

RECAP. 2020. Low volume road design manual: Section B design. UKAID-DFID for Department of Rural Road Development, Ministry of Construction. Government of Myanmar. https://www.research4cap.org/ral/DRRD-2020-LVRRDesignManual-SectionB-Ch5t011-AsCAP-MYA2118A-200623-compressed.pdf.

SEACAP. 2008. Low volume rural road standards and specifications: Part: I: Classification and geometric standards. SEACAP 3, UKAID-DFID for Ministry of Public Works and Transport. Lao PDR. SEACAP 3. Low volume rural roads: Classification and geometric standards. GOV.UK. www.gov.uk.

SEACAP. 2009. Rural road standards and specifications classification, geometric standards and pavement options. Final project report. DFID for Royal Government of Cambodia. https://www.gov.uk/research-for-development-outputs/seacap-19-03-rural-road-standards-and-specifications-classification-geometric-standards-and-pavement-options-final-project-report.

TRL. 1988. *Overseas Road Note 6. A Guide to Geometric Design. Overseas Unit.* UK: Transport Research Laboratory.

Chapter 10

Earthworks

INTRODUCTION

In order to comply with horizontal or vertical geometric guidelines and thus permit reasonable access for users, rural road alignments in hilly or mountainous areas may require the construction of cut or embankment earthworks. Embankments may also be required in low lying areas to raise alignments above flood levels or as approaches to bridges. These earthworks should be designed to minimise the risk of slope failure by implementing designs and construction procedures that are compatible with the engineering properties of the excavated soil–rock or the placed fill, whilst at the same time considering the impact of these earthworks on existing slopes or foundations.

The interaction of route alignment and the geometry or instability of the natural slopes may be such that construction to recognised safe angles is not an economical or engineering feasibility. Engineered or bio-engineered stabilisation or protection may have to be considered, particularly in areas of identified natural hazard or where significant potential climate impacts have been identified. If temporary Low Volume Rural Road (LVRR) closures and debris clearance can be tolerated and dealt with through maintenance management, then less earthworks may be a preferred option on economic grounds.

EXCAVATION

Principles

Excavation is the process of loosening and removing rock or soil from its original position and transporting it to a fill or waste deposit. Excavated materials capable of being compacted to form a stable fill are used for construction of embankments, subgrades and shoulders, or as backfill for structures. Selected rock material may be processed for use as aggregate or erosion protection. Unsuitable and surplus excavated material is disposed of as 'spoil'. If the material forming the bottom of a cutting is not suitable as the foundation for the road pavement, it may be necessary to remove it and replace it with satisfactory imported subgrade material. Excavated topsoil is preferably stockpiled for later use on earthwork side slopes to promote vegetation growth.

In an ideal situation there should be a focus on achieving an 'earthwork balance', whereby the volume of excavated materials is equal to the volume of fill required, thus removing the requirement either for spoil disposal or for additional borrow areas. Although in many cases this target of balanced earthworks is not achievable, there should still be a target to reduce excess spoil or excess borrow to a minimum, on both cost and environmental impact grounds. In this context, the innovative use of marginal earthwork materials and approaches to cut-slope stabilisation should be a priority for LVRR designers.

Excavation

Selection of equipment for excavation and moving of earth should be based on the nature of the material, haulage distances, climatic conditions, the skill of the equipment operators and economic considerations. Excavation and moving of earth can be carried out through a wide range of methods using various mixes of labour and equipment. In LVRR works there are likely to be cost constraints on use of expensive equipment, and there should be a focus on the use of 'appropriate technology', which may either Labour Based, equipment based, or a combination of both, rather than an inflexible adherence to a pre-conceived approach. The concept of appropriate technology is considered in more detail in Chapter 22.

On equipment-based projects, bulldozers are normally used for very short hauls and for spreading dumped materials. For moderate hauls motorscrapers may offer an appropriate option, although not commonly used on LVRR construction in LMICs. For moderate to long, or over the public roads, trucks working in partnership with front-end loaders, or power shovels may be the most economical solution. A motor grader may be used for shaping ditches, trimming slopes, and shaping the cross section of the road.

In the case of rock excavation, drilling and blasting may be necessary with materials then transported by trucks filled by front-end loaders. Methods of rock excavation are determined by rock mass strength, structure and the power of available plant. Pettifer and Fookes (1994) presented excavation modes as being based on the inter-relationships between intact rock material strength and a rock fracture index (or rock blockiness), whilst the *Caterpillar Handbook* (2019) proposes a more straightforward relationship between rock mass seismic velocity and excavation using a range of bulldozers with ripper tynes (see Figures 10.1 and 10.2). In LVRR projects, there may well be a constraint on the availability of heavy plant for

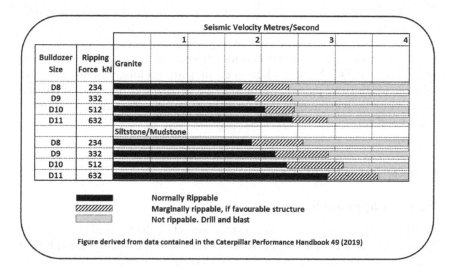

Bulldozer Size	Ripping Force kN	Seismic Velocity Metres/Second				
			1	2	3	4
		Granite				
D8	234					
D9	332					
D10	512					
D11	632					
		Siltstone/Mudstone				
D8	234					
D9	332					
D10	512					
D11	632					

Normally Rippable
Marginally rippable, if favourable structure
Not rippable. Drill and blast

Figure derived from data contained in the Caterpillar Performance Handbook 49 (2019)

Figure 10.1 Excavatability and seismic velocity

(a)   (b)

Figure 10.2 Laterite caprock being excavated by ripping. (a) D8 bulldozer fitted with ripping tynes. (b) Detail of triple tyne ripping arrangement.

excavation, and the use of drill and blasting in rock-like materials may be more prevalent than in larger road projects.

Structural excavation includes the excavation of material to permit the construction of culverts, foundations for bridges, retaining walls, etc. Suitable materials taken from structural excavations may be used either in backfilling around the completed structure or in other parts of the construction site. Both machine and hand methods are used for structural excavation, but more hand labour is generally required in this operation than in other types of excavation.

When sufficient material for the formation of embankments is not available from excavations along the alignment, additional suitable material may be taken from borrow pits. Use of borrow-pit material brings with it the need to comply with environmental considerations, concerning the condition in which borrow pits should be left when they are no longer being used (see Chapter 6). Borrow-pit excavation may be carried out using labour-based methods or by back-acting shovels or front-end loaders, and the material loaded and transported on trucks.

EMBANKMENT DESIGN

General approach

Embankments are used in road construction when the vertical alignment of the road has to be raised above the level of the existing ground. Embankment design is based primarily on two related elements: the character of the available fill materials and geotechnical nature of the foundations. Embankment slopes and cross sections should be designed taking into account both elements as well as the geometric alignment requirements on height and other issues such as climate impact, as outlined in Chapter 17.

High embankments impose a heavy load on the underlying foundation soil. On some soils, this may result in excess settlement, and if the foundation soil is extremely weak, an embankment foundation slip failure may occur, if no preventative measures have been taken. The settlement characteristics of soil profiles vary considerably from minimal-problem well-drained granular soils to geotechnically difficult soft clay or organic soils.

Most types of soil and broken rock can be used for construction of embankments, but fine to medium granular materials with low plastic fines are generally preferable (AASTHO classification A-1, A-2-4, A-2-5 and A-3). More plastic materials may create construction problems in wet weather. Highly expansive clays and organic soils should not be used as fill (UDSI 1988).

Earth embankments are constructed by placing relatively thin layers of material. On equipment-based LVRR projects, the material is usually dumped at the required location by trucks, and spread by bulldozers or graders, with a maximum thickness of loose soil usually 250–300 mm. The soil is then thoroughly compacted to the required specification before the next layer is placed. During the construction operation the embankment should be kept well drained, especially when material of high plasticity has to be used.

Typical slope angles for embankment fill on sound foundations are presented in Table 10.1 (Ingles 1985). Side slopes are normally protected from erosion by establishing a cover of vegetation and in extreme cases by rock-fill, riprap or concrete.

Table 10.1 Suggested Fill Slope Gradients

Fill materials	Embankment side-slope gradients (V:H) for various heights			
	<5m	5–10m	10–15m	15–20m
Well graded sand, gravels, sandy or silty gravels.	1:1.5–1:1.8	1:1.5–1:1.8	1:1.8–1:2.0	
Poorly graded sand.	1:1.8–1:2.0	1:1.8–1:2.0	1:1.8–1:2.0	
Weathered rock spoil.	1:1.5–1:1.8	1:1.5–1:1.8	1:1.5–1:1.8	1:1.8–1:2.0
Sandy soils, hard clayey soil and hard clay.	1:1.5–1:1.8	1:1.5–1:1.8	1:1.8–1:2.0	–
Soft clayey soils.	1:1.8–1:2.0	1:1.8–1:2.0	–	–

For embankments >3 m in height detailed geotechnical investigation and analysis may be required. Fill placed near or against a bridge abutment or foundation, or fill that can impact on a nearby structure, may also require specific stability analysis.

Before constructing a fill slope on side-long ground, it is necessary to terrace or step the formation to prevent a possible slip surface from developing at the interface between the fill and the natural ground. The potential for failure along a deeper surface in the ground beneath the fill should be considered where layering in the soil–rock mass beneath the fill dips parallel to the ground slope.

Soft soils foundations

Soft soils, if not recognised and carefully investigated, can cause problems of instability and unacceptable settlement of embankments. The aim of the embankment designer should be to ensure that any settlement is complete prior to the placement of the road pavement. In the case of problem soils, this may require the use of special construction techniques. Research in Bangladesh (Mott Macdonald 2017) and Indonesia (WSP International 2002) where road embankments can be required on up to more than 20 m of soft or organic soil, identified the following commonly applied options through a literature review and discussions with local stakeholders:

Earthwork

- Excavate and replace/displacement.
- Use of side berms.
- Surcharge.
- Staged construction.
- Use of light material.

Ground improvement

- Sand compaction pile.
- Sand drain (with surcharge).
- Prefabricated vertical drains (with surcharge).
- Geo-textile basal reinforcement.
- Cement or lime columns.

The constrained budgets of most LVRR programmes demands that the many geotechnically attractive solutions may have to be rejected on cost grounds in favour of more pragmatic options. Some pragmatic LVRR earthwork options for constructing embankments on weak or problem spoils are outlined below and illustrated in Figure 10.3.

Replacement: The weak or problem soil is removed, either partly or completely, and replaced by suitable material. The economic limits to full removal are around 3–4 m depth of weak material. The removal of the deposit solves stability and settlement problems, as the embankment can be founded on firmer ground and settlement will be greatly reduced. In the partial excavation of soft materials, the remaining soft deposit is later consolidated. If necessary, surcharging is provided to accelerate settlement so that most of the settlement will be completed during construction.

The excavation and replacement method may need a considerable quantity of suitable backfill material which, ideally, should be available within acceptable economical hauling distance. Therefore, this method would be most suitable in sections of alignment where fill material is available from nearby cutting areas. Granular, free-draining material (sand, gravel or a mixture of sand and gravel) should be used as fill material when filling is to be done below the water table. Acceptable cohesive soil can be used when the excavation is dry, and the fill material can be compacted, as normal, in lifts.

In partial excavation, a layer of free-draining material may be required as a drainage blanket at the base of the fill to speed up consolidation of the remaining soft layer during construction.

Counterweight berms: The principle of counterweight berms is to add weight to the toe of the embankment to increase the resistance against slip or lateral spreading. When used in front of an approach fill to a bridge this will increase stability and reduce lateral pressure on the substructure. This option is very effective in solving stability problems with soft inorganic soils but will not solve the long-term settlement problem that may be particularly associated with organic materials.

Surcharging: Surcharging involves placing temporary additional load onto the proposed embankment to increase primary settlement. The load applied should be sufficient such that the settlement during the construction period is equal to the total expected settlement from the embankment less the allowable post construction settlement. When the desired settlement has

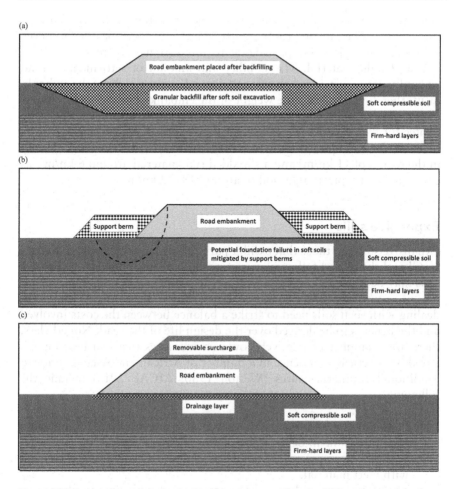

Figure 10.3 Low-cost options for embankments over soft soils. (a) Excavate and replace. (b) Use of support berms. (c) Removable surcharge.

been achieved, the surcharge is removed. The effectiveness of this method depends on the following factors:

- Thickness of the soft soil.
- Permeability of the soft soil.
- The presence of drainage layers.
- Available construction time.
- Shear strength of the soft soil.

Staged construction: This method involves constructing the embankment in series of stages with a pause between stages to allow the soil under the embankment to consolidate and gain the required strength to proceed with the next lift. The rate and timing of the stages must be controlled to allow sufficient consolidation to provide the required strength increase.

Use of light material: The stability and amount of settlement of road embankments constructed on soft soil depend on the weight of the embankment; therefore, reducing the weight of the embankment will reduce stress in the subsoil and reduce excessive settlement and instability. By using lighter fill material than ordinary fill the weight of embankment will be reduced; for example, a volcanic pumice material will have bulk density in the region of 11 kn/m³ and a shredded tyre material around 6 kn/m³, as compared to a typical sandy soil of around 18–22 kn/m³.

Expansive soils

Expansive soils are those that exhibit particularly large volumetric changes (swell and shrinkage) following variations in moisture contents. Expansive soils may be thick and laterally widespread, which makes the implementation of countermeasures costly, particularly for LVRRs. Any measures for dealing with such soils need to strike a balance between the costs involved and the benefits to be derived over the design life of the road. Nonetheless, there are a number of relatively low-cost measures that can be adopted, based on practical experience in a number of African and Asian countries. Traditional countermeasures (Weston 1980; AfCAP 2013) include the following:

- Pre-wetting (2–3 months) to induce attainment of the Equilibrium Moisture Content before constructing the pavement.
- Partially or completely removing the expansive soil and replacement with inert material.
- Modifying or stabilising the expansive soil with lime to change its properties.
- Increasing the height of the fill (surcharge) to suppress heave.
- Minimising or preventing moisture change using waterproofing membranes.

Compaction

Compaction increases the density of a material by expelling air from the voids in the material, thereby bringing the particles into more intimate contact with each other and improving the shearing resistance of the soil mass. Soils in embankments and subgrades are usually compacted using dedicated compaction equipment. The result of compaction work depends primarily on the moisture content of the soil, the type of the soil, the compaction equipment used and the energy applied (Parsons 1992).

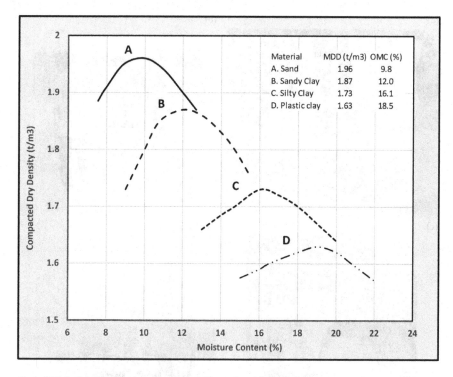

Material	MDD (t/m3)	OMC (%)
A. Sand	1.96	9.8
B. Sandy Clay	1.87	12.0
C. Silty Clay	1.73	16.1
D. Plastic clay	1.63	18.5

Figure 10.4 Compaction curves for a range of soil types.

For most soil types, the Maximum Dry Density (MDD) is achieved at a particular moisture content, the so-called Optimum Moisture Content (OMC) as illustrated in Figure 10.4. Because of the moisture–density relation, water must be added to dry soils, and overly wet soils must be aerated before compaction. However, in dry areas, where it is difficult to provide the large amount of water needed to bring the moisture content of a dry soil to the optimum level, it may be better to compact the soil in the dry state rather than adding an insufficient amount of water to the soil. In rainy weather, it may be necessary to replace an overly wet soil with more suitable material or to modify it with lime (see Chapter 8).

Compaction equipment options are varied, and the choice of compactor should ideally be a function of the material type. It is appreciated, however, that there may be economic constraints on the ideal use of differing compactors in some LVRR situations in LMICs, particularly when employing local contractors or community groups. Figure 10.5 illustrates a range of compacting plant.

It is possible to compact a soil to a higher density then theoretically required by increasing the amount of compactive effort, and this option is

Figure 10.5 Range of compaction plant for rural roads. (a) Plate compactor. (b) Pedestrian compactor. (c) Three-wheel 8 tonne deadweight roller. (d) Pneumatic tyred roller. (e) 2 tonne vibrating roller. (f) 8 tonne vibrating rollers working together.

Table 10.2 Typical Material Bulking or Shrinkage Factors

	Relative volume		
Condition	Rock fill	Sand & gravel	Silty clay
In situ	1.0	1.0	1.0
Loose stockpiled	1.75	1.2	1.4
Compacted	1.4	0.9	0.85

sometimes recommended as a means of increasing strength and sustainability. However, the over-compaction must not exceed a critical limit, particularly in fabric-sensitive materials, such as some residual soils. In such cases, the soil and particle fabric may breakdown and the fill become remoulded and weakened. In general, repetitions of roller passes are effective only up to a certain limit, normally between 8 and 16 passes, depending on the soil and compactor type. When constructing embankments with marginal materials or with materials whose compaction characteristics are not fully defined, compaction trials are strongly recommended.

Compaction requirements are commonly specified using an end-product specification, for example 90%–98% Modified MDD (ASTM T180). In some LVRR environments, with local contractors or inexperienced supervisors, the use of a 'Method Specification' has distinct advantages. In this case, the compaction target is defined in terms of the easily counted machine passes. This may be backed up by in situ density testing, including the effective use of the Dynamic Cone Penetrometer (DCP) measuring defined penetration rates. (See Chapter 8 – Construction Materials and Chapter 19 – Construction Supervision.)

Most soil or rock materials change volume between their in situ state, their loose (haulage) state and their final compacted state. It is important to take this volume change into consideration when computing excavation, haulage and fill quantities, see Table 10.2.

CUT FORMATION

General approach

Where possible, LVRR cut-slopes are generally designed empirically on precedent or modified precedent principles; that is, based on past experience with similar soil and rock materials in similar environments (Cook et al. 1992). Critical cuts or those greater than around 5 m in height may require a more detailed engineering geological assessment depending on the complexity of the ground conditions. This would include an assessment of the strength of the soil–rock materials and their mass structure. (See Chapter 7.)

Table 10.3 Suggested Cut-Slope Gradients

Soil–rock classification		Slope gradients (V:H) for various cut heights		
		<5m	5–10m	10–15m
Hard rock (without adverse structure)		1:0.3–1:0.8	1:0.3–1:0.8	1:0.3–1:0.8
Soft rock		1:0.5–1:1.2	1:0.5–1:1.2	1:0.5–1:1.2
Sand	Loose, poorly graded	1:1.5	1:1.5	1:1.5
Sandy soil	Dense or well graded	1:0.8–1:1.0	1:1.0–1:1.2	–
	Loose	1:1.0–1:1.2	1:1.2–1:1.5	–
Sandy soil, mixed with gravel or rock	Dense, well graded	1:0.8–1:1.2	1:0.8–1:1.2	1:1.0–1:1.2
	Loose, poorly graded	1:1.0–1:1.2	1:1.0–1:1.2	1:1.2–1:1.5
Cohesive soil		1:0.8–1:1.2	1:0.8–1:1.2	1:0.8–1:1.2
Cohesive soil, mixed with rock or cobbles		1:1.0–1:1.2	1:0.8–1:1.2	–

Cuttings in strong homogenous rock masses can often be excavated to very steep angles, depending on whether adverse geological structure is present. However, in weaker, fractured, or weathered soil-rock profiles, it is necessary to use shallower slopes. In heterogeneous slopes, where both weak and hard rock occur, the appropriate cut-slope angle can be determined on the basis of the location, nature and structure of the different materials making up the soil–rock mass, the interfaces between them and the variations in permeability between the different horizons.

The slope angles indicated in Table 10.3 have been provided as a general guide for LVRRs (Ingles 1985). Note that these angles cannot be applied without due consideration of the actual ground conditions. Individual countries or organisations may have similar or slightly different basic guidelines.

One of the most cost-effective ways to decide upon a suitable cut-slope is to survey existing cuttings in similar materials along other roads or natural exposures in the surrounding areas (Bulman 1968; Cook et al. 1992). Generally, new cuttings can be formed at the same slope as stable existing cuttings, if they are in the same material with the same overall structure. In rock excavations, persistent joint, bedding or foliation surfaces may determine the final cut-slope profile.

Excavation of rock slopes should be undertaken in such a way that disturbance due to blasting is minimised. In larger rock cuts the use of pre-splitting blast techniques has distinct advantages in terms of producing regular, more easily supported faces. Blasting should also be undertaken in a manner to produce material of such size that allows it to be placed in embankments without significant prior processing (Harber et al. 2011).

Cut-slope profiles can be single-sloped or benched. Single-sloped profiles are usually cut in uniform soil or rock materials or excavations less than 5–10m. Benched slopes are generally used in deeper cuts or where layered soil-rock profiles are encountered. The construction of benches can be advantageous in terms of overall stability, intercepting falling debris, control the

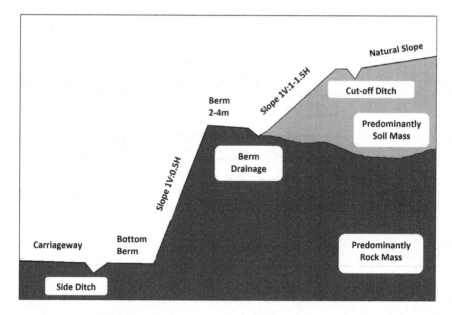

Figure 10.6 Key components of a cut-slope (Slope drains preferably lined trapezoidal).

flow of water and access for future slope maintenance. Although there is no fixed rule regarding the dimension of benches which are generally between 2 and 3 m wide, an important consideration is the width required for equipment access for future slope maintenance. In weaker materials, rain runoff should be encouraged to drain along the bench to an end or intermediate discharge, such as a cascade drain, rather than directly over the slope-face. Maintenance of these drains is important to prevent water accumulating on the bench. Figure 10.6 illustrates some of the important aspects of cut-slope formation.

Cut to fill

Cut to fill cross sections are a combination of excavation into hillside above the alignment and placement of the excavated fill on the "down" side. Although the cut-fill option is attractive in terms of cut-fill balance and is a common situation in many hill or mountainous access routes, it also the frequent cause of road failure, unless adequate design and construction precautions are adopted against adverse features, as illustrated in Figure 10.7.

Vital requirements for an adequate cut to fill design are:

- Suitable cut-slope excavation and drainage.
- Fill Fill sections are keyed-in by excavation of steps in the natural slope.

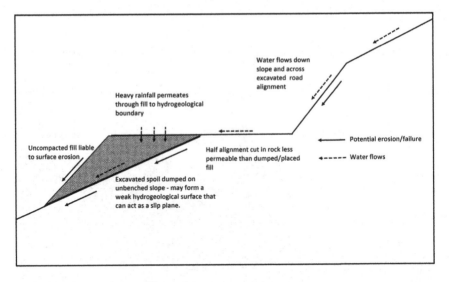

Figure 10.7 Risk in the cut-fill situation.

- Adequate drainage to prevent pore pressure build-up or lubrication of the cut-fill interface.
- Specification for compaction of fill in layers and not simply dumped over the alignment edge.
- Specification of complete removal of vegetation and organic material prior to construction.
- Prohibiting of embankment construction on loose spoil material derived from earlier excavations.
- Prevention of erosion on slopes immediately below the embankment.
- Ensure safe management of rainwater run-off away from cut-fill area.

TROPICALLY WEATHERED SOILS AND EARTHWORKS

Chapter 7 highlighted the potential challenges posed by some tropically weathered soil-rock materials and masses in terms of testing and geotechnical characterisation. Related challenges exist as regards designing and constructing earthworks in tropically weathered soil–rock terrain. These are summarised in Table 10.4 (Brand 1985).

On the positive side, the occurrence of chemical bonding and, or, the existence of pore suction in partially saturated materials can allow steeper angles of cut, provided due account is made of the points raised in Table 10.4.

Table 10.4 Earthwork Challenges in Tropically Weathered Masses

Characteristic	Potential problem
Cut-slopes	
Gradational and variable weathering boundaries.	Detailed Ground Model difficult to predict leading to challenges for cut-slope design.
Relict rock structure carried through into soil materials.	Planes of weakness contained within already weakened weathered soil materials reducing mass stability.
Erodibility of some residual soils.	Eroded slope faces can lead to larger scale slumps and face failures.
Embankments.	
Fabric-sensitive materials.	Soil fabric can break down if over-compacted. Potential problems on wet earth haul roads with material breaking down to a totally remoulded slurry.
Non-standard mineralogy.	Drying out of some minerals (Allophane and Halloysite) at standard laboratory temperatures gives very different moisture contents from natural air-dried samples and also leads to changes in particle size distributions.
Non-standard index property correlation.	Empirical relationships between index properties derived for sedimentary soils may not hold true for some residual soils leading challenges with regard to behaviour prediction as a possible fill.

SLOPE PROTECTION AND STABILISATION

Earthwork and natural roadside slope instability

Roadside instability in natural slopes, cuts and embankments has the potential to disrupt traffic flow and create access and safety problems for road users, particularly during rainy seasons. Intense rainfall is a frequent trigger for slope failure due to a combination of surface erosion, additional weight and a reduction in slope mass shear strength.

Erosion can take place on unprotected fill slopes adjacent to river channels, especially downstream of culverts, bridges and roadside turnouts. Once slope erosion is initiated, it can develop rapidly. In addition, uncontrolled runoff can erode roadside drains and road pavement.

Roadside instability may involve one or more of the following general situations.

- Above alignment erosion or failure of cut-slopes.
- Below alignment erosion or failure of embankment slopes.
- Failure in natural ground.

Figures 10.8 and 10.9 illustrate common instability scenarios, and Table 10.5 lists common types of erosion and slope failure. Cruden and Varnes (1996) provides a detailed classification of slope failure types.

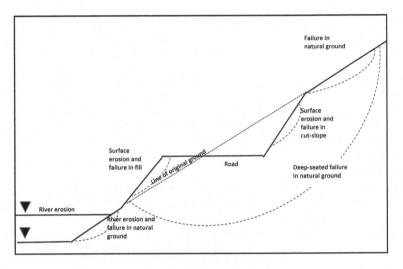

Figure 10.8 Typical soil mass roadside slope failures in hill terrain.

Figure 10.9 Typical soil-rock mass failures. (a) Joint and bedding controlled planar failure. (b) Joint controlled rock toppling failure. (c) Joint controlled rock wedge failure. (d) Soil-weathered rock mass slump failure.

Table 10.5 Common Types of Erosion and Slope Failure

Mechanism	Description	Depth
Surface erosion.	Rills and gullies form in weak, unprotected surfaces. Most common on fill and tipped spoil. Erosion should be expected on all bare or freshly prepared slopes.	Usually in the top 0.1 m, but can become deeper if not controlled.
Gully erosion.	Surface erosion that is established on the slope continues to develop and grow bigger.	Usually in the top 0.5 m, but can become deeper if not controlled.
Shear failure (translational landslide, planar or debris slide).	Mass slope failure on a shallow slip plane parallel to the surface. This is a common type of landslide, slip or debris fall. It may occur on any scale.	May be either shallow or several metres deep.
Shear failure (rotational landslide).	Mass slope failure on a deep, curved slip plane, with the toe of the debris rising slightly. Not common in structurally controlled tropical soil–rock profiles, although rotational landslides can occur in thick residual soil.	Often more than 1.5 m deep, even in a small failure.
Slumping or flow of material when very wet.	Slumping or flow where material is poorly drained or has low cohesion between particles and liquefaction is reached. These sometimes appear afterwards like planar slides but are due to flow rather than sliding.	Frequently 0.5 m or less below surface.
Structurally controlled rock failure.	Failures in fresh to moderately weathered rock masses controlled by mass structures; bedding planes, faults or joints – either singly or in combination. The geometry of the planes in relation to each other and the cut face orientation is important in conjunction with the discontinuity condition.	Can be to a wide range of scales.
Rock or debris fall or collapse.	Collapse due to failure of the supporting material. This normally takes the form of a rock fall where there is a weakness or fracture in a rock mass, or where a weaker band of material has eroded to undermine a harder band above.	0.5–2 m in road cuts; deeper in natural cliffs.
Debris flow.	In gullies and small, steep river channels (bed gradient usually more than 15°), debris flows can occur following intensive rain-storms. This takes the form of a rapid but viscous flow of liquefied mud and debris. The depositional area may cover a broad area below the outlet of the channel.	The flow depth is usually 1–2 m deep.

General approaches to slope stabilisation and protection

Slope stabilisation and erosion control can be addressed through a range of methods to reduce the causes of failure and improve stability. It is important to select affordable methods that are relevant to the nature of the slope problem to be solved, the materials involved, the climatic regime, the extent of the slope instability and, not least, the budget constraints.

For LVRRs, a combination of bio-engineering, low-cost retaining walls such as gabions, dry-stone and mortared masonry walls, and surface drainage structures is a cost-effective approach to slope stabilisation (Fookes et al. 1985). This is in addition to minor remedial work associated with regular maintenance activities such as debris clearance, trimming, and the removal of overhangs. Common LVRR slope stabilisation or protection options are summarised in Figure 10.10. Before selection of an option, appropriate site investigations may be required to define the slope problem within the overall geotechnical environment. In addition, there are well known stabilisation

General Option	General Road-side Earthwork Problem									
	Earthfall	Rockfall-Topple	Rock Slide	Debris Slide	Soil Slide (R)	Soil Slide (T)	Debris Flow	Earth Flow (Soil Creep)	River Erosion	Slope erosion
Removal	1	1	2	2	2	2	2			
Realignment	1	1	2	2	2	2	2	2	1	
Earthwork	1	1	2	2	2	2				
Drainage			2	1	1	1	1	2		1
Retaining wall			2	2	2	1	2	1	1	
Revetment Wall			2	2	1	1	2	2		1
Bioengineering				2	2	2	2	1	2	1
Check Dams				1			1			2
Tied-back wall			1	1	1	1	2	2	1	
Pile wall			2	1	1	1	2	2	1	
Buttress	1	1	2	2						
River Training			2	2	2	2			1	
Anchors-Bolts	1	1	2		2	2				
Catch works	2	2	2	2						
Surface Protection	2	2	2							1

1   Principal option to be considered for problem solution
2   Secondary option

Figure 10.10 General options for LVRR slope protection works.

methods that involve more substantial engineering works, including anchoring, piling and deep subsurface drainage, but these are not commonly used on LVRRs and require specialist geotechnical input. The Nepal Department of Roads manual on roadside slopes (DoR 2007) provides sound general guidance on slope stabilisation in the LVRR context.

Retaining and revetment walls

Retaining walls must be designed to withstand the pressure exerted by the retained material attempting to move forward down the slope due to gravity. The lateral earth pressure behind the wall depends on the angle of internal friction and the cohesive strength of the retained material, as detailed in standard geotechnical texts, as for example by Das (2020). The wall must also withstand pressure due to material placed on top of the fill behind the wall ("surcharge").

Groundwater behind the wall that is not dissipated also exerts a horizontal hydrostatic pressure on the wall and must be considered in the design. Dissipation of ground water is normally achieved by constructing horizontal drains behind the wall with weepholes. These must be adequately maintained throughout the design life of the wall.

The following paragraphs briefly summarise the key points with respect to retaining walls in the LVRR environment, and general options for retaining and revetment walls are summarised in Figures 10.11 and 10.12.

Gravity walls depend on their weight to resist pressures from behind the wall that tend to overturn the wall or cause it to slide. Gravity walls are normally designed with a slight 'batter' to improve stability by leaning the wall back into the retained soil. The foundations should be wide enough to ensure that excessive pressure is not applied to the ground.

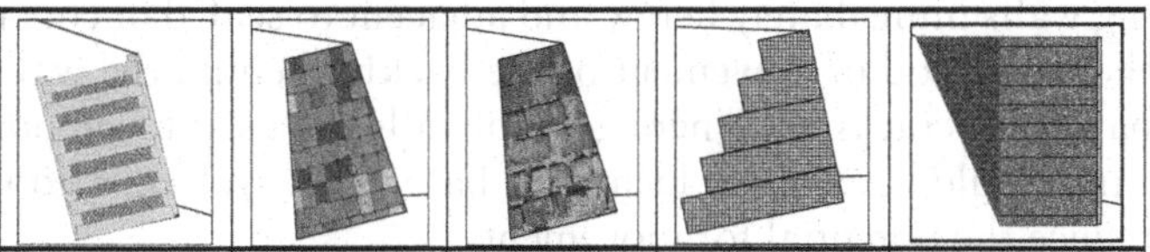

Typical Dimensions	Crib (RCC)	Dry Masonry	Cemented Masonry	Gabion[1]	Reinforced Earth[1]
Top width	1-2m	0.6-1.0m	0.5-1.0m	1m	0.75H
Base width(m)	0.4-0.6H	0.5-0.7H	0.5-6H	06-0.75H	0.75H
Height (m)	4-12m	1-4m	4-8m	1-10m	3-12m
Front batter (V:H)	4:1	3:1	10:1	4:1 - 6:1	3:1
Foundation dip (V:H)	1:4				Horizontal
Foundation depth	0.5-1.0m	0.5m	0.5-1.0m	0.5m	0.5m
1 Specialist organisations may have a wide range of design options to suit specific situations.					

Figure 10.11 Typical retaining wall dimensions for LVRR slope support.

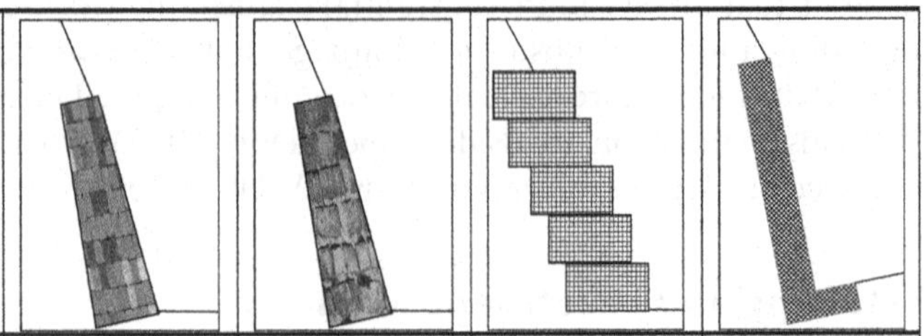

Typical Dimensions	Dry Masonry	Cemented Masonry	Gabion	Concrete
Top width	0.5m	0.5m	2m	0.5
Base width(m)	0.3-0.35H	0.25H	2m	0.25H
Height (m)	3-6m	1-8m	1-6m	1-12m
Back batter (V:H)	3:1	3:1 to 5:1	3:1 to 5:1	3:1
Foundation dip (V:H)	1:3 to 1:5	1:3	1:5	1:3
Foundation depth	0.5m	0.5m	0.5-1.0m	0.5m

Figure 10.12 Typical revetment wall dimensions for LVRR slope protection.

Gabion walls are built from gabion baskets wired together, as shown in Figure 10.13. A gabion basket is made up of steel wire mesh in a shape of rectangular box. It is strengthened at the corners by thicker wire and by mesh diaphragm walls that divide it into compartments. The wire should be galvanised and sometimes PVC coated for greater durability. The baskets have a double or triple twisted hexagonal wire mesh, which allows the gabion wall to deform without the box breaking or losing its strength. Gabion walls are most cost effective where they can employ locally available rock and labour. Gabion structures, commonly used for walls of up to 6 m high, are usually preferred where the foundation conditions are variable, the retained soils require drainage, and continued slope movements are anticipated (Agostini 1987; ICIMOD 1991).

Because of their inherent flexibility, Gabions are not preferred as retaining walls immediately below and adjacent to sealed or concrete roads due to the likelihood of movement of the backfill behind the wall and subsequent pavement cracking. Where gabion walls are used to support a sealed road, care should be taken to locate the base of the wall on a good foundation, to reduce the potential for movement.

Gabion walls have the following advantages:

1. Gabions can be easily stacked in different ways, with internal or external indentation to improve the stability of the wall.
2. They can accommodate some movement without rupture.
3. They allow free drainage through the wall.
4. The cross section can be varied to suit site conditions.
5. They can take limited tensile stress to resist differential horizontal movement.

Figure 10.13 Gabion basket wall.

Their disadvantages include:

1. Gabion walls need large spaces to fit the wall base. (This base width normally occupies about 40%–60% of the height of the wall.)
2. The high degree of permeability can result in a loss of fines through the wall.
3. For road support retaining walls, this can result in potentially problematic settlement behind the wall, although this can be prevented using a geo-textile (filter fabric) between the wall and the backfill.

Gabion mattresses are similar in nature to gabion walls but have a much flatter shape (Figure 10.14) and are commonly used to provide stream or riverbank erosion protection, frequently in conjunction with gabion retaining walls.

Dry-stone walls are constructed from stones without any mortar to bind them together. The stability of the wall is provided by the interlocking of the stones. The great virtue of dry-stone walls is that they are free-draining. They also avoid the high cost of cement mortar. The durability of dry-stone walls depends on the quality and amount of the stone available and the quality of the work. In a slope management situation, they are useful as revetments for erosion protection and as a means of supporting soil against very shallow movement. Dry-stone walls should not exceed 5 m in height.

As with gabion walls and dry-stone walls, a mortared masonry wall design uses its own weight and base friction to balance the effect of earth

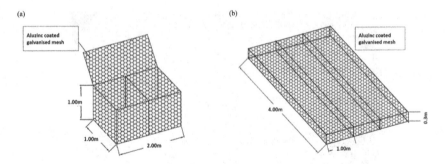

Figure 10.14 General gabion dimensions. (a) Gabion box. (b) Gabion mattress.

pressures. Masonry walls are brittle and cannot tolerate large settlements. They are especially suited to uneven founding levels but perform equally well on a flat foundation. Mortared masonry walls tend to be more expensive than other gravity wall options. If the wall foundation is stepped along its length, movement joints should be provided at each change in wall height so that any differential settlement does not cause uncontrolled cracking in the wall.

Mortared masonry walls require the construction of weep holes to prevent build-up of water pressure behind the wall. Weep holes should be of 75 mm diameter and placed at 1.5 m centres with a slope of 2% towards the front of the wall. A filter of lean concrete or geo-textile should be placed at the back of the weep holes to permit free drainage of water.

Slope drainage

Slope stability is greatly influenced by hydrology, either by the erosive impacts of surface water or the changes in pore pressure resulting from rainfall infiltration and concentration within the slope mass. Water may decrease pore suction in the underlying soil or increase pore water pressure, thereby reducing the effective strength and hence the stability of the slope. The construction of surface and subsurface drainage is a vital aspect of slope stabilisation. Common LVRR earthwork drainage options are summarised below.

Cut-off drains are used to reduce surface runoff at the crest of a cut-slope or slope failure. In order to reduce the likelihood of continuing slope movements breaching the drain, they are sometimes located many tens of metres above the failure crest. The potential problem with cut-off drains is that unless they are regularly maintained, they can create their own instability problem. It is not recommended that cut-off drains be constructed unless regular maintenance can be assured. Low cut-off bunds are an alternative solution.

Herringbone (or chevron) drains are constructed herringbone fashion on slope faces to collect surface seepages and surface runoff. They are often quite shallow (about 1 m deep) but can be much deeper. Care needs to be taken to ensure that the construction of the drain does not lead to further instability and to ensure that the drain can still function in the event of minor downslope movements.

Counterfort drains are used to depress a high water table. These drains are constructed at right angles to the toe of the slope and are often dug to a depth of 3 m or more at intervals of 3–10 m, depending on the permeability of the subsoil.

Horizontal (or sub-horizontal) drains are used to intercept groundwater and seepage at depth. They require the use of plastic pipes and specialist drilling equipment that may not always be available, and they are not easy to install.

Lined channels or cascades are likely to be necessary if a watercourse or gully is a direct cause of the instability in the first place. A lined channel may be necessary to divert an existing watercourse from the failed area or to train the watercourse within defined limits. The lining itself may be impermeable (mortared masonry and/or concrete) or permeable (gabion). The structure may comprise cascades, chutes and check dams. As a general rule, gabion structures are preferred, since they are flexible and allow water ingress provided they are located below the wet season groundwater table.

BIO-ENGINEERING

General

Bio-engineering can be broadly defined as the use of vegetation, either alone or in conjunction with engineering structures, and non-living plant material, to reduce erosion and shallow-seated instability on slopes. Key references, on which this section is based, are available for detailed guidance derived from recent research and practical application; these include:

Hearn, (2011). Slope Engineering for Mountain Roads. Geological Society of London Special Publication No 24. 301 pp. Contains key chapters of slope stability, slope stabilisation and bio-engineering.

Howell. (1999), Roadside Bio-engineering: Site Handbook and Reference Manual. Department of Roads, Kathmandu. A comprehensive study of bio-engineering options in mountainous terrain Nepal.

ICEM. (2017). A series of documents prepared for Ministry of Agriculture and Rural Development (Vietnam) and ADB and based on a research and demonstration sections in Northern Vietnam. Detailed design of bio-engineering options; guidance on their use; outline BoQ and Final Report.

Scott Wilson (2008). A collection of manuals, documents and training materials based on a DFID-funded South East Asian Community Access Programme (SEACAP).

TRL (1997) 'Principles of low-cost road engineering in mountainous regions, with special reference to the Nepal Himalaya', Overseas Road Note 16, Crowthorne: Transport. Contain basic outline designs on bio-engineering and associated "hard" engineering options.

In bio-engineering applications, there is an element of slope stabilisation as well as slope protection in which the principal advantages are:

1. Vegetation cover protects the soil against rain splash and erosion, and prevents the movement of soil particles down slope under the action of gravity.
2. Vegetation increases the soil infiltration capacity, helping to reduce the volume of runoff.
3. Plant roots bind the soil and can increase resistance to failure, especially in the case of loose, disturbed soils and fills.
4. Plants transpire considerable quantities of water, reducing soil moisture and increasing soil suction.
5. The root cylinder of trees holds up the slope above through buttressing and arching.
6. Tap roots or near vertical roots penetrate into the firmer stratum below and pin down the overlying materials.
7. Surface runoff is slowed by stems and grass leaves.

An important associated advantage of the use of most bio-engineering techniques is the added benefit they can bring to local communities. This is not only terms of employment during construction and maintenance, but also in terms of the use to which some of the vegetation can be put to. For example, vetiver grass may be used as cattle fodder, roofing materials or as an ingredient in traditional medicine (Booth et al. 2008).

In summary, vegetation is important in the control of erosion and shallow forms of instability (1–2 m depth at most). However, it is also important to appreciate that the beneficial effects may be insignificant under extreme conditions of rainfall or drought.

Key factors

The main factors to be addressed when selecting a particular species for use in bio-engineering works are summarised as follows:

1. The plant must be of the right type to undertake the bio-engineering technique that is required. The possible categories include:
   • A grass that forms large clumps.
   • A shrub or small tree that can be grown from woody cuttings.
   • A shrub or small tree that can grow from seed in rocky sites.

- A tree that can be grown from a potted seedling.
- A large bamboo that forms clumps.
2. The plant must be capable of growing in the location of the site.
3. There is no single species or technique that can resolve all slope protection problems.
4. It is always advisable to use local species that don't invade and harm the environment, and were able to protect the slope from eroding in the past.
5. Large trees are suitable on slopes of less than 3H:2V or in the bottom 2 m of slopes steeper than 3H:2V, maintaining a line of large trees at the base of a slope can help to buttress the slope and reduce undercutting by streams. Due account must be taken, however, of the potential damage that may be caused by the uprooting of trees due to high winds during storm events.
6. Grasses that form dense clumps generally provide robust slope protection in areas where rainfall is intense. They are usually best for erosion control, although most grasses cannot grow under the shade of a tree canopy.
7. Shrubs (i.e. woody plants with multiple stems) can often grow from cuttings taken from their branches. Plants propagated by this method tend to produce a mass of fine, strong roots. These are often better for soil reinforcement than the natural rooting systems developed from a seedling of the same plant.
8. In most cases, the establishment of full vegetation cover on unconsolidated fill slopes may take one to two rainy seasons. Likewise, the establishment of full vegetation on undisturbed cut-slopes in residual soils and colluvial deposits may need 3–5 rainy periods. Less stony and more permeable soils have faster plant growth rates, and drier locations have slower rates.
9. Most plant roots cannot be expected to contribute to soil reinforcement below a depth of 500 mm; the exception is vetiver grass, which can grow down to 2–2.5 m.
10. Plants cannot be expected to reduce soil moisture significantly at critical periods of intense and prolonged rainfall.
11. Grazing by domestic animals can destroy plants if it occurs before they are properly grown. Once established, plants are flexible and robust. They can recover from significant levels of damage (e.g. flooding and debris deposition).

Preparation

Before bio-engineering treatments are applied, the site must be properly prepared. The surface should be clean and firm, with no loose debris. It must be trimmed to a smooth profile, with no vertical or overhanging areas.

Table 10.6 Recommended General Bio-engineering Procedures

Site characteristics	Recommended techniques
Cut-slopes	
Cut-slope in soil, very highly to completely weathered rock or residual soil, at any grade up to 1H:2V.	Grass planting in lines, using slip cuttings.
Cut-slope in colluvial debris, at any grade up to 1H:1V (steeper than this would need a retaining structure).	
Landslide head scarps in soil, at any grade up to 1H:2V.	
Roadside lower edge or shoulder in soil or mixed debris.	
Cut-slope in mixed soil and rock or highly weathered rock, at any grade up to about 1H:4V.	Direct seeding of shrubs and trees in crevices.
Trimmed landslide head scarps in mixed soil and rock or highly weathered rock, at any grade up to about 1H:4V.	
Fill slopes	
Fill slopes and backfill above walls without a water seepage or drainage problem; these should first be re-graded to be no steeper than 3H:2V.	Brush layers (live cuttings of plants laid into shallow trenches with the tops protruding) using woody cuttings from shrubs or trees.
Debris slopes underlain by rock structure, so that the slope grade remains between 1H:1V and 1H:1.5V.	Palisades (the placing of woody cuttings in a line across a slope to form a barrier) from shrubs or trees.
Other debris-covered slopes where cleaning is not practical, at grades between 3H:2V and 1H:1V.	Brush layers using woody cuttings from shrubs or trees.
Fill slopes and backfill above walls showing evidence of regular water seepage or poor drainage; these should first be re-graded to be no steeper than about 3H:2V.	Fascines (bundles of branches laid along shallow trenches and buried completely) using woody cuttings from shrubs or trees, configured to contribute to slope drainage.
Large and less stable fill slopes more than 10 m from the road edge (grade not necessarily important, but likely eventually to settle naturally at about 3H:2V).	Truncheon cuttings (big woody cuttings from trees).
The base of fill and debris slopes.	Large bamboo planting; or tree planting using seedlings from a nursery.

Figure 10.15 Examples of bio-engineering roadside slope protection. (a) Vetiver grass slips prior to planting. (b) Vetiver grass after 8 months on road embankment. (c) Brush layer just after planting on road embankment. (d) Brush 6 months after planting on road embankment. (e) Live poles and riprap for embankment protection. (f) Combination of concrete frames and grass on cut-slope.

The object of trimming is to create a semi-stable slope with an even surface, as a suitable foundation for subsequent works.

Trim soil and debris slopes to the final desired profile, with a slope angle of between 30° (2H:1.15V) and 60° (1H:1.5V). (In certain cases the angle will be steeper, but review this carefully in each case). Trim off excessively steep sections of slope, whether at the top or bottom. In particular, avoid slopes with an over-steep lower section, since a small failure at the toe can destabilise the whole slope above.

Remove all small protrusions and unstable large rocks. Eradicate indentations that make the surrounding material unstable by trimming back the whole slope around them. If removing indentations would cause an unacceptably large amount of work, excavate them carefully and build a buttress wall. Remove all debris from the slope surface and toe to an approved tipping site. If there is no toe wall, the entire finished slope must consist of undisturbed material.

Recommended techniques

Table 10.6 provides the different types of bio-engineering techniques recommended for various kinds of slopes and soil materials for both cut and fill situations, as illustrated in Figure 10.15.

REFERENCES

Agostini, R. 1987. *Flexible Gabion Structures in Earth Retaining Works*. Bologna, Italy: Maccaferri.

Booth, D., A. Adinata and R. Dewi. 2008. Vetiver systems for community development and poverty alleviation in Indonesia. National Workshop on Vetiver System at Cochin, Kerala, India.

Brand, E. W. 1985. Geotechnical engineering in tropical residual soils. In *Proceedings 1st International Conference on Geomechanics in Tropical Lateritic and Saprolitic Soils, Brasilia*, vol. 3.

Bulman, N. 1968. A survey of road cuttings in Western Malaysia. In *Proceedings of the First Southeast Asian Conference on Soil Engineering, Bangkok*, pp. 219–300.

Caterpillar. 2019. *Caterpillar Performance Handbook*. https://www.warrencat.com/performance-handbook/.

Cook, J. R., P. Beaven and A. Rachlan. 1992. Indonesian slope inventory studies. In *Proceedings of the 7th Conference Road Engineering Association. of Asia & Australasia, Singapore*. https://transport-links.com/research-archive/indonesian-slope-inventory-studies-seventh-reaaa-conference-singapore-22-26-june-1992/.

Cruden, D. M. and D. J. Varnes. 1996. Chapter 2 Landslide types and processes. In *Landslides. Investigation and Mitigation*. Special report no. 247, TRB.

Das, B. M. 2020. *Principles of Geotechnical Engineering*. 10th edition. CENGAGE.

DoR (Department of Roads). 2007. Roadside geotechnical problems: A practical guide to their solution., Gov. of Nepal, Kathmandhu.

Fookes, P. G., M. Sweeney, C. N. D. Manby and R. P. Martin. 1985. Geological and geotechnical engineering aspects of low-cost roads in mountainous terrain. *Engineering Geology* 21:1–152.

GEO, Hong Kong. 2000. *Highway Slope Manual.* Geotechnical Engineering Office, Civil Engineering Department. Gov. of Hong Kong.

Harber, A. J., I. M. Nettleton, G. D. Matheson, P. McMillan and A. J. Butler. 2011. Rock engineering guides to good practice: road rock slope excavation. TRL Project Report, PPR556. TRL Ltd for Transport Scotland.

Hearn, G (ed). 2011. *Slope Engineering for Mountain Roads. Geological Society of London Special Publication No 24:*301 pp.

Howell, J. 1999. Roadside bio-engineering: site handbook and reference manual. Department of Roads, Kathmandu. A comprehensive study of bio-engineering options in mountainous terrain Nepal.

ICIMOD. 1991. *Mountain Risk Engineering Handbook. II Application.* Nepal: International Centre for Integrated Mountain Development.

Ingles, O. 1985. *Translation of Japanese highway design manual, part 1: earthworks.* ARRB research report 51. Special report 247. Washington: TRB.

Mott MacDonald. 2017. Ground Improvement for Khulna Soft Clay Soil. Final report. Research for Community Access Partnership (ReCAP) report. Reference BAN2083A, for UKAID-DFID. https://www.gov.uk/research-for-development-outputs/ground-improvement-for-khulna-soft-clay-soil-final-report.

Parsons, A. W. 1992. *Compaction of Soils and Granular Materials. A Review of Research at the Transport Research Laboratory.* HMSO.

Pettifer, G. S. and P. Fookes. 1994. A revision of the graphical method for assessing the excavatability of rock. *Quarterly Journal of Engineering Geology and Hydrogeology* 27:145–164.

Roughton International. 2000. Guidelines on borrow pit management for low-cost roads. DFID KaR project report (Ref. R6852). https://www.gov.uk/research-for-development-outputs/guidelines-on-materials-and-borrow-pit-management-for-low-cost-roads.

Scott Wilson. 2008. Slope maintenance site handbook. South East Asian Community Access Partnership (SEACAP) 2. UKAID-DFID Report to Ministry of Public Work and Transport, Lao PDR. https://www.gov.uk/research-for-development-outputs/seacap-21-slope-maintenance-site-handbook.

TRL. 1997. *Overseas Road Note 16. Principles of Low-Cost Road Engineering in Mountainous Regions, with Special Reference to the Nepal Himalaya.* Crowthorne: Transport. https://www.gov.uk/research-for-development-outputs/principles-of-low-cost-road-engineering-in-mountainous-regions-with-special-reference-to-the-nepal-himalaya-overseas-road-note-16.

U.S. Dept of Interior (UDSI). 1998. *Earth Manual.* 3rd edition. Bureau of Reclamation.

Weston, D. J. 1980. *Expansive roadbed treatment for southern Africa. RR288. CSIR, South Africa.*

WSP . 2002. Indonesian, 1. Road embankments on soft soils geoguide 1, occurrence & general nature of soft soils. For Institute of Road Engineering, Government of Indonesia.

Chapter 11

Hydrology and drainage

INTRODUCTION

One of the most important aspects of the design of a road is the provision made for protecting the road from the effects of rainwater, surface water and groundwater. From a safety viewpoint, water on the road pavement slows traffic and contributes to accidents from hydroplaning and loss of visibility from splash and spray. If water is allowed to enter the structure of the road, the strength and deformation resistance of the pavement, surface or subgrade will be weakened and roads much more susceptible to damage by traffic. When roads fail, it is often due to inadequate drainage or poor maintenance of the drainage system. Hydrology triggered failures can occur spectacularly as, for example, when cuttings collapse, or when embankments and bridges are carried away by flood water (Hald et al. 2004).

This chapter summarises key aspects of the hydrology of drainage and its input into hydraulic design. Specific options for road drainage assets are discussed elsewhere in this book:

- Earthwork drainage in Chapter 10.
- Culverts, fords, drifts and small bridges in Chapter 12.
- Pavement drainage in Chapter 13.

Long-established standard approaches to engineering hydrology as detailed, for example, by Garg (2010) and Wilson (1990) are adopted, but adjusted where required by the specific needs of Low Volume Rural Road (LVRR) design within a climate change context.

A drainage system has six main functions:

- To convey rain and storm water from the surface of the carriageway to outfalls.
- To control the level of the water-table in the subgrade beneath the carriageway.

DOI: 10.1201/9780429173271-11

- To intercept ground and surface water flowing towards the road.
- To control water running down the faces of earthworks.
- To lower groundwater pressures within earthworks.
- To convey water across the alignment of the road in a controlled fashion.

There are two main steps in a general procedure for road drainage design (Figure 11.1), and they are:

1. Hydrological analysis to estimate the quantity and flow of water necessary to be dealt with by the drainage structures.
2. Hydraulic design to determine details of drains and structures to carry the estimated flows.

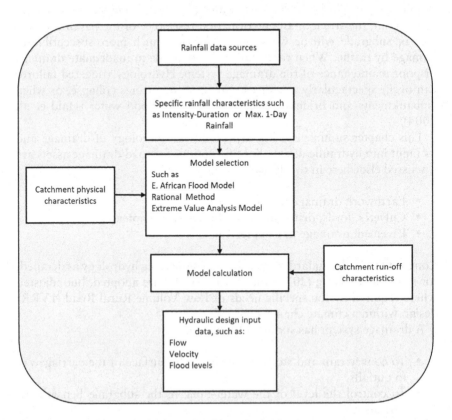

Figure 11.1   General steps from rainfall data to hydraulic design.

HYDROLOGICAL DATA

Rainfall

Although some regional and global patterns of rainfall can be identified, rainfall can be unpredictable and highly variable both in time and space. Topography can create microclimates, which will significantly influence local rainfall intensity. The rainfall volume and intensity are the basic input to many flood estimation methods, and data requirements vary from the average annual rainfall to detailed knowledge of variations in rainfall intensity during a rainstorm with a selected recurrence interval. Large variations in monthly rainfall maxima and minima can occur, even within countries, because of physical setting. This is illustrated in Figure 11.2 which compares rainfall patterns in locations in Myanmar less than 400 km apart but separated by a significant mountain chain.

The extreme events that cause floods are often rainstorms with a duration of between half an hour to a few hours depending on the catchment size. For the larger catchments which have long response times, rainstorms of longer duration with less intensity will be most critical. For small catchment areas with short response time, short-duration high-intensity rainstorms will be most critical. Hence, knowledge of rain event duration as well as intensity is important (Wilson 1990).

Rainfall is recorded using daily rainfall or autographic gauges. An autographic gauge records both time and accumulated rainfall either graphically or digitally (WMO 2018). The accuracy of measurements is influenced by the wind and any obstructions in the vicinity of the gauge. The wind effect may underestimate the actual rainfall by up to 10%–15%, depending on wind speed and exposure.

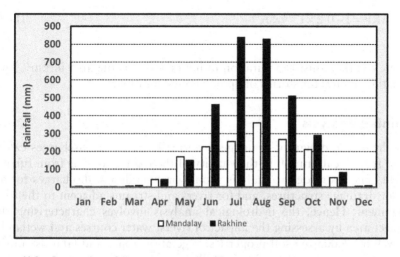

Figure 11.2   Regional rainfall variations within Myanmar.

Measurements made by the above methods will represent point-rainfall, that is, the rainfall at a particular location. However, most often the total quantity of rain falling on a particular area is needed to assess discharges. The conversion of point-rainfall to area-rainfall for a selected standard duration (year, season, month, day, or fraction of a day) can be done by various methods. If data from several rain gauges are available in the catchment under study, then the following methods can be applied:

1. Averaging: A simple average of the amounts of rain at the different gauges for the period of interest.
2. Theissen method: A weighted average; neighbouring stations are connected on a map with lines, and perpendiculars are raised in the middle of each connecting line; each station is represented by an area bounded by the perpendiculars.
3. Isohyets: Curves representing the same amount of rainfall. Areas between isohyet curves can be measured, and the total rainfall calculated by multiplying the areas by the mean value of the higher and the lower isohyets.

The isohyet method is particularly useful in mountain areas, where they can be drawn to consider the effects of topography. High-intensity short-duration storms normally only affect areas of a limited size (area $<10\,\mathrm{km^2}$). For areas larger than this, reduction factors have to be applied to rainfall run-off modelling. In general, the area reduction factor applied to the conversion between point-rainfall and area-rainfall is dependent on both the area and the duration. An example of an area reduction factor (ARF) is

$$ARF = 1 - 0.2 \times t^{-1/3} \times A^{1/2} \tag{11.1}$$

where t is the rainstorm duration in hours and A is the area in $\mathrm{km^2}$. This equation applies for storms of up to 8 hours duration.

Rainfall analysis

The hydraulic design of drainage is normally based on discharges calculated from local rainfall in the vicinity of the road as well as from further afield. The hydrological analysis should provide design discharges for all major drainage structures, and for rivers and streams adjacent to the road alignment. Hence, the hydrological analysis involves characterising the project area by assessing the impact of rivers, water courses and wetlands, as well as evaluating soil properties, vegetation zones and land use in the catchment area, together with rainfall.

The high variability associated with the rainfall requires a probabilistic approach to describing its characteristics. The worst rainstorm occurring in, say, 50 years has a 50-year interval or 50-year 'return period'. The 'recurrence interval' (T) is defined as the average number of years between a rainstorm of a given size, and the recurrence interval is the inverted probability of occurrence. It should not be inferred that the storm occurs regularly at T-year intervals, but that it has a probability of occurrence of $1/T$ in any year. It can be shown that the probability of the occurrence of the T-year event within a period of T years is 63%.

When analysing a raw set of rainfall data, the most widely applied tool is an Extreme Value Analysis (EVA). This method is also applicable to other data series, such as river discharge data. EVA is a statistical process for dealing with the extreme deviations from the median of probability distributions; hence, it is suited for dealing with climate storm events (ADB 2020). It seeks to assess, from a given ordered sample of a given random variable, the probability of events that are more extreme than any previously observed. For hydrology, EVA is used to estimate the probability of an unusually large flooding or drought events, such as the 100-year flood. It is used for both rainfall and discharge data.

The analysis is normally performed for an annual series of data. For a selected duration of rainfall, for instance, the 24-hour, 2-hour, 1-hour or 30-minute rainfall, the highest value is determined from each calendar year. This series is termed an 'annual maximum series'.

This approach can result in the selection of two rainfalls within the same wet season (e.g. one in December and one in January) and sometimes one from another wet season. This problem occurs when the wet season extends over the new year, for example from October to April. This can be overcome by taking the 'hydrological year' instead of the calendar year, so that one peak is selected for each wet season. The annual records are listed in order of descending values, and a rank is assigned to each of them, the first having rank '1', the second rank '2', and so on. Next, the 'non-exceedance probability', or recurrence interval/return period for each value can be calculated by:

$$F = 1(1 - 1/T)^N \tag{11.2}$$

where F is the non-exceedance probability, T the recurrence interval, N the number of years in the series.

A distribution curve is fitted to the series of annual rainfall extremes. The best fit is usually obtained by using the 'Gumbel distribution' (Gumbel 1954). The Gumbel Method can eliminate the skew and linearise the distribution, so that standard linear regression techniques can be used to find the line of best fit for the relationship between storm rainfall and recurrence interval. The straight-line relationship can then be extrapolated, with caution, to predict the magnitude of longer recurrence interval events. Worked examples of Gumbel distribution relevant to extreme rainfall estimation for hydraulic structures are included in the ADB (2020) manual on climate change adjustments for the design of roads.

A meaningful statistical analysis requires at least 10 years of reliable records, and, even then, extrapolated values need to be used. If records are shorter than 10 years, the 'station-year' technique may be applied. This technique mixes records from several stations, assuming that the records are homogeneous and that the events are independent. Records from several stations are then lumped together, considered as a single record, and analysed accordingly. The results are then considered valid for all the stations that were sources of data. The station-year technique may be appropriate for rainstorm analyses but should not be used in an analysis of discharge data.

Intensity–Duration–Frequency (IDF) curves for rain stations show the average rainfall intensity variation during the length of storm, with a particular return period (RP). Examples of (IDF) curves for different return periods (RPs) are shown in Figure 11.3.

It is unusual for adequate rainfall data to exist for the development of a full set of rainfall intensity curves for a location of interest. Normally, only rainfall data for a fixed duration, such as the 24-hour rainfall, is available, and no information on the time distribution exists. In such cases, it may be possible to use a generalised relationship and determine the appropriate constants to use in this. The rainfall is converted to intensity by dividing by the duration.

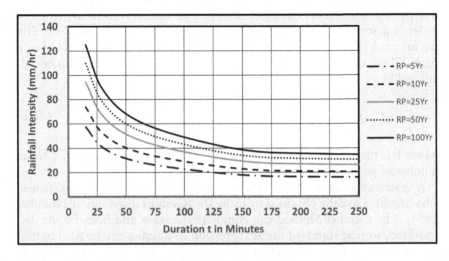

Figure 11.3 Typical intensity–duration curves (location in central Laos).

Data challenges

Cost constraints, non-availability of detailed maps and other basic data, together with difficult logistics limit the quality and extent of hydrological surveys. The following challenges often occur with hydrological assessment for rural roads in LMICs:

- Rain gauge and discharge station networks are of low density.
- Quality and accessibility of data or record keeping can be poor.
- Few rainfall intensity measurements are available.
- Only short time-series data exist.
- Flood estimation practices are not standardised.
- Land-use changes, high-intensity rainfall and erosion cause dynamic changes in the drainage systems, making predictions difficult.

The nature of hydrology investigations made will depend on the data availability and the possibilities for acquisition of supplementary data. The most comprehensive hydrological analysis is carried out using numerical rainfall run-off models such as MIKE 11 (https://www.dhigroup.com/download/mike-2017/) which has been under continuous development since 1972 (Hald et al. 2004). The Mike 11 model simulates surface flow, interflow and groundwater base flow, by taking into account time-varying storage capacity, irrigation and groundwater pumping. Numerical simulations have proved to be a reliable design tool.

Information surveys

The information and hydrological data sets that ideally should be collected are as follows:

1. Topographic maps, with scale depending on project scope, but 1:25,000 or 1:50,000 will usually be adequate for identification of the geometric parameters of the catchments.
2. Air photos, photo mosaics or satellite images from which land use can be studied when land-use maps are not available.
3. Soil and vegetation maps or general descriptions.
4. Water use in the project area, dams, reservoirs, abstractions for irrigation and other factors that may affect the run-off pattern.
5. Rainfall data, in terms of maximum intensities, as well as general climatologic information.
6. Discharge gauging station data, in terms of annual flood discharges, and relation to stage discharge for the purpose of evaluating the accuracy of the flood flow data.

Further data collection will have to take place in the field as part of the road site investigation. At stream sites where bridges or major culverts are to be constructed, data on the following should be collected:

1. Stream cross-section area.
2. Bed material (in order to estimate the 'Manning roughness').
3. Longitudinal slope of stream bed.
4. Details of historic floods obtained from local residents (to estimate peak flood discharges).

Likely future changes in the run-off regime due to climate change should be evaluated. Evidence should be noted of morphological changes, building up or degradation of riverbed levels, erosion patterns, debris size and controls, hydraulic controls, existing drainage structures, evidence of scour, stability of river banks, and land-use changes.

Flood discharge estimation

The run-off in rivers and streams is the result of several hydrological processes. In general, run-off comprises the surface run-off, the interflow and the groundwater flow. The surface run-off is the direct result of the excess rainfall, while interflow is the water moving through the soil near the surface. The groundwater component is the contribution from the aquifers through which the stream passes. During a major flood, the surface run-off is predominant, and usually only the contribution from the excess rainfall has to be taken into account. The variation in discharge over time, as a result of the run-off processes, depends on a number of characteristics of the rainfall, the catchment and the stream. The most significant parameters are the rainfall intensities and their geographical distribution, the catchment area size, the shape and slope of the catchment, the soil infiltration capacity, the vegetation, the slope and roughness of the stream, and the storage capacity of the catchment.

A plot of the discharge variation over time is called a 'hydrograph', and the peak of this hydrograph is the maximum discharge. For small catchments, maximum discharge occurs when rain falling on the remotest part of the catchment reaches the stream or drainage system. At this point in time, called the 'time of concentration', the whole catchment contributes to the discharge. For larger catchments, the run-off will be delayed before reaching the aquifer. The shape of the hydrograph varies significantly depending on the rainfall intensities, durations and on the catchment characteristics. The most peaked hydrographs are found in mountain catchments, where no vegetation attenuates the surface run-off and where the infiltration capacity is very small. Lower peaks are found in low-relief catchments with extensive vegetation and good infiltration capacity. These catchments serve as reservoirs and will delay the run-off significantly. A typical mountain catchment hydrograph is shown in Figure 11.4.

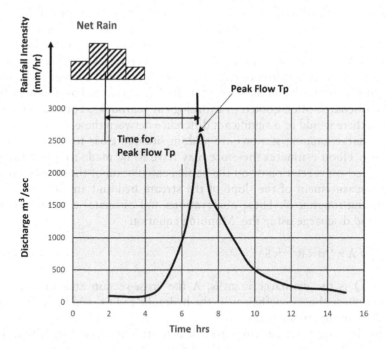

Figure 11.4 Typical mountain catchment hydrograph.

Different methods for estimation of flood discharges have been developed. The methods vary in complexity from 'black-box' estimates of maximum flood peaks to detailed reconstruction of hydrographs. The methods may be divided into three categories:

- Direct flow data.
- Run-off modelling.
- Regionalised flood formulae.

The three methods are listed in order of decreasing reliability and also decreasing data requirements. In all cases of flood assessment, estimates must be checked against historic evidence, local experience and practice, research, and earlier studies. In spite of considerable efforts, flood estimation is still an uncertain science, and allowances must be made accordingly when engineering works are being designed.

Direct flow data

The most common form of stream flow measurement comprises monitoring of the water levels at a suitable cross-section. The relationship between water level and discharge at the location is established through a series of discharge measurements (stage–discharge curve). Extrapolations of this curve will give estimates of the peak flood discharge for different flood water levels. An Extreme Value Analysis of the annual maximum series will yield estimates of floods for different recurrence intervals. It is important to distinguish between the momentary peak and the maximum discharge averaged over a day, which is the discharge often quoted in hydrological yearbooks. For smaller catchments, there would be a significant difference between these values.

Discharge gauging stations are seldom situated close to the location of interest. Flood estimates therefore may have to be made from levelling of marks left on vegetation or on the banks, calculation of the cross-sectional area, measurement of the slope of the stream bed and an estimate of the 'Manning roughness'. These observations and estimates combine to yield the flood discharge using the 'Manning equation':

$$Q = A \times 1/n \times R^{2/3} \times S^{1/2} \tag{11.3}$$

where Q is the discharge in m^3/s, A the cross-section area in m^2, n the Manning roughness coefficient, R the hydraulic radius, and S the longitudinal slope of the stream bed.

This method for estimating a peak discharge is termed the 'slope-area method'. Roughness changes with variation in factors such as vegetation and water level. Thus, large seasonal changes in the roughness must be expected, and the choice of Manning roughness coefficient should reflect the flooding situation. Table 11.1 presents some typical Manning Roughness coefficients (AfCAP 2013).

Table 11.1 Manning Roughness Coefficients (n)

Stream characteristics	Ranges of values of n
Streams in upland areas	
1. Gravels, cobbles and boulders with no vegetation	0.030–0.050
2. Cobbles and large boulders	0.040–0.070
Streams on plains	
1. Clean straight bank	0.025–0.033
2. Same as 1 but with some weeds and stones	0.030–0.040
3. Winding watercourse, some pools and shoals but clean banks	0.035–0.050
4. As 3 but straighter river with less clearly defined banks	0.040–0.055
5. As 3 but with some weeds and stones	0.035–0.045
6. As 4 but with stony sections	0.045–0.060
7. River reaches with weeds and deep pools	0.050–0.080
8. Very weedy river reaches	0.080–0.150
9. Overbank flooded stream flowing across grass	0.030–0.050
10. Overbank flooded stream flowing through light bush	0.040–0.080

Another way of assessing flood estimates, when no gauging station is available at the location of interest, is by carrying out a series of discharge measurements to obtain the short-term statistical distribution. Simultaneous discharge measurements at an old discharge gauging station will provide short-term statistics for the same period, and the correlation between the two locations can be determined. Long-term data will be available from the gauging station, and it is possible to construct a long-term statistical distribution, which can then be transformed to the location of interest provided that a correlation exists.

The most widely used rainfall run-off relationship for ungauged areas is the 'rational method'. The method applies constant rainfall over the entire catchment and is thus most suitable for small catchments of sizes up to, say, a few square kilometres. The consequence of applying the rational method to larger catchments is an over-estimate of the discharges and, thus, a conservative design. The area reduction factor (ARF) can be included to account for spatial variability over the catchment to compensate for the over-estimates. The basic form of the 'rational' equation is

$$Q = (C \times I \times A)/3.6 \tag{11.4}$$

where Q is the flood peak discharge at catchment exit (m³/s), C the rational run-off coefficient; I the average rainfall intensity over the whole catchment (mm/h) for a duration corresponding to the time of concentration and A the catchment area (km²).

The time of concentration is defined as the time required for the surface run-off from the remotest part of the drainage catchment to reach the location being considered. The time of concentration can be calculated by the 'Kirpich formula' (Hald et al. 2004):

$$T_c = \left[\left(0.87 \times L^2 \right) / (1000 \times S) \right]^{0.385} \tag{11.5}$$

where T_c is the time of concentration in hours, L the main stream (km), and S the average main stream (m/m) slope.

Having determined the time of concentration, the corresponding rainfall intensity can then be obtained from the intensity–duration curve for the selected recurrence interval or return period.

The run-off coefficient, C, combines many factors influencing the rainfall run-off relationship, that is, topography, soil permeability, vegetation cover and land use. The run-off coefficient can be estimated using Table 11.2. The selection of the correct value for the run-off coefficient presents some difficulty because of the wide range of parameters reflected. The value can vary from one time to another depending on changes, especially in soil moisture conditions. Thus, the choice of run-off coefficient should be accompanied by field observations and by an evaluation of sensitivities of discharge estimates.

Table 11.2 Run-off coefficient for the Rational Method

Components	Runoff Coefficients			
Relief (terrain) Cr	0.28-0.35 Steep, slopes >30%.	0.20-0.28 Hilly, slopes 10-30%.	0.14-0.20 Rolling, slopes 5-10%	0.08-0.14 Flat, slopes 0-5%.
Soil infiltration Ci	0.12-0.16 No effective soil cover.	0.08-0.12 Slow, clayey soils; poorly drained.	0.06-0.08 Normal; well drained soils, sandy soils.	0.04-0.06 Deep sand; very light, well-drained soils.
Vegetal cover Cv	0.12-0.16 Bare or very sparse cover.	0.08-0.12 Less than 20% of drainage area has good cover	0.06-0.08 Fair to good; about 50% of area in good grassland or woodland.	0.04-0.06 Good to excellent; about 90% in good grassland, woodland.
Surface Storage Cs	0.10-0.12 Negligible; few surface depressions and shallow, drainageways.	0.08-0.10 Well-defined system of small drainageways.	0.06-0.08 Normal; storage lakes and ponds and marshes	0.04-0.06 Much surface storage, large floodplain storage, ponds or marshes

Note: The total runoff coefficient based on the 4 runoff components is **C** = Ci+ Ci+ Cv+ Cs

Catchment areas are determined from regional maps or aerial photographs. Varying topography and widely spaced contours on large-scale mapping sometimes make it difficult to fix catchment boundaries.

The rational method assumes that:

1. The design storm produces a uniform rainfall intensity over the entire catchment.
2. The relationship between rainfall intensity and rate of run-off is constant for a particular catchment.
3. The flood peak at the catchment exit occurs at the time when the whole catchment contributes to the discharge.
4. The coefficient, C, is constant and independent of rainfall intensity.

CLIMATE CHANGE AND HYDROLOGY

Key issues

Climate change is now a generally accepted reality. There is a consensus that climate change will continue to occur over the coming decades and that it will have a significant impact on rural road network sustainability (World Bank 2015). The principles of climate change as they relate to the overall resilience of LVRRs are discussed in Chapter 17. It is however important in the following section to recognise the relationship between climate change, hydrology and drainage.

The changes in rainfall amounts, intensities and patterns will have an increasingly significant impact on the hydrological environments and how roads perform in terms of service delivery. It is important, therefore, that rainfall changes within the design life of roads assets are reflected in the data sets used to define the hydrological parameters used in hydraulic drainage design.

The standard approach to drainage design using historical records does not take into account possible effects of climate changes. Drainage provision that was satisfactory in the past may well not be adequate in the future. A fundamental issue should, therefore, be recognised: hydrology inputs to hydraulic design can no longer be based solely on historical data, but must take into account assessed future climate data sets. It is likely to be highly misleading to base the hydraulic design for a bridge, for example, with a 50–100 years design life on rainfall data sets from the previous decades.

General data sources

Global Climate Change Models (GCMs) are used by the IPCC (Intergovernmental Panel on Climate Change) to define alternative greenhouse gas emission levels with different atmospheric concentration future trajectories known as Representative Concentration Pathways (RCPs). The projections from RCPs are then used as inputs to GCMs in generating climate change projections. The Representative Concentration Pathways (i.e. RCP2.6, RCP4.5, RCP6.0 and RCP8.5) represent assumptions on future climate change mitigation; RCP2.6, for example, represents a very strong mitigation scenario, whereas RCP8.5 assumes little if any mitigation (IPCC 2014).

The World Bank Climate Change Knowledge Portal (CCKP) is a hub for climate-related information, data, and provides online access to comprehensive data related to climate change and development. Climate data aggregations are currently available at national, sub-national and watershed scales (World Bank 2021).

Other generally available sources include:

- The Southern Africa Development Community (SADC), Regional Climate Services Centre (RCSC), http://csc.sadc.int/en/climate/historical-reference/climdex-indices.
- The Koninklijk Nederlands Meteorologisch Instituut (KNMI) Climate Change Atlas can be accessed through website: http://climexp.knmi.nl/plot_atlas_form.

Table 11.3 Scenarios Used in IPCC 2013

RCP scenarios	Project CO_2 atmospheric concentration by 2100	Climate policy
RCP 2.6	421	Mitigation
RCP 4.5	538	Stabilisation
RCP 6.0	670	Stabilisation
RCP 8.5	936	No policy application

Specific country data on climate change projections and guidance may be available through relevant ministry websites.

Changing data and standards

The design of climate resilient drainage infrastructure is based on knowing how much water the drainage system needs to deal with during its design life. This is essentially in turn based on knowledge of climate change hydrology and change in rainfall related to storm return periods.

The accuracy and availability of climate change hydrology data for use in LVRR design are highly variable between countries and dependant on, for example, the amounts and level of downscaling from regional models undertaken or funded by relevant ministries. For most LVRR projects there will be limited budget available for specific project downscaling or modelling. The use of data such as that from the World Bank CCKP usually provides sufficient data to make sensible projections when allied with locally available climate information.

The appropriate selection of climate change data sets to be applied to hydrological parameters will be a function of:

1. The selection climate period in terms of years: this will be a function of the design life of the road asset(s) in question from, for example, 10–20 years for a LVRR pavement to 50–100 years for a small bridge.
2. The RCP scenario to be used. General guidance on this is shown in Table 11.3 and is related to the amount of risk to be accepted and may be governed by National Standards or Guidance. For example the Vietnamese Ministry of Natural Resources and Environment (MoNRE 2016) guidance states that RCP 8.5 scenario data should be used for all long-term structures, but that RCP 4.5 data may be used for non-long-term structures.
3. Design standards. Different design standards will normally apply to different classes of road and type of drainage component being considered. Country-specific design manuals will often contain recommended design flood recurrence intervals, for example, the South Sudan Low Volume Roads Design Manual (AfCAP 2013).

Key data sets that cover the road asset design life in this context are:

- Overall rainfall pattern change.
- Maximum daily rainfall changes.

Figure 11.5 presents an example of the general rainfall information that can be obtained at province or district level from the World Bank CCKP for a specific future period (2080–2099) and for a specific RCP value (8.5). The CCKP use a wide range of climate models in their data analyses, which normally produces a wide range of results; hence it usually presents the option of a range (10%–90% confidence levels) and median values. The latter are usually used, although some national guidance procedures require the use of specific models which may show higher percentage changes than the median value. Figure 11.6 illustrates a typical range of model outputs that require careful selection.

Maximum 1-day or 5-day rainfall changes are used as basic of hydrological climate change input data. Typical figures in Table 11.4 indicate the sensitivity of maximum rain levels to the selection of scenario and return period, and the need for pragmatic selection of data to avoid unnecessary or unjustifiable costs.

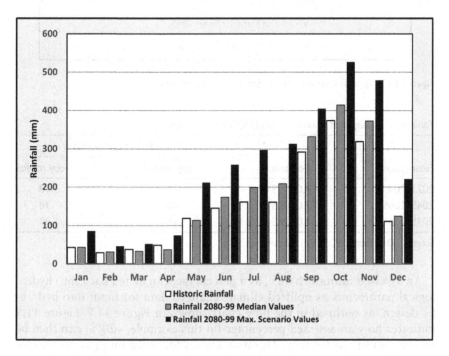

Figure 11.5 Example of current and future rainfall – Gai Lai province Vietnam.

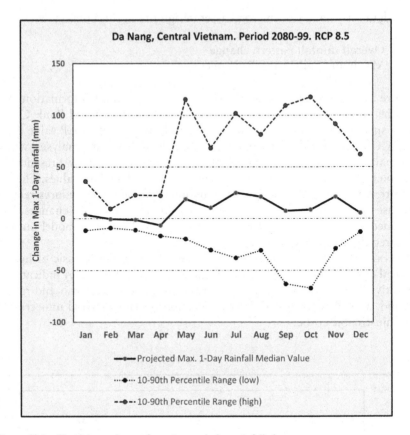

Figure 11.6   Typical envelope of maximum 1-day rainfall changes.

Table 11.4   Example of Maximum Daily Rainfall Variations

		Percentage increases	
Future years	Scenario	1-day rainfall	5-day rainfall
2025–2045	RCP 8.5	17	18
2045–2065	RCP 8.5	40	38
2045–2065	RCP 4.5	28	23

Source: Data from IMC (2018), Tanzania.

An assessed rainfall change, as a percentage, can be fed back into hydrological parameters as uplifted climate changed data for input into hydraulic design, as outlined in the flow chart shown in Figure 11.7. Figure 11.8 indicates how an assessed percentage (in this example, +20%) can then be used to adjust the Intensity-Duration curves for input into parameters for hydraulic design.

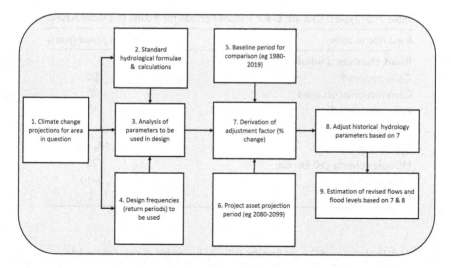

Figure 11.7 Use of a climate change adjustment factor for hydraulic design.

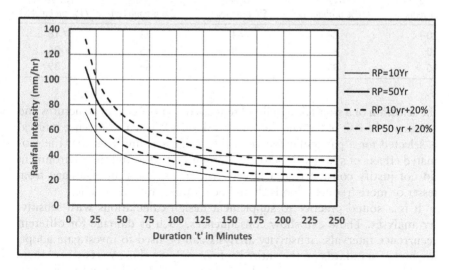

Figure 11.8 A typical uplift adjustment of IDF curves for assessed climate change.

HYDROLOGY AND RISK

The design discharge for a drainage facility depends on the selected flood frequency or rainfall event return period (RP). By selecting a large recurrence interval and a correspondingly large design flood, the probability of having such a flood occurring and the risk of damage is reduced, but costs of structures to accommodate such large floods are increased. Conversely,

Table 11.5 Typical Climate Event Return Periods for Roads and Road Assets

Road type or asset	Return period (years)
Road class (as a whole)	
Collector road	50
Community access road	25
Agricultural road	15
Drainage asset	
Small bridges	50–100
Multiple culverts (>0.9 m dia)	25
Small culverts	5–10
Ditches	2

Table 11.6 Probability of Design Standards Being Exceeded during Design Life

Design life in years	Return period (years) and probability of design flow being exceeded (%)			
	10 years (10% probability)	20 years (5% probability)	50 years (2% probability)	100 years (1% probability)
20	88	64	33	18
30	96	78	45	26
50	99	92	64	39

the selection of a small design flood reduces the initial cost of structures and increases the risk of damage from larger floods. When a recurrence interval is selected for a particular location, the designer is implying that the estimated effects of a larger flood on life, property, traffic and the environment do not justify constructing a larger structure at the time. Damage from lesser or more frequent floods should be minimal and acceptable.

It is a sound practice to supplement design calculations with sensitivity analyses. These can show consequences, such as damage for different recurrence intervals. Sensitivity analysis can be used to investigate adopting higher standards on the entire route or only on those parts where there would be severe consequences. The analysis can also indicate where it might be reasonable to accept slightly lower standards, if the cost of achieving the preferred standards is too great.

Risk can be defined in terms of the recurrence interval or the probability of a flood of stated magnitude being exceeded in any 1 year. Drainage structures are designed to accommodate a flow of a given return period or probability of occurrence. For example, for rural roads in the tropics it is usual to design small culverts for a 1 in 10 years (or 10% probability of occurrence event) but bridges for 1 in 50 or 100 years (or 2%, or 1%, probability of occurrence, respectively), as shown in Table 11.5.

Over any period of time, there is a given probability that the design flow will be equalled or exceeded during the lifetime of the structure, Table 11.6. The probability that drainage capacity has been exceeded can be calculated as:

$$P = 1 - (1 - 1/T)^L \tag{11.6}$$

where P is the probability, T the recurrence interval or return period in years, and L the lifetime in years.

Lower standards are adopted for culverts than say bridges, for example, because these are smaller structures and cheaper to repair if damage occurs from an under-design event. To build culverts to accommodate longer return period events would increase capital costs beyond economic levels. The selection of the design flood recurrence interval involves an evaluation of the risk of disruption or damage to the road, that is, possible loss of life, property damage, the interruption of traffic, and the economic consequences. This should be balanced with the capital and operational cost of providing and maintaining drainage. However, a theoretical risk-based approach to design can still present problems. For example, the 50-year flood can be exceeded at one location with minimal damage, but occurrence of the same frequency flood elsewhere might be disastrous. The combining of theoretical risk with local knowledge, site investigation and good engineering practice is essential.

REFERENCES

ADB. 2020. *Manual on Climate Change Adjustments for Detailed Engineering design of Roads Using Examples from Vietnam.* http://dx.doi.org/10.22617/TIM200147-2.

AfCAP. 2013. *South Sudan low volume road design manual; volume 2 cross drainage and small structures. DFID for Ministry of Roads and Bridges, Government of South Sudan.* https://www.gov.uk/research-for-development-outputs/south-sudan-low-volume-roads-design-manual-volume-1-2-and-3.

Garg, S. K. 2010. *Hydrology and Water Resources Engineering.* Khanna Publishers. ISBN 8174090614.

Gumbel, E. J. Statistical. 1954. *Theory of Extreme Values and Some Practical Applications, National Bureau of Standards Applied Mathematics, Series 33.* US Department of Commerce.

Hald, T., J. Hassing, M. Høgedal and A. Jacobsen. 2004. Chapter 10 Hydrology and drainage. In *Road Engineering for Development.* 2nd edition. Robinson R. and Thagesen B., Eds. Taylor and Francis. ISBN 0-203-30198-6.

IMC Worldwide. 2018. Building climate resilience into the improving rural access, Tanzania (IRAT) PART B, climate adaptation strategies and plans for the pilot District (Bahi), Tanzania. Report for DFID.

IPCC. 2014. Climate change 2014: Synthesis report. Contribution of Working Groups I, II and III to the fifth assessment report of the intergovernmental panel on climate change IPCC, Geneva, Switzerland, 151 pp. https://www.ipcc.ch/report/ar5/syr/.

MoNRE 2016. *Climate Change and Sea Level Rise Scenarios for Viet Nam*. Ministry of Natural Resources and Environment. Vietnam. http://imh.ac.vn/files/doc/KichbanBDKH/CCS_SPM_2016.pdf.

Wilson, E. M. 1990. *Engineering Hydrology*. 4th edition. Macmillan Education Ltd.

WMO. 2018. *Guide to Hydrometeorological Practices*. Geneva: World Meteorological Organization. https://library.wmo.int/doc_num.php?explnum_id=5541.

World Bank. 2015. *Moving toward Climate-Resilient Transport: The World Bank's Experience from Building Adaptation into Programs*. Washington DC: World Bank Transport & ICT. http://hdl.handle.net/10986/23685.

World Bank. 2021. *User Manual. Climate Change Knowledge Portal; (CCKP)*. Washington, DC: World Bank Group. https://climateknowledgeportal.worldbank.org/.

Chapter 12

Cross-drainage small structures

INTRODUCTION TO CROSS-DRAINAGE

It is essential for sustainable access and mobility that that cross-drainage structures are located along the road alignments where water courses would otherwise be cut. Water courses do not distinguish between low and high-volume traffic roads. The engineering approaches to protecting a road from the effects of water are similar and largely independent of traffic, although selection of options may vary on economic grounds. Hence water crossing structures can form a major construction cost component of a road that, depending on the road category, surface type and topography, may account for a large percentage of the total route cost.

Many rural roads can be satisfactorily passable for the great majority of their length but, following rain, may then be impassable where they are traversed by watercourses, or at low points in the alignment where adequate drainage structures are absent, as illustrated in Figure 12.1. Once a road has been constructed, the all-season passability and maintenance cost are closely linked to the quality of the cross-drainage provision.

General guidelines on rural road structures are widely available (TRL 2000; IRC 2000; ERA 2016), while Larcher et al. (2010) focus on Low Volume Rural Roads (LVRR) structures with the provision of standard designs and standard Bills of Quantity. Individual countries may have specific structure guidance, for example USAID (2006).

Because of their importance to sustainable access, it is appropriate to highlight key issues relating to the planning, design, construction and maintenance of culverts, drifts, fords and small bridges in this book. For the purposes of this chapter discussion of small structures covers the following rural road components:

- Culverts, single and multiple.
- Fords, drifts and causeways, also known as Low-Water Crossings (LWCs).
- Small single-span bridges.

DOI: 10.1201/9780429173271-12

(a)   (b)

Figure 12.1 Access hotspots at stream/river crossings. (a) Potential stream crossing hotspot - dry season. (b) Access hotspot - wet season.

CROSS-DRAINAGE TYPES

Simple culverts

Culverts are required at small river or stream crossings where they convey water directly under the road. In addition, culverts may also be required as follows:

1. Where roads traverse low lying areas, especially those which have been developed for irrigated agriculture.
2. To enable transfer of water through raised embankments, which otherwise could act as dams retaining flood waters.
3. In steep terrain where relief culverts are required to ensure that side drainage capacity is not over loaded.

Even though relief culverts may transmit very small flows, it is always recommended that the minimum allowable size for any culvert is installed at 0.8–1.0 m in diameter, as smaller culverts can easily become blocked with debris and may prove difficult to clear manually. Table 12.1 presents current recommendations on culvert spacing, irrespective of the occurrence of permanent or transient water courses.

Barrel or box culverts may have single or multiple openings (see Figure 12.2). Barrel culverts usually consist of a concrete or steel pipe, while box culverts are generally constructed using reinforced concrete, brick or masonry. In general, box culverts are easier to construct using a local labour and construction methods, and are more easily checked for quality. However, most countries make precast concrete pipes of up to about 1 m

Table 12.1 Minimum Recommended Relief Culvert Spacing

Road gradient (%)	Culvert interval (m)
12	40
10	80
8	120
6	160
4	200
2	250–500

(a)    (b)

Figure 12.2   Common LVRR culvert types. (a) pipe culvert. (b) box culvert.

diameter, and these may be cost-effective, provided that they can be satisfactorily transported, handled, bedded, placed and backfilled.

Reinforced-concrete box culverts are normally cast in place, although smaller sizes may be precast. Corrugated galvanised steel pipes, often known by the trade name 'Armco', are often available in large diameters and are easy to handle and place (Figure 12.3). Annual inspections and clearing of debris are required to counter corrosion that could occur to the metal pipes.

Most culverts have an upstream headwall and terminate downstream with an end wall (or downstream headwall). Headwalls direct the flow into the culvert, while end walls provide a transition from the culvert to the outlet channel. Both protect the embankment or roadbed from erosion by flood water. Straight headwalls placed parallel to the roadway are used mainly with smaller pipe culverts. Culvert headwalls are normally supplemented with wingwalls at an angle to the embankment. Most headwalls and wingwalls are concrete, although masonry or brick are also used. In all cases, a cut-off wall (toe-wall) extending below the level of expected scour should be incorporated in the design of the outlet. Often, a paved apron

Figure 12.3   'ARMCO' culvert construction.

inlet extending beyond the cut-off wall is a wise addition; Figure 12.4 illustrates key components.

For larger volumes of water, multiple pipes or boxes can be used in parallel under the road. Multiple pipes can also be used where the planned embankment height is insufficient to cover a single pipe of sufficient diameter (see Figure 12.5).

Locations of culverts should be selected carefully. The alignment of a culvert should generally conform to the alignment of the natural stream. The culvert should, if possible, cross the road at a right-angle to minimise cost. However, skew culverts located at an angle to the centre-line of the road may be needed in some instances. The slope of the culvert should generally conform to the existing slope of the stream. To avoid silting or erosion, the slope of the culvert should not be less than 1%.

The following issues influence the determination of culvert size:

- Headwater depth.
- Tailwater depth; outlet velocity.
- Culvert flow with 'inlet control'.
- Culvert flow with 'outlet control'.

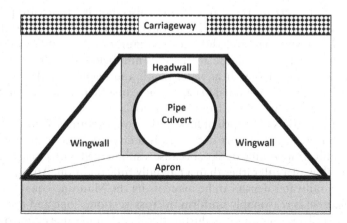

Figure 12.4   Culvert nomenclature.

Figure 12.5   Pipe culvert.

Culverts generally constrict the natural flow of a stream and cause a rise in the upstream water level. The height of the water at the culvert entrance is termed headwater elevation, and the total flow depth in the stream measured from the culvert inlet invert is termed headwater depth. The headwater elevation should not be so high that it can damage both the culvert and the road, and interrupt the traffic. It may also damage upstream property and create a risk to human life. Headwater depth is a function of the discharge, the culvert size and the inlet configuration, and its elevation for the design discharge should be at least 500 mm below the edge of the road shoulder. A ratio between headwater depth and height of culvert opening equal to 1:1.2 is recommended in cases where insufficient data are available to predict the flooding effect from high headwater.

Tailwater depth is the depth of flow in the downstream channel measured from the invert at the culvert outlet. Tailwater depth can be an important factor in culvert hydraulic design, because a submerged outlet may cause the culvert to flow in a full rather than a partially full condition. An approximation of the tailwater depth can be made using the Manning equation, if the outlet channel is reasonably uniform in cross-section, slope and roughness. However, tailwater conditions during floods are sometimes controlled by downstream obstructions or by water conditions. A field inspection should always be made to check on features that may influence tailwater conditions, but considerable experience is then needed to predict tailwater depth.

The outlet velocity, measured at the downstream end of the culvert, is usually higher than the maximum natural stream velocity. This higher velocity can cause stream bed scour and bank erosion for a limited distance downstream. There are two major types of culvert flow: with inlet control or with outlet control. A culvert operates with inlet control when the flow capacity is controlled at the entrance by the following factors:

- Culvert type (shape of barrel/opening).
- Type of culvert inlet.
- Culvert cross-sectional area.
- Headwater depth.

When a given culvert operates under inlet control, the headwater depth determines the culvert capacity, with the barrel usually flowing only partially full. For culverts flowing under inlet control, wingwalls improve the hydraulic characteristics.

When a culvert operates under outlet control, the flow capacity is determined by the same factors as under inlet control, but, in addition, the performance depends on roughness of the inner surface of the culvert (barrel roughness), longitudinal slope of the culvert (barrel slope), tailwater depth or critical depth. Culverts operating under outlet control may flow full or partly full, depending on various combinations of the determining factors.

Fords and drifts

Fords, sometimes referred to as drifts, are Low-Water Crossings (LWCs) that can offer a desirable low-cost alternative to bridges for stream crossings on low volume roads where road use and stream flow conditions are

Figure 12.6 Drift with thin concrete surfacing.

appropriate. Ideally, they should be constructed at a relatively narrow, shallow stream location and should be in an area of bedrock or coarse granular soil for good foundation conditions. A ford can be narrow or broad but should not be used in deeply incised drainages that require a high earthwork or excessively steep road approaches (Keller and Sherar 2003).

LWCs are commonly used in areas with highly variable flows, such as dry streams subject to floods. High, short-duration peaks followed by long intervals of very low or no flow is most conducive to LWCs where short-term traffic interruptions during floods are tolerable. LWCs, in general, may also be a cost-effective option for crossing shallow wide rivers that are dry for the majority of the year or have low permanent flows and can be closed for short periods of time without serious consequences.

The simplest watercourse crossing is a ford, as illustrated in Figure 12.6. Fords utilise suitable existing channel bed, and they are appropriate for shallow, low-velocity, or seasonal water courses. Fords that have a concreted or partially concreted surface may be termed drifts. Fords may be appropriate for local traffic of up to about 50 vehicles per day. Gravel or cobbles can be used to line the bottom of the ford to provide a firm footing for vehicles. Fords should normally only be used for watercourses that are not subject to flash floods, which may cause them to be washed away, although in many instances repair or replacement may be inexpensive and adequately undertaken by local communities.

(a)   (b)

Figure 12.7   Causeways. (a) Low-level causeway. (b) Vented causeway.

If a ford is not able to perform effectively, it may be upgraded with concrete or masonry slabs and approaches. This type of crossing may be termed a causeway (see Figure 12.7a). A causeway is suitable as a crossing for watercourses that are normally passable or are seasonally prone to floods. Causeways are appropriate for similar traffic volumes to fords, although they may be able to carry larger vehicles. Attention must be paid to providing suitable upstream cut-off to prevent seepage and erosion under the structure, and downstream measures to avoid erosion from turbulence.

Where the water is running most of the year, the causeway can be provided with openings to permit normal water flow to pass below road level and reduce the frequency and depth of over-topping during flash floods. This type of crossing is termed a 'vented causeway' (see Figure 12.7b). Vented fords with culverts that are small relative to the bankfull channel are said to have a low Vent Area Ratio (VAR). A vent opening that approximates or exceeds the size of the bankfull channel is said to have high a VAR.

Potential benefits of low-water crossings

LWCs are generally less expensive to construct and maintain than small bridges or large multi- bore culverts (Table 12.2). Designs are generally less complicated, construction is quicker, and fewer materials are involved. Simple LWCs like unvented fords are useful in naturally unstable channels, such as alluvial fans and braided streams or in channels with extreme flow variations. Because they obstruct flows less than most culverts, they are less likely to cause flow diversions or accelerations, both of which can exacerbate a channel's inherent tendency toward instability. They can also be inexpensive to reconstruct in a new location if the channel does shift location.

Table 12.2 Advantages and Disadvantages of Low-Water Crossing Structures

Advantages	Disadvantages
Structures designed for over-topping.	Can be dangerous to traffic, particularly non-motorised traffic (NMT), during high-flow periods.
Non-vented structures less likely than culverts to be damaged by debris or vegetation plugging.	Have periodic or occasional traffic delays during high-flow periods, especially for NMT.
Typically, less expensive structures than large culverts or bridges.	Are not well-suited to deeply incised drainage channels.
Less susceptible than other structures to failing during flows higher than the design flow.	Are typically not desirable for high-use or high-speed roads.
Good for provision of climate 'shock-event' resilience, and where large amounts of sediment and debris are expected after a large storm event.	Can be considered as having an environmental impact through blocking passage of fish.

LWCs are very useful in catchments where substantial mobilisation of rock and woody debris is expected and are most suited for rural roads with low-to-moderate traffic speeds. Unimproved fords may only be driven over at low speeds of less than 15–30 km/h. Vented fords with a broad, smooth dip and gentle transitions may be suitable for speeds up to 50–70 km/h.

Small bridges

Where water channel conditions are too wide to install culverts, or there is a risk of large debris being washed down the water course and lodging in culverts, a small bridge is the preferred option. Small bridges usually comprise of a single span over the natural river bed, unlike a box culvert which has a concrete base, as illustrated in Figure 12.8. Small single-span bridges may be constructed of appropriate combinations of concrete, masonry, wood or steel. Pre-fabricated, short-span steel truss bridges (Bailey Bridges) have advantages in the rural road context.

There are many appropriate options for small bridge structures and materials, so long as they are structurally designed to withstand anticipate traffic loads and are suitable for the governing and hydraulic and ground conditions. Standard designs can be found for many simple bridges as a function of bridge span and loading conditions. Larger or more complex structures should be specifically designed by a structural engineer (Keller and Sherar 2003; USAID 2006).

The size of the bridge span will generally be dependant on the width of the channel to be crossed and the bridge invert be adjusted to the maximum flow level design discharge, unless additional height is required

Figure 12.8 Short-span reinforced concrete rural bridge.

for passage of water transport. The maximum flow level will often be chosen based on known or estimated future high flood levels (HFLs) in the channel at the crossing point, taking climate change predictions into account. Small bridge capacity can be estimated based on the hydrology principles discussed in Chapter 11. Modelling software is available, but it may often not be considered a cost-effective method for design of small rural bridges.

A further option is the use of 'floodable' or low-water bridges that could be described as LWCs with elevated decks and a natural stream bed bottom. Low-water bridges generally have greater capacity and are able to pass higher flows underneath the driving surface than most vented and unvented fords. Low-water bridges are designed and constructed with the expectation that they will be under water for short periods at higher flows, and suitable protection to mitigate over-topping erosion is included in their design (see Figure 12.9).

Figure 12.9 Floodable bridge.

KEY DESIGN CROSS-DRAINAGE ISSUES

Selection

Key issues to be addressed include:

- Basic design framework.
- Appropriate design life of the structure.
- Physical siting.
- Traffic and loading requirements.
- The hydrology and flood regime at the structure location.
- Economic and social use of available local resources: materials, skills and enterprises.
- Climate resilience levels.
- Safety.
- Construction and maintenance regimes.

Basic design framework

The design process for cross-drainage structure should be compatible with that of the LVRR Project Cycle as outlined in Chapter 3. Figure 12.10 presents a summary flow chart of the overall design process.

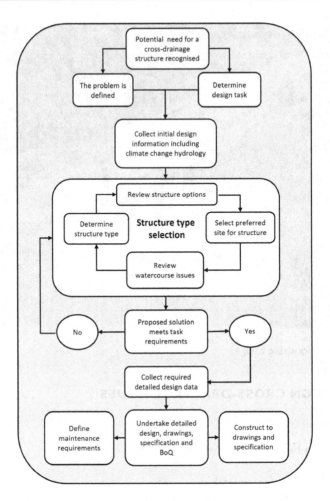

Figure 12.10 The structure selection and design process.

There is a logical sequence of work that should be undertaken in the selection and design of road cross-drainage structures. The two initial activities that should be carried out are to identify the problem and determine the design criteria for the structure. Design data may then be collected to enable the preliminary design to be carried out. It is common practice in LVRR projects to undertake an alignment walkover survey of actual or potential water crossings. This may be carried out as part of more general walkovers, as described in Chapter 7. Desk study or field surveys may need to be carried out to ascertain catchment details and characteristics. Runoff assessment should be based on available current and future climate data and checked with local stakeholders to tap into knowledge on the occurrence and extent of previous high-flow incidences.

It is suggested that a review of structural options is initially undertaken followed by an appraisal of potential construction sites. The water flow characteristics of the watercourses should then be considered before a selection of the most appropriate structures is made. It is likely that an iterative preliminary design loop will be needed to assess different potential structures and construction sites. It may be that short-period impassability may be considered for vented or standard drifts, if full climate resilient serviceability would lead to too expensive a structure. This should be balanced against local traffic demand and alternative access options.

Following completion and checking of the preliminary design, detailed design of the structure can then be undertaken. It will also be necessary to review the options for construction materials that may be available. Larcher et al. (2010) provide detailed guidance on the entire process of small structures provision, as well as data collection, processing and design development.

Appropriate design life of the structure

The design life of a structure is the length of time that the structure can be expected to be operational without reconstruction or replacement of failed elements. It assumes that, throughout the life of the structure, regular routine and periodic maintenance arrangements will be carried out. When determining a structure's design life the factors that should be taken into account are:

- Expected life spans for different structure elements and materials.
- Expected initial and recurrent costs for the design life options.
- Pragmatic assessment of future maintenance probability.
- Future changes in the use of the road (e.g. increased traffic volumes or loadings).
- Flood return periods (taking account of climate change).
- Economic, social and life consequences of possible structure failure.
- Influence of climate change on future life of the structure, risk and consequences of failure.

The design life of the road itself (i.e. the length of time before the road will become obsolete or require substantial improvement or rehabilitation) should also be taken into account. Typical design life periods for rural road assets are as follows:

- Pavement and associated drainage: 10–20 years (depending on pavement type).
- Minor structures (culverts, causeways): 15–25 years.
- Major structures (bridges): 50–100 years.

Physical siting

The foundation conditions and topography at a cross-drainage structure site will influence the type of structure selected. In some circumstances, the road alignment may need to be adjusted to take advantage of better site conditions.

Traffic

As with pavement design, careful consideration must be given to the types of vehicles that may use the road, their dimensions, their numbers and loading. If a road is close to existing or potential quarries, industrial or logging areas, it is likely that heavily loaded or probably overloaded vehicles will use the route and its structures. It can be extremely difficult to effectively control such loading against legal loading limits, especially on rural routes.

An important resource-related decision is whether a structure should be provided to accommodate one or two lanes of traffic.

The hydrology and flood regime at the structure location

Rainfall and catchment data need to be collected and reviewed, as per guidance within Chapter 11. Data availability will vary considerably depending on location and information management capability. An assessment must be made regarding the most adverse storm flow conditions for the selected design period. Most countries will have their own standards regarding recommended storm return periods, but Table 11.2 in Chapter 11 provides typical information for a range of road assets.

Some special design considerations

Lessons to be learnt from surveys of existing small structures on networks in various regions indicate that insufficient attention is often paid to some basic issues regarding small structures. Some of these deficiency issues are summarised below:

1. Water concentration: Roads interfere with the natural surface and groundwater flows and care must be taken to manage the collection, management and dispersal of water flows to minimise concentration issues and erosion possibilities to the road infrastructure and land downstream of the structure.
2. The watercourse hydraulic vertical alignment: A common misjudgement is to give the road alignment priority over the natural watercourse gradient. Many culverts are installed at the wrong level as the

road vertical alignment has been given priority. Once disturbed, the natural water flow will work determinedly to regain its natural vertical alignment. This can lead to erosion or silting and blocking of badly sited structures.

3. The watercourse horizontal alignment: In terms of horizontal alignment, location is an important consideration. Where there is a choice in the selection of the position of a watercourse crossing, it is desirable that it should be located:

   - On a straight reach of the watercourse, away from bends.
   - As far as possible from the influence of tributaries or main rivers.
   - On a length of water course with well-defined banks.
   - At a site that makes straight approach roads feasible.
   - At a site that makes a right-angle crossing possible.

   See summary options in Figure 12.11.

4. Outfall arrangements: The details of cut-offs to prevent water seepage under the structure and outfall erosion prevention details are vitally important. Due to the flow concentration effects of structures, the outfall arrangements and erosion prevention considerations are important.

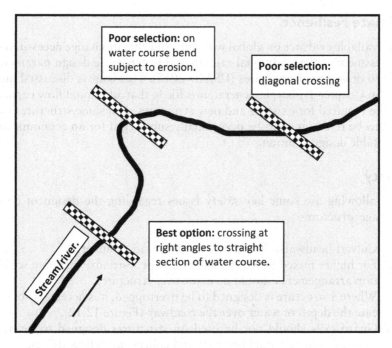

Figure 12.11 Horizontal alignment issues for water crossings.

Local resources

Most national standards and specifications for small structures focus on the use of concrete. This is understandable, as the component materials and skills are widely available in most developing regions. However, appropriate use of local resources for structures such as masonry, brick and wood will often produce the lower cost structures in both initial and Whole Life-Cycle Cost terms. It is certainly inadvisable to uncritically apply standards, practices and 'rules of thumb' derived from developed economies for use in LMIC structures.

For economic and social benefits, it is sensible to aim at maximising the local resource content of structures with respect to materials use, local enterprise, local employment and skills. Labour and credit availability and cost differences mean that the most appropriate solutions will probably be very different between high wage, developed economies and rural locations in developing regions. Such a strategy will mean that subsequent maintenance of the structure will be cheaper and more easily facilitated, rather than relying on imported materials and skills.

Larcher et al. (2010) provide comprehensive guidance on all aspects of appropriate technology structures, to include use of local materials such as masonry (mortared, dry jointed or gabion), fired clay brick and sustainably sourced timber, as well as concrete.

Climate resilience

The available evidence on global warming and climate change necessitates a reassessment of the traditional rainfall and watercourse design parameters used to design road structures (USAID 2015). This issue is discussed more fully in Chapter 17, but, in general, it is likely that additional flow capacity will be required for existing and new structures, and some structures may need to be re-designed to be occasionally submerged for an economic and acceptable design solution.

Safety

The following are some key safety issues regarding the design of cross-drainage structures:

1. Culvert headwalls should not obstruct road shoulders.
2. For higher mixed flows of vehicles and pedestrians, adequate separation arrangements should be made over structures.
3. Where a structure is designed to be overtopped, it is necessary to indicate the depth of water over the roadway (Figure 12.12).
4. Guard rails should not be used on structures designed to be overtopped, as these would obstruct and collect storm flow debris.
5. Site-specific guidance should be available on the safe depths of water for passage of vehicles, cyclists and pedestrians.

Figure 12.12   Depth markers on a causeway.

MAINTENANCE REGIME

The maintenance regime for a structure is just as important as design and construction, if the asset is to achieve its intended design purpose and life and provide sustainable provision of access; see Figure 12.5 for an example of poor maintenance impacting fitness for purpose. The maintenance of cross-drainage structures should be an integral part of overall road or network asset management arrangements, as discussed further in Chapters 20 and 21. There are, however, a number of maintenance aspects that should be appreciated for structures in particular, either individual structure or a large number within road network, as summarised below:

1. Structures may not need significant periodic maintenance for periods of many months or sometimes even years.
2. Deterioration or damage to a structure can progress slowly (e.g. corrosion or erosion) or suddenly (e.g. in a flood or vehicle accident).
3. The need for repairs may not be obvious to road users or through casual observation from the roadway. However, the deterioration can progress, if not checked, to result in the need for major works at great cost and requiring substantial unplanned resource mobilisation.

4. The resources for maintenance and repair of a typical structure are required intermittently, not continuously.

5. It is usually most efficient to provide maintenance resources only when the structure requires maintenance or repair works.

REFERENCES

ERA. 2016. *Manual for Low Volume Roads. Part Explanatory Notes and Design Standards for Small Structures*. Ethiopian Roads Authority. https://www. research4cap.org/index.php/resources/rural-access-library#.

IRC (Indian Roads Congress). 2000. *Project Preparation Manual for Bridges, Special Publication 54. First revision*. https://archive.org/details/gov.in.irc.sp.054.2018/ page/8/mode/2up.

Keller, G. and J. Sherar. 2003. *Low Volume Road Engineering Best Practices*. USDA Forest Service. https://pdf.usaid.gov/pdf_docs/PNADB595.pdf.

Larcher, P., R. C. Petts and R. Spence. 2010. Small structures for rural roads. A practical planning, design, construction and maintenance guide. Report by for DFID, published by ADB. https://www.gov.uk/research-for-development-outputs/ small-structures-for-rural-roads-a-practical-planning-design-construction-maintenance-guide.

TRL. 2000. *Overseas Road Note 9: A Design Manual for Small Bridges*. TRL Ltd. ORN9: A guide to small bridge design for highway engineers – GOV.UK. www. gov.uk.

USAID. 2006. *Bridge Design Manual*. Ministry of Roads and Bridges, Government of S Sudan.

USAID. 2015. *A guide for USAID Project Managers. Bridges. Incorporating Climate Change Adaptation in Infrastructure Planning and Design*. https://pdf.usaid.gov/ pdf_docs/pbaah356.pdf.

Chapter 13

Low volume rural road pavement design principles

INTRODUCTION

Aims

A Fit-for-Purpose pavement is central to achieving the goal of sustainable rural access and network connectivity. To achieve this goal, the designer must have sufficient knowledge of the foundations of the available materials, the traffic, and the local Road Environment in order to satisfactorily meet road task requirements. Good practice pavement design and management depends on a number of factors, ideally addressed within a Whole Life Cycle (or a 'cradle to grave' approach), within which the primary steps are:

- Design: including Road Environment impact factors.
- Construction: to meet design requirements, including quality and drainage.
- Maintenance: to maintain pavement integrity.

Pavement components

Despite major differences in detail, all road pavements are similar in their load-bearing aims and structural make-up and comprise as listed below and illustrated in Figure 13.1:

1. A stable foundation on which to construct a pavement, commonly termed the subgrade.
2. A structural component to take and spread traffic loading over the foundation; this may be split into base and sub-base layers.
3. A suitable surface on which traffic can safely travel.

In simple rural road options, such as earth or gravel roads, two or more these components may be combined, while in higher volume roads each of the above components may be subdivided further.

DOI: 10.1201/9780429173271-13

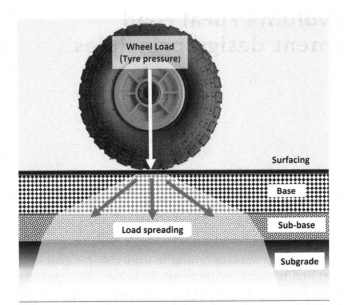

Figure 13.1 General pavement components and purpose.

Background

Multilateral Development Banks (MDBs) and bilateral aid agencies such as UKAID-DFID and KfW have for some decades been developing cost-effective, locally focussed approaches to the design of Low Volume Rural Roads (LVRRs) and their pavements in the increasing recognition that traditional international procedures were not proving to be appropriate. Good practice for LVRR pavements, founded on lessons learnt from this work, is summarised by Cook et al. (2013) and places emphasis on the following:

- Fitness for purpose within the Road Environment (as discussed in Chapter 6).
- Whole-Life costing and budgeting.
- Phased design.
- Local solutions for local challenges.

In the Introduction to this book, the upper limit for LVRR pavements in terms of traffic was defined as being around 3 million ESA. This is especially important as regards pavement design, where traffic loading becomes an important consideration. Even within the LVVR envelope there are significant differences to consider in design, depending on the relative importance of traffic to other road environmental issues as impact factors. Very low trafficked road pavement options, particularly those with minimal commercial vehicles, can be selected and designed with an emphasis on mitigating surface deterioration rather than structural failure, thus providing

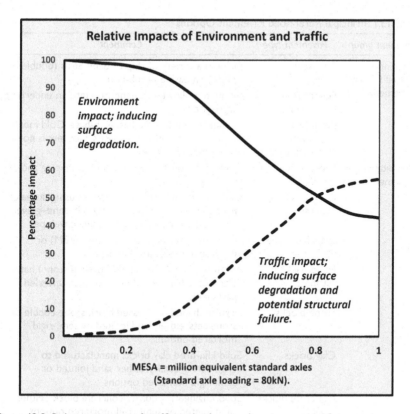

Figure 13.2 Relative impacts of traffic and other road environmental factors.

a potential saving in pavement material costs. Figure 13.2 outlines the general impacts of traffic and the local environment on the modes of pavement deterioration in low volume roads.

KEY PAVEMENT DESIGN ISSUES

Rural road pavements options

A wide range of pavement options have been proven and established for use on LVRRs, each of which has its own advantages and disadvantages depending on its purpose and the environment in which it is to be used (Henning et al. 2005; Cook et al. 2013; World Bank 2020). Pavements can be classified as either unsealed, sealed flexible, rigid or semi-rigid. For purposes of clarity in this book they have been broadly grouped and discussed as bituminous sealed (flexible) pavements (Chapter 14), concrete (rigid) pavements, and other non-bituminous pavements including semi-rigid options, (Chapter 15). Table 13.1 summarises common options under these headings.

Table 13.1 Principal Rural Road Pavement Options

Pavement group	Pavement type	Comment
Bituminous sealed flexible pavements	Thin seals	Includes a variety of chip or Sand Seals (double or single) on base and sub-base.
	Penetration macadam	Bitumen is allowed to penetrate into an underlying aggregate layer.
	Pre-mixed aggregate-bitumen carpet.	Carpet laid on flexible base-sub-base. Cold mix is an increasing option for LVRRs. Hot mix is not a common LVRR option.
Non-bituminous pavements	Earth or natural surface	'Pavement' construction limited to 'shaping and draining'.
	Gravel wearing course	Common default option for LVRRs using unsealed naturally occurring graded clay–silt–sand–gravel materials within a defined grading envelope.
	Crushed stone	Unsealed water-bound macadam (WBM) or dry-bound macadam (DBM).
	Hand-packed stone	Irregular shaped stone (hand-packed stone) hand placed and hammered-in to form an unsealed pavement.
	Stone blocks	Regular shaped or dressed block stone cobble or stone sets; either sand jointed or semi-rigid mortared options.
	Clay bricks	Solid kiln-fired clay bricks manufactured to engineering quality; either sand jointed or semi-rigid mortared options.
	Concrete blocks	Solid, engineering quality concrete blocks; either sand jointed or semi-rigid mortared options.
Concrete options	Concrete slabs	Reinforced or non-reinforced slabs (continuous paving unusual in the LVVR context); usually overlying a granular sub-base.
	Geocells	Concrete placed within strong, lightweight, three-dimensional polyethylene honeycomb like structure.

Note: Options may be applied individually, or in combination, within a pavement structure.

Fitness for purpose

Fitness for purpose can be considered as the over-arching driver of appropriate pavement design. Roads must deliver a level of service in line with their identified purpose or task, which includes compatibility with the nature of the traffic (people and freight as well as the vehicles) that will pass along them while minimising costs over the whole life of the road. To be Fit for Purpose, a pavement design should also be:

1. Compatible with the construction materials that are readily available within appropriate specifications.

2. Within the construction capacities of the contractors and labourers who will build them.
3. Within the means of communities or local organisations to maintain them.

Whole Life Cycle and costs

As with other elements of LVRR design, a life-cycle approach can provide the framework for presenting guidance, act as a route map to the appropriate levels of information required for making knowledge-based pavement decisions, and at the same time provide a link to Whole Life Costs, Figure 13.3.

The consideration of all present and expected future costs, including maintenance, involved with an investment in rural road infrastructure should be an integral part of the pavement design process. Whole-Life Costing (WLC) of LVRR pavements is, therefore, a preferred approach to comparing road design options. Minimising WLC for road assets steers road authorities toward looking for the best use of available construction and maintenance budgets that best preserves the value of the road or network infrastructure.

WLC is the process of assessing all costs associated with a road over its intended or design lifetime. The aim is to reduce the sum of these values to obtain the minimum overall expenditure on the road asset while achieving an acceptable level of service from the investment. Usually, an assessment

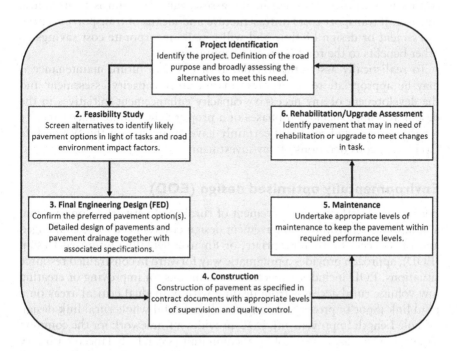

Figure 13.3 The pavement life-cycle.

of the residual value of the asset at the end of the assessment period is included. There are two basic approaches to the assessment of whole-life costs for rural road pavements that can each reflect discrete objectives and may result in different conclusions depending on the local circumstances. These can be characterised as follows:

1. Whole- Life Costs for the Road Asset (Whole-Life Asset Costs).
2. Whole- Life Transport Costs.

Whole-Life Asset Cost (WLAC) assessment aims to define the costs of construction, and maintenance of a particular road and pavement over a selected assessment period. The principal cost components are the initial investment or construction cost and the future costs of maintaining (or rehabilitating) the road over the assessment period selected (for example, 12 years from construction). Whole-Life Asset Cost (WLAC) is the cost of the road for the road asset owner or manager.

Since the purpose of the road is to cost-effectively transport the local road users, Whole Life Transport Costs assessment will, in addition, include a component for the savings in Vehicle Operating Costs (VOC) for the road users under the various investment and maintenance strategies. This component can be substantial on higher traffic rural roads. Other socio-economic factors may also be included in the assessment. The aim is to minimise the overall transport costs (infrastructure and means of transport) over the assessment or design lifetime and will usually incorporate cost savings or other benefits to the road users and community.

To realistically assess the likelihood of effective future maintenance it may be appropriate to incorporate maintenance capacity assessment and the development of any necessary capacity enhancement initiatives in the design and implementation phases of a project. The maintenance capacity for the constructed road will certainly have an impact on the Whole Life Cycle Cost considerations of any investment.

Environmentally optimised design (EOD)

For more cost-effective improvement of rural road networks, conventional assumptions regarding road pavement design criteria need to be challenged and the concept of an appropriate, or Environmentally Optimised Design (EOD), approach provides a pragmatic way forward in constrained resource situations. EOD includes a spectrum of solutions for improving or creating low volume rural access – from dealing with individual critical areas on a road link (Spot Improvements) to providing a total whole rural link design (Whole Length Improvement). EOD provides a framework for the common situation where aspirations of local communities can be balanced with very limited budgets (Roughton 2008; Cook et al. 2013). The application of

EOD to rural road networks has a significant advantage in terms of providing cost-effective climate strengthening; see Chapter 17.

Whole Length Improvement applies the principle of adapting road designs to suit environments to individual segments of the road alignment. This allows differing pavement options to be selected in response to different impacting factors along an alignment. Most rural roads are designed using standard national designs along their entire length. However, this can be expensive and sometimes does not meet the needs of the users. The EOD approach for each road or road section is intended to meet their specific Road Environment conditions and optimises the application of available investments resources along a route.

Spot Improvement is an extreme expression of EOD that permits available budget resources to be concentrated on specific priority areas to the exclusion of other areas of lower priority. Spot Improvement is discussed separately in Chapter 16.

Appropriate use of locally available materials

The appropriate use of locally available materials is a fundamental issue in the design and construction of sustainable LVRR pavements and their surfaces. Important factors to be considered are:

- Material location, quality and quantity.
- Variability.
- Behaviour characteristics.
- Suitability for road task.
- Processing requirements.
- Quality Control on excavation and delivery.
- Environmental impacts.

The strategic options for addressing an apparent lack of suitable local materials may be overcome by:

1. Adapting the specification and road design to suit local materials; or
2. Adapting or modifying the materials to suit a realistic specification.

The issues around the use of marginally acceptable local constructions materials are discussed in detail in Chapter 8 based on summaries in Austroads (2018) and Cook et al. (2000).

Equilibrium Moisture Content

Equilibrium Moisture Content (EMC) occurs where the moisture conditions under a sealed pavement reach a state of stable equilibrium with the moisture regime of the local environment at some stage after construction.

The conditions are generally those found towards the centre of the pavement, beyond about 1–2 m from the edge of the pavement. However, fluctuations in moisture conditions can result from relatively rapid weather or climate impact changes in shoulder moisture content, which in turn can potentially cause weakening in outer wheel path materials and pavement strength in comparison with the stable inner pavement governed by EMC conditions.

There are clear cost advantages in terms of material strength in being able to design a pavement based on a stable EMC. The concept of EMC may not be fully applicable to many single-lane (3–4 m carriageway) rural roads.

PAVEMENT DESIGN FRAMEWORK

A phased approach

A two-phase selection and design approach to LVRR pavement, as shown in Figure 13.4, is compatible with the phases of the Whole-Life Pavement Cycle and comprises:

1. Phase I: General assessment of appropriate pavement option(s) compatible with the Road Environment and budget constraints – at the Feasibility Stage (FS) of the Pavement Cycle.
2. Phase II: Detailed design of the selected pavement components compatible with engineering standards and requirements – at the Final Engineering Design (FED) stage of the Pavement cycle.

Road Environment compatibility

The principal criteria in the detailed road design process are traditionally based around traffic loading, material properties, and subgrade strengths. In the case of rural roads, this traditional approach requires some modifications, as indicated previously by Figure 13.2. It is now appreciated that additional factors should be considered that cumulatively can be described as the 'Road Environment'. Factors important to the Road Environment can be broadly grouped as shown in Figure 13.5. This important concept has been discussed in detail in Chapter 6 (Figure 6.1).

- Environment factors.
- Road task factors.
- Operational environment.
- Local resources.

Phase I pavement option selection

The aim of Phase I is to identify one or more pavement options that are likely to be compatible with the Fit-for Purpose design targets. Initially, in

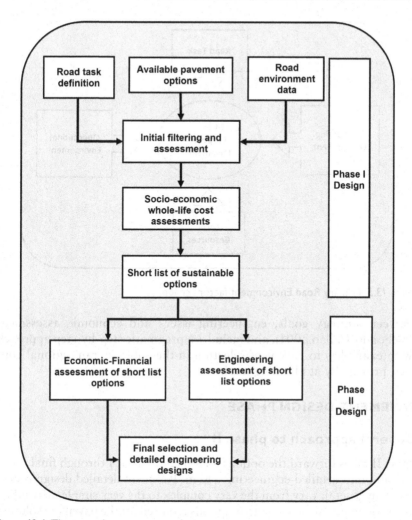

Figure 13.4 The two-phase pavement design process.

many LMIC situations, this will involve a fundamental decision on whether to opt for an unsealed or a paved solution.

Engineering-related issues that influence the choice of pavement options can be grouped and assessed as shown in Table 13.2 that presents a series of checks or filters that can guide the road designer towards selecting appropriate options.

In a wider context, The World Bank 'Pave or No Pave' document (2020) draws on the previous pavement assessment work (Henning et al. 2005; Intech-TRL 2006; Cook et al. 2013) to provide strategic guidance through a multi-criteria framework for initial pavement option selection in the Sub-Saharan regional environment. It provides useful links

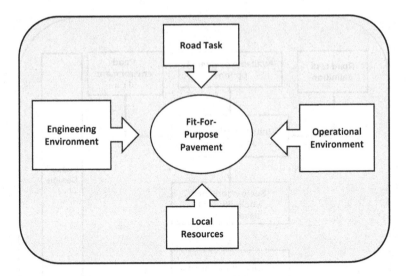

Figure 13.5 Primary Road Environment factors groups.

between strategy goals, engineering issues and economic assessment (Archondo-Callao 2001), and includes a pragmatic step-by-step approach (with examples) to pavement selection in the Sub-Saharan regional context, principally at planning level.

PAVEMENT DESIGN PHASE

General approach to phase II

Phase II takes forward the options identified in Phase I through final selection and into detailed engineering design. Pavement detailed design procedures, in general, vary from the very complex to the very simple, depending on the data requirements or assumptions required for the pavement solution in question. For practical and pragmatic reasons, the procedures adopted for LVRR pavement design should be simple and appropriate for the technical resources available in the majority of LMICs and complex mechanistic approaches are not recommended.

Many LMICs now have some form of road or pavement design manual that is applicable to LVRRs. These are predominantly based around the use of design catalogues or charts that enable local road engineers to follow a straightforward design process summarised in Figure 13.6.

The extrapolation of catalogues or charts for use outside their region or country of origin should be undertaken with caution, and users must take into account any assumptions or restrictions inherent in their development. Comparison of Road Environment factors (Chapter 6) between countries or regions can be useful in the context of assessing the suitability of using specific catalogues or charts.

Table 13.2 Initial Pavement Option Selection Checklist

	Flexible pavement single seal	Flexible pavement double seal	Flexible: pre-mix	Flexible pavement: Pen-Mac	Unsealed: engineered natural surface	Unsealed: gravel surface	Unsealed: macadam	Unsealed: hand-packed stone	Dressed stone/cobble stone	Fired clay brick pavement	Concrete slabs	Concrete blocks	Concrete geocell
Traffic regime (see Table 13.3)													
Light traffic	✓	O	X	O	✓	✓	✓	✓	✓	✓	O	O	O
Moderate traffic	O	✓	O	✓	X	✓	✓	✓	✓	✓	✓	✓	✓
Heavy traffic (overload risk)	X	O	✓	O	X	X	X	O	✓	O	✓	✓	✓
Construction regime													
High labour content	O	O	O	O	✓	✓	✓	✓	✓	✓	✓	✓	O
High machinery input	✓	✓	✓	✓	O	✓	X	X	X	X	O	X	X
Construction cost													
Low	O	O	X	X	✓	✓	✓	✓	O	O	X	O	X
Moderate	✓	✓	✓	✓	O	O	✓	O	✓	✓	O	✓	O
High	O	O	O	O	O	O	O	O	✓	O	✓	O	✓
Maintenance management regime													
Poor	O	O	O	O	O	X	X	O	✓	✓	✓	✓	✓
Fair	✓	✓	✓	✓	O	O	O	✓	✓	O	O	O	O
Good	O	O	O	O	✓	✓	✓	✓	O	O	O	O	O
Erosion regime (see Table 13.4)													
A: low erosion regime	✓	✓	✓	✓	✓	✓	✓	✓	✓	✓	✓	✓	✓
B: moderate erosion regime	✓	✓	✓	✓	X	O	O	✓	✓	✓	✓	✓	✓
C: high erosion regime	O	✓	✓	✓	X	X	X	O	✓	O	✓	✓	✓
D: very high erosion regime	X	O	✓	O	X	X	X	X	✓	X	✓	✓	✓

✓, option recommended; O, option possible, depending on other criteria; X, option not recommended.

Rural road pavement design methods

This section aims to provide a brief background to the practical design of LVRR pavements and summarise some of the options available. Although some LMICs have derived very simple empirical pavement designs based around road classification and location, it is more normal to base the designs around anticipated traffic, foundation (subgrade) strength and one or more physical environmental factors, such as climate and topography. Principal LVRR pavement design procedure options are listed and summarised in Table 13.5 along with key references in Table 13.6. Information in Table 13.5 is largely drawn from a review undertaken by RECAP (2020) supplemented by recommendations from Austroads (2017), and ARRB

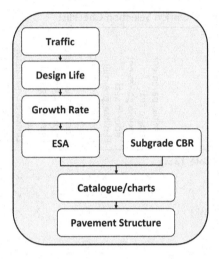

Figure 13.6 Simplified pavement design steps.

Table 13.3 Indicative Traffic Regime

Indicative category	Traffic description
Light	Mainly non-motorised, pedestrian and animal modes, motorbikes & less than 25 motor vehicles per day, with few medium/heavy vehicles. No access for overloaded vehicles. Typical of a rural road with individual axle loads up to 4 tonne.
Moderate	Up to about 100 equivalent vehicles per day including up to 20 medium goods vehicles, with no significant overloading. Typical of a rural road with individual axle loads up to 6 tonne.
High	Between 100 and 300 equivalent motor vehicles per day. Accessible by all vehicle types including some heavy and multi-axle (3 axle +) trucks, construction materials, industrial and forestry haulage routes. Specific design methodology may need to be applied.

Table 13.4 Definition of Erosion Regime

Road alignment longitudinal gradient	Annual (monsoonal) rainfall (mm)			
	<1,000	1,000–2,500	2,500–4,000	>4,000
Flat (<1%)	A	A	B	C
Moderate (2%–6%)	A	B	B	C
High (6%–8%)	B	C	C	D
Very high (>8%)	C	C	D	D

A = low; B = moderate; C = high; D = very high.

Areas prone to regular erosive flooding should be classed as 'High Risk' irrespective of rainfall.

Table 13.5 General Methods of Pavement Design for LVRRs

Ref.	Method	Description	Comment
A	The AASHTO method	A method that is based on the California Bearing Ratio (CBR) test, which is used to quantify the strength of the subgrade soil and the strength of any existing unbound pavement layers. Based on this, a structural number (SN) of the pavement for the estimated traffic loading is derived or assumed. The structural number for the various layers involved in the pavement are input into an experimentally derived formula along with various environmental factors to derive materials characteristics and layer thicknesses.	Rarely used directly for LVRR design.
B	Overseas Road Note 31	This is a CBR based catalogue method with series of charts based the assessment of subgrade strength for the structural design of layers to resist ranges of traffic loading, derived using a similar approach to the ASSHTO method.	Not commonly used directly for LVRR design because of the predominantly high traffic ranges. A revised Road Note 31 (2023) has extended traffic ranges and included concrete options.
C	Low traffic ORN 31 chart derivates.	There are a number of CBR-Traffic chart systems similar ORN31 but focussed much more on lower volume or lighter traffic and derived or modified through additional research, specific Road Environments, specific pavement types or local experience.	Many LMICs use their own versions of a catalogue or chart-based system, see Table 13.6. Austroads.
D	ARRB Rural Road Guidelines. Austroads Pavement Design.	These related guidelines present design chart plots based around deriving thickness of cover over foundation levels based on traffic and e strength (CBR) going down to very low traffic levels. The approach involves deriving total pavement thickness and allowing flexibility in the selection of layers.	The ARRB sealed and unsealed guidelines offer the option of solutions to 90% or 95% confidence limits. The Austroads guidance is aimed at main roads but with sections on low volume roads.
E	DCP-based (TRL)[a]	The TRL-DCP method derives layer thicknesses and strengths based on empirical relationships between DCP cone penetration resistance (mm/blow) and CBR strength adjusted for moisture condition. The methodology is included within a specific DCP Design Manual using free-to-download software.	The DCP-CBR relationship is also commonly used in conjunction with methods 1–4 to derive in situ CBR values. Purely DCP-based systems have a major disadvantage for alignment with significant cuts and fills that are deeper than the DCP length (+/− 1.6 m).

(Continued)

Table 13.5 (Continued) General Methods of Pavement Design for LVRRs

Ref.	Method	Description	Comment
F	DCP (DN)[a]	The DCP-DN does not convert the resistance to penetration to CBR but instead uses the resistance directly to undertake the pavement design. The DCP-DN design method is empirical in nature and currently based on measurements and limited monitoring observations on a range of soil types and environmental conditions prevailing in Southern Africa.	The DCP-DN was derived principally for the scenario of upgrading gravel to sealed roads and offers some advantages in reduced layer thickness, within its proven environment. Has similar cut and fill terrain disadvantages as above.

[a] The moisture condition of the ground at time of DCP testing must be noted and a correction made, if necessary, to take account of differences between this and the assessed design condition.

Table 13.6 Example References to General Methods of LVRR Pavement Design

Method	Sample reference	Country/region
A	American Association of State Highway and Transportation Officials (AASHTO) (1993). AASHTO guidelines for design of pavement structures. American Association of State Highway and Transportation Officials. Washington, DC.	International
B	TRL (1993). ORN 31 (4th Edition). A guide to Structural Design of Bitumen Surfaced Roads in Tropical and Sub-tropical countries.	International
	TRL (2023). RN 31 (5th Edition) A guide to Structural Design of Surfaced Roads in Tropical and Sub-tropical regions (includes concrete options).	International
C	Gourley and Greening (1999). Performance of low volume sealed roads: results and recommendations from studies in Southern Africa.	East and Southern Africa
	SADC (2003). Guideline on low volume sealed roads.	Southern Africa
	Gov. of Myanmar (2020). Low volume road design manual: Section B design.	Myanmar
	Gov. of Lao PDR (SEACP 2008) Low volume rural road standards and specifications: Part II: pavement options and technical specifications.	
	MPW Liberia (2019). Manual for low volume roads. Part B: Materials. Pavement design and construction.	S E Asia
		West Africa
D	ARRB (1995). Sealed Roads Manual: Guideline to good practice.	Australia
	ARRB (2000). Unsealed Roads Manual: Guideline to good practice.	
	Austroads (2017). Guide to Pavement Technology. Part 2 Pavement structural design.	Australia

(Continued)

Table 13.6 (*Continued*) Example References to General Methods of LVRR Pavement Design

Method	Sample reference	Country/region
	Austroads (2009). Guide to Pavement Technology. Part 6, Unsealed Pavements. Sydney, Australia.	Australia, New Zealand
E	TRL (2006). UK DCP 3.1 User manual.	International
F	Paige-Green and van Zyl (2019). A Review of the DCP-DN Pavement Design Method for low volume sealed roads: Development and Applications.	East and Southern Africa
	MOWTC Tanzania (2016). Low volume roads manual: Part D, Chapter 14. Structural pavement design.	Tanzania

Full reference details are included at the end of the chapter.

(1995, 2000). Further detail on the design of specific pavement types is contained in Chapter 14 (bituminous pavements) and Chapter 15 (non-bituminous pavements).

Pavement design catalogues and charts

Many of the above approaches depend on the knowledge that if the traffic level and subgrade strength are known, it is straightforward to present the structural designs for each pavement type as a matrix with traffic level as one component and subgrade strength as the other input component. Figure 13.7 illustrates the key elements of typical catalogue or chart approaches; with key steps as follows:

1. Figure 13.7a: Austroads/ARRB approach. Assess design traffic (1), move vertically to meet the appropriate CBR strength (2) and read across (3) to note the total required thickness of base/sub-base/ capping layers above the subgrade. This approach allows significant flexibility in thicknesses of the various layers and materials – provided a minimum base layer is maintained (usually 100 mm).
2. Figure 13.7b: This style of chart is very common across a range of guidance documents. The key steps are: select appropriate subgrade CBR (1) and design traffic value (2) and read off (3) the recommended thicknesses of base, sub-base and capping layer (if required). These charts are simpler to use than above, but do not necessarily have their flexibility of choice.

Material characteristics normally accompany design charts such as these, as discussed further in Chapters 14 and 15.

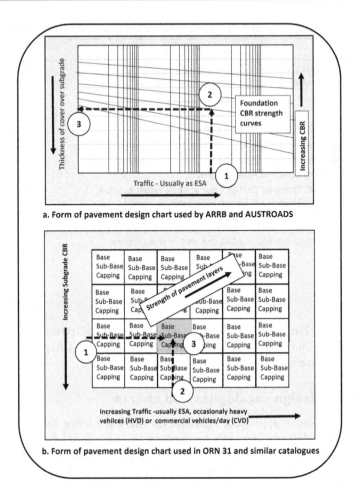

a. Form of pavement design chart used by ARRB and AUSTROADS

b. Form of pavement design chart used in ORN 31 and similar catalogues

Figure 13.7 Typical design charts.

The role of subgrade and its assessment

The updated Overseas Road Note 31 (TRL 2023) describes the subgrade as the foundation layer for the pavement and notes that the assessment of its working condition as a critical element of the pavement design process. The subgrade foundation may comprise a number of different elements:

1. Native (or In Situ) subgrade. In situ soil or rock, for example at the base of a cutting, that after appropriate preparation may act as a subgrade. Existing pavement layers or compacted fill on which a road rehabilitation or upgrade is proposed may be considered as an in situ subgrade component of the foundation if they are of sufficient quality.

2. Imported Subgrade. Material imported from borrow or earthwork excavation, and then placed and compacted to act either as an element within the pavement foundation or as a capping layer.
3. Capping Layer. An additional layer placed over in situ subgrade or imported subgrade when they are of insufficient strength for the proposed pavement design requirements.
4. Sub-foundation. Soil or rock materials, imported or in situ, below the normally defined Foundation or Material Depth.

The foundation components perform the following tasks during the construction and during the design life of the pavement:

- Ensure a uniform load-bearing layer evenly for the full width of the carriageway.
- Create a platform to ensure the quality of above sub-base compaction.
- Facilitates drainage of overlying pavement layers.
- Facilitates the access during the construction process that does not damage the road foundation.

Subgrade fitness for purpose to perform its tasks may be considered a combination of its material characteristics, its moisture condition, its geometry and its working environment, which includes the stresses to which it is subjected during the construction process and pavement design life. Assessment of subgrade as an integral part of the pavement design must take the above into account (TRL 2023; Austroads 2017; SANRAL 2013).

Pavement foundation investigations should fit in with and be complementary to the phases of project investigations for alignment, materials, earthworks and structures, as described in Chapter 7. In terms of pavement foundations, in particular, the following are the key investigations aims:

1. Define a general Ground Model along the length of the alignment in terms of topography, soil-rock profiles, hydrology and potential geotechnical hazards in the sub-foundation.
2. Define the design life working condition of the proposed subgrade in terms of strength (CBR%) under equilibrium moisture or reasonable worst case moisture conditions, if required.
3. Define the geotechnical properties of proposed subgrade materials.
4. Identify any weak materials or materials with problems within the zone of influence.
5. Identify any problem materials below the zone of influence (foundations) that may have an impact on the performance of the subgrade or pavement.

The design of the pavement foundation investigations for rural roads will be governed by the nature of the road project and its general environment

Table 13.7 Subgrade Investigation Scenarios

Scenario	Description	Investigation-testing implications	Investigation-testing implications
		Pavement on new alignment	Pavement on existing alignment
A	Alignment at or near existing surface	Test pit sampling and testing. DCP-CBR testing for in situ conditions is relevant.	Test pit sampling and testing. DCP-CBR testing for in situ conditions is relevant.
B	Alignment on embankment (>1 m)	Test pitting for subgrade sampling irrelevant. Testing on material from cut or borrow required. DCP-CBR testing only relevant after embankment fill is placed and compacted.	Test pit sampling and testing of existing embankment. DCP-CBR testing for in situ condition of embankment is relevant.
C	Alignment in cut (>1 m)	Test pitting for subgrade sampling only relevant for material within 2 m of surface. In situ testing irrelevant. Drilling may give indication of in situ 'native' subgrade condition.	Test pit sampling and testing of existing roadbed material. DCP-CBR testing for in situ condition is relevant.

and is particular by the relationship between the vertical alignment and the existing ground level, as illustrated in Table 13.7.

PAVEMENT DRAINAGE

General requirement

The following sections are concerned primarily with the basic principles of pavement drainage. Other aspects of rural road drainage are covered in Chapter 8 (earthwork drainage), Chapter 11 (hydrology) and Chapter 12 (cross drainage structures).

Effective pavement drainage is a critical issue in LVRR design and construction but one which, while emphasised in manuals and guidelines, is all too frequently poorly addressed in practice (Cedergen 1997; Rolt et al. 2002). This challenge is accentuated in low volume roads that may be constructed from natural, often unprocessed materials that tend to be moisture sensitive.

It is impossible to guarantee that roads will remain waterproof throughout their lives, hence it is important to ensure that if any layer of the pavement, including the subgrade, consists of material that is seriously weakened by the presence of water, then it must be facilitated to drain away quickly. Austroads (2017) summarises the requirements for pavement drainage as being to:

- Provide local lowering of the water table (drainage of subgrade).
- Cut off water ingress to the subgrade or pavement from water-bearing strata.
- Drain specific pavement layers.
- Control surface run-off.
- Achieve a combination of some or all of the above.

External drainage

The role of external drainage is to keep water away from the road, and its pavement and its design should include consideration of drainage ditches, surfacing, shoulders, cross section and appropriate alignment, with links to culverts as considered in Chapter 12.

Unsealed roads are, of course, particularly vulnerable to water ingress and consequent deterioration. However, even paved roads are not fully impermeable because of, for example, oxidation and cracking of bitumen seals or infiltration through concrete slab joints. The following measures should be considered to minimise ingress of water into the pavement:

- Bituminous seals and concrete slab joints seals should be well maintained.
- Use a dense well-graded, highly compacted materials in base and sub-base.
- Prevent ingress of moisture from the pavement edges by sealing the shoulder.
- Raise embankment levels.
- Line side drains where appropriate.

It is commonly stated in road design manuals, in relation to external drainage design, that the road must be raised above the level of existing ground such that the crown height of the road (i.e. the vertical distance from the bottom of the side drain to the finished road level at the centre-line) is maintained at a minimum height Hw, as shown in Figure 13.8. However, this should be considered only a partial solution and height Hf, height of finished road

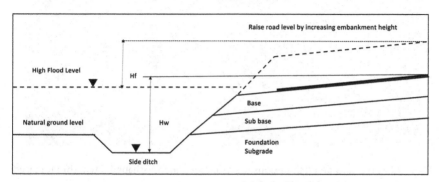

Figure 13.8 Road crown heights.

level above flood level, should also be considered. Hf and Hw are normally required to be a minimum of between 0.5m and 0.65m, depending on local specifications. This flood level should be assessment in line with the principles of future climate and flood assessment in Chapter 17.

Gourley and Greening (1999) emphasised the cost benefits of sealing shoulders and they recommended a sealed shoulder width such that the outer wheel track is more than 1.5 m from the edge of the sealed area. The simplest way to achieve this is by extending the carriageway seal through to the shoulders, preferably up to the side drain edge.

The road surface must be constructed with a sufficient camber or cross-fall to shed rainwater quickly (see Chapter 9: Geometric design).

Longitudinal drainage

Ditches and associated turnouts and intercepting ditches may be used to provide open road-side drainage. Ditches are channels provided to remove the run-off from the road pavement, shoulders, and cut and fill slopes. The depth of the ditch should be sufficient to remove the water without risk of saturating the subgrade. Ditches may be lined or provided with scour checks to control erosion. Unlined ditches should preferably have side slopes not steeper than 2 to 1 horizontal to vertical. Side drains are typically constructed in V-shape rectangular or trapezoidal cross section (see Figure 13.9). Unlined V-shaped are amenable to maintenance by motor or towed grader, whereas labour will find it easier to construct and maintain rectangular or trapezoidal sections with the aid of templates. Rectangular and trapezoidal cross sections are also less susceptible to erosion.

(a)   (b)

Figure 13.9 Typical side ditch options. (a) Concrete lined V-shaped ditch. (b) Mortared stone lined trapezoidal ditch.

The amount of erosion control and maintenance can be minimised to a great extent by using the following controls:

- Gentle side slopes in erodible material.
- Regular turnouts and intercepting ditches.
- Ditch checks or drop structures.
- Protective lining of ditches on gradients 6-8%.

Turnouts or mitre drains are short, open, skew ditches used to remove water from the road-side ditches or gutters, as illustrated in Figure 13.10. Use of turnouts reduces the necessary size of the side ditches, minimises the side drain velocity of water and thereby the risk of erosion. The interval between turnouts depends on run-off, permissible velocity of water and slope of the terrain. To prevent the flow through turnouts from generating soil erosion at the outlet, the discharge end of the turn-out should be fanned out. Consultation with adjacent landowners and users can minimise the opportunities for conflict due to erosion or siltation issues.

It is more important to maintain a minimum longitudinal gradient for pavement side ditches to avoid stormwater spreading over the pavement. Vegetation encroachment along the pavement edge may also impede the run-off of water, if the cross-fall or road gradient is flat. Where the longitudinal gradient of the road is close to zero, the depth of side drains may have to be varied to obtain sufficient gradient of the ditch. Longitudinal gradients should preferably not be less than 0.2% in very flat terrain. Their hydraulic capacity is often designed to contain a 5- or 10-year frequency storm run-off (see Chapter 11).

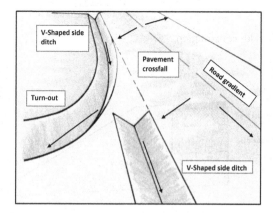

Figure 13.10 Ditch turnout.

Internal and subsurface drainage

Internal drainage is concerned with the movement of water that enters the road pavement despite an efficient and functioning external drainage system. When permeable base/sub-base materials are used, particular attention must be given to their drainage. Ideally, the base and sub-base should extend right across the shoulders to the drainage ditches in full-width construction, as shown in Figure 13.11. In addition, cross-fall is needed to assist the shedding of water across shoulders and into the side drains.

Trench (or boxed-in) type of cross section should be avoided, if possible, as pavement layers may be confined between continuous impervious shoulders, as shown in Figure 13.12. This type of construction has the undesirable effect of trapping water at the pavement–shoulder interface and preventing flow of the water into drainage ditches, which, in turn, facilitates damage to the shoulders and eventual pavement failure, even under light trafficking. It may be too costly for some rural road projects to extend the base and sub-base material across the shoulder, in this case drainage channels, or grips, at 3–5 m intervals should be cut through the shoulder to a depth of 50 mm below sub-base level. These

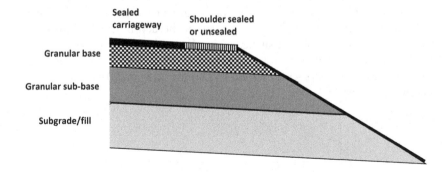

Figure 13.11 Full-width construction.

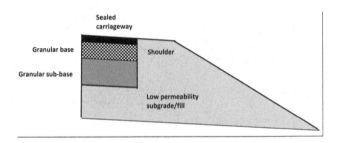

Figure 13.12 Boxed-in construction.

channels should be backfilled with material of base quality but which is preferably more permeable than the base itself and should be given a fall of 1 in 10 see Figure 13.13. Maintenance procedures must ensure the continued functioning of these grips.

A permeability inversion should be avoided in pavement construction. This occurs when the permeability of the pavement and subgrade layers decreases with depth. Under infiltration of rainwater, there is potential for moisture accumulation at the interface of the layers. The creation of such a perched water table often leads to rapid lateral wetting under a seal. This may lead to base or sub-base saturation in the outer wheel track, and result in potholing and potential failure of the base layer when trafficked. A permeability inversion often occurs at the interface between sub-base and subgrade, since many subgrades are of cohesive and relatively impermeable fine-grained materials. Under these circumstances, a conservative fully soaked design approach may be necessary. Figure 13.14 summarises typical permeabilities for pavement construction materials.

Subsurface drainage systems such as French drains, underdrains, edge drains, wells, interceptor drains, filter layers or drainage blankets can be placed in or under the pavement to enhance pavement drainage in high water table areas (Rolt et al. 2002). However, because of budget constraints, their use in rural roads may be limited to high-risk areas.

Figure 13.13 Typical grips into granular base layer.

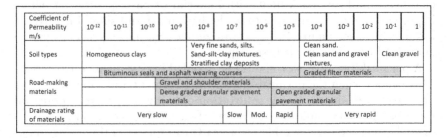

Coefficient of Permeability m/s	10^{-12}	10^{-11}	10^{-10}	10^{-9}	10^{-8}	10^{-7}	10^{-6}	10^{-5}	10^{-4}	10^{-3}	10^{-2}	10^{-1}	1
Soil types	Homogeneous clays				Very fine sands, silts. Sand-silt-clay mixtures. Stratified clay deposits				Clean sand. Clean sand and gravel mixtures,			Clean gravel	
Road-making materials			Bituminous seals and asphalt wearing courses						Graded filter materials				
				Gravel and shoulder materials									
				Dense graded granular pavement materials					Open graded granular pavement materials				
Drainage rating of materials	Very slow					Slow	Mod.	Rapid		Very rapid			

Figure 13.14 Typical materials permeability values.

REFERENCES

AfCAP. 2013. South Sudan low volume road design manual; volume 2 cross drainage and small structures. DFID for Ministry of Roads and Bridges Government of South Sudan. https://www.gov.uk/research-for-development-outputs/south-sudan-low-volume-roads-design-manual-volume-1-2-and-3.

American Association of State Highway and Transportation Officials (AASHTO). 1993. *AASHTO Guidelines for Design of Pavement Structures*. Washington, DC: American Association of State Highway and Transportation Officials.

Archondo-Callao, R. S. 2001. Roads Economic Decision model (RED) economic evaluation of low volume roads. Africa Region Findings & Good Practice Info briefs; no. 179. World Bank. https://openknowledge.worldbank.org/handle/10986/9820.

ARRB. 1995. *Sealed Roads Manual: Guideline to Good Practice*. Australia: ARRB Transport Research Ltd.

ARRB. 2000. *Unsealed Roads Manual: Guideline to Good Practice*. Australia: ARRB Transport Research Ltd. https://austroads.com.au/publications/pavement/agpt06.

Austroads. 2009. *Guide to Pavement Technology Part 6: Unsealed Pavements*. AGPT 02-09. Sydney, NSW: Austroads. https://austroads.com.au/publications/pavement/agpt06.

Austroads. 2017. *Guide to Pavement Technology Part 2: Pavement Structural Design*. AGPT 02-17. Sydney, NSW: Austroads. https://austroads.com.au/publications/pavement/agpt02/media/AGPT02-17_Guide_to_Pavement_Technology_Part_2_Pavement_Structural_Design.pdf.

Austroads. 2018. *Appropriate Use of Marginal and Non-standard Materials in Road Construction and Maintenance*. Technical report AP-T335-18. https://austroads.com.au/publications/asset-management/ap-t335–18.

Cedergren, H. 1997. *Seepage, Drainage, and Flow Nets*. 3rd edition. John Wiley and Sons.

Cook, J. R., E. C. Bishop, C. S. Gourley and N. E. Elsworth. 2000. Promoting the use of marginal materials TRL Ltd DFID KaR Project PR/INT/205/2001 R6887. https://www.gov.uk/research-for-development-outputs/promoting-the-use-of-marginal-materials.

Cook, J. R. and C. S. Gourley. 2002. A framework for the appropriate use of marginal materials. World Road Association (PIARC)-Technical Committee C12, Mongolia. http://transport-links.com/wp-content/uploads/2019/11/1_796_PA3890.pdf.

Cook, J. R. and R. C. Petts. 2005. Rural road gravel assessment programme. SEACAP 4, module 4, final report. DFID Report for MoT, Vietnam. https://www.gov.uk/research-for-development-outputs/seacap-4-rural-road-gravel-assessment-programme-final-report.

Cook. J. R., R. C. Petts and J. Rolt. 2013. Low volume rural road surfacing and pavements: A guide to good practice. Research report for AfCAP and UKAID-DFID. https://www.research4cap.org/index.php/resources/rural-access-library.

Gourley, C. S. and P. A. K. Greening. 1999. Performance of low volume sealed roads: Results and recommendations from studies in Southern Africa. TRL project report PR/OSC/167/99. project record 6020. Transport Research Laboratory, Crowthorne. UK. https://www.gov.uk/research-for-development-outputs/collaborative-research-programme-on-highway-engineering-materials-in-the-sadc-region-volume-1-performance-of-low-volume-sealed-roads-results-and-recommendations-from-studies-in-southern-africa.

Gov. of Myanmar. 2020. *Low Volume Road Design Manual: Section B Design.* UKAID-DFID for Department of Rural Road Development, Ministry of Construction. https://www.research4cap.org/ral/DRRD-2020-LVRRDesignManual-SectionB-Ch5t011-AsCAP-MYA2118A-200623-compressed.pdf.

Henning, T., P. Kadar and R. Bennet. 2005. Surfacing alternatives for unsealed roads. World Bank TRN-33, Washington. https://documents1.worldbank.org/curated/en/473411468154454358/pdf/371920Surfacing0Alternatives01PUBLIC1.pdf.

Hindson, J. 1983. *Earth Roads: A Practical Guide to Earth Road Construction and Maintenance. Intermediate Technology Publications,*

Intech-TRL. 2006. SEACAP 1 Rural road surfacing trials final report. UKAID-DFID report for MoT, Vietnam. https://www.gov.uk/research-for-development-outputs/seacap-1-rural-road-surfacing-research-rrsr-final-report-volume-1-main-report.

JKR Malyasia. 2012. *Design Guide for Alternative Pavement Structures, Low Volume Roads.* Malaysia: Road and Geotechnical Engineering Unit, Public Works Dept.

MoWTC (Ministry of Works, Transport and Communication). 2016. *Low Volume Roads Manual. Part D Design.* https://www.research4cap.org/index.php/resources/rural-access-library

MPW Liberia. 2019. *Manual for Low Volume Roads. Part B: Materials. Pavement Design and Construction.* UKAID-DFID for Ministry of Public Works Liberia. https://www.gov.uk/research-for-development-outputs/manual-for-low-volume-roads-liberia.

Paige-Green, P. and G. D. van Zyl. 2019. A review of the DCP-DN pavement design method for low volume sealed roads. Development and applications. *Journal of Transportation Technologies* 9:397–422. https://doi.org/10.4236/jtts.2019.94025.

Rolt, J. R. 2004. Chapter 15 Structural design of asphalt pavements. In *Road Engineering for Development.* 2nd edition. Robinson R. and Thagesen B., Eds. Taylor and Francis. ISBN 0-203-30198-6.

Rolt, J. R. 2007. Behaviour of engineered natural surface roads. SEACAP 19 technical paper 2.1. UKAID-DFID for MRD, MPWT, Cambodia. https://www.research4cap.org/ral/Rolt-TRL-Cambodia-2007-Technical+Paper2.1+ENS-SEACAP19-v100203.pdf.

Rolt, J., C. S. Gourley and J. P. Hayes. 2002. Rational drainage of road pavements. TRL report PR/INT/244/2002. Transport Research Laboratory, Crowthorne, Berkshire, UK.

Roughton International. 2008. Local resource solutions to problematic rural road access in Lao PDR. SEACAP 17 Rural access roads on route no. 3. Module 2 – Completion of construction report. https://www.research-4cap.org/ral/Roughton-LaoPDR-2008-Module2+Construction+Report-SEACAP17-v080918.pdf.

RECAP. 2020. Rural Road Note 01: A guide on the application of pavement design methods for Low Volume Rural Roads. UK Aid. https://assets.publishing.service.gov.uk/media/5fa193578fa8f57896ad026a/Roltetal-TRL-2020-RuralRoadNote01PavementDesignMethodsLVRR-ReCAP-GEN2166B-200710.pdf.

SADC. 2003. Chapter 5 Pavement design, material and surfacing. In *Low Volume Sealed Roads Guideline*. SADC-SATCC. https://www.gtkp.com/assets/uploads/20100103-131139-9347-SADC%20Guideline-Part2.pdf.

SANRAL. 2013. Chapter 10 Pavement design. In *South African Pavement Engineering Manual*. South African National Road Agency Ltd.

TRL. 1993. *Overseas Road Note 31. A Guide to Structural Design of Bitumen Surfaced Roads in Tropical and Sub-tropical Countries*. 4th edition. UK: TRL Ltd for DFID. https://www.gov.uk/research-for-development-outputs/orn31-a-guide-to-the-structural-design-of-bitumen-surfaced-roads-in-tropical-and-sub-tropical-countries.

TRL. 2006. *UK DCP 3.1 User Manual. Measuring Road Pavement Strength and Designing Low Volume Sealed Roads Using the Dynamic Cone Penetrometer*. UK: TRL Ltd, for DFID. https://www.gov.uk/research-for-development-outputs/uk-dynamic-cone-penetrometer-dcp-software-version-3-1.

TRL. 2019. Development of guidelines and specifications for low volume sealed roads through back analysis. Draft final report. ReCAP reference number: RAF2069A. https://www.gov.uk/research-for-development-outputs/development-of-guidelines-and-specifications-for-low-volume-sealed-roads-through-back-analysis-phase-3-final-report.

TRL. 2023. *Road Note 31 (5th Edition). A guide to Structural Design of Surfaced Roads in Tropical and Sub-tropical regions*. TRL Ltd for UK. Aid. https://transport-links.com/download/road-note-31-a-guide-to-the-structural-design-of-surfaced-roads-in-tropical-and-sub-tropical-regions-online-edition/

SEACAP. 2008. Low volume rural roads standards and specifications; Part III. SEACAP 3, UKAID-DFID for MPWT Lao PDR. https://assets.publishing.service.gov.uk/media/57a08bbb40f0b64974000d28/Seacap3-LVRR3.pdf.

World Bank. 2020. *To Pave or Not to Pave. Developing a Framework for Systematic Decision Making in the Choice of Paving Technologies for Rural Roads*. Mobility and Transport Connectivity Series. https://openknowledge.worldbank.org/handle/10986/35163.

Chapter 14

Bituminous pavements

INTRODUCTION

The detailed design of bituminous sealed pavements follows on from the general principles and initial design selection process described in the previous chapter. This chapter concentrates on bitumen surfaced flexible pavements, and Chapter 15 deals with non-bituminous options. Bituminous surfaces consist of selected aggregates bound together by a bituminous binder and comprise a range of different types, from thin surface dressings to thicker layers of asphalt mix, for use in Low Volume Rural Road (LVRR) design.

The Scottish engineer John Loudon McAdam developed the stone macadam road surface in the 1810s (Skempton et al. 2015). This technique was later improved by the addition of tar derived as a by-product of town gas production from coal to produce an even more durable road surface (tarmacadam). Tar as a road surface sealer and binder was then later superseded by bitumen products derived from petroleum processing, and today bitumen sealers and binders are the almost universal technique used in 'flexible' road surfacing. Technically, bitumen is a highly viscous liquid or semi-solid sealer/binder; hence the 'flexible' terminology. Figures 14.1 and 14.2 illustrate the principal components of bitumen sealed pavements; not all pavement layers may be required for some LVRRs.

Durable bituminous seals are critical for achieving the intended design life of the pavement. These surface options fulfil a variety of functions that, collectively, preserve the integrity of roadway and any pavement layers, and improve the functionality of the road in service. The design of a particular type of bituminous surface treatment is usually project specific and related to such factors as traffic volume, climatic conditions, available type and quality of materials, and sub-surface characteristics.

In the last 2–3 decades there have been significant advances in sealed pavement design developed through UKDFID, World Bank or ADB funded research programmes, and outcomes from this work have been included in recent country specific or region-specific rural road design guides and manuals. This chapter provides information on characteristics of bituminous pavement options appropriate for rural roads. An overview of the various

DOI:10.1201/9780429173271-14

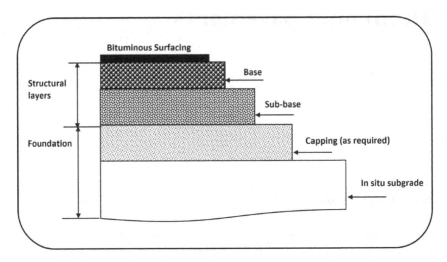

Figure 14.1   Sealed pavement elements.

Figure 14.2   Profile of a typical sealed pavement.

types of bitumen seal is followed by summary descriptions of the structural design options, with a focus on practical catalogues and charts for use in low volume rural road design.

SURFACING OPTIONS

Bitumen options

The general types of bitumen used in LVRRs are penetration grade, cutback or emulsion option, as outlined in Table 14.1.

Bitumen emulsions are two-phase systems consisting of a dispersion of bitumen droplets in water containing an emulsifier. After spraying, or placement, the water migrates from the emulsion, leaving a residue of bitumen—a process known as 'breaking'. This can take from as little as 10 minutes to several hours depending whether the emulsion is a 'rapid setting' or 'slow setting', depending on its designed composition. It is simple to check whether the emulsion has broken, as the colour of the emulsion changes from brown to black. Commonly, bitumen emulsions are available in two classes; cationic and anionic. Acidic aggregates, such as those from granite and quartzite, are negatively charged, providing good adhesion to the positively charged bitumen in a cationic emulsion. Conversely good adhesion is achieved between anionic emulsions and positively charged aggregates such as dolomite and limestone.

Bitumen use

In road surfacing, bitumen bonds aggregate material together into a cohesive layer to serve as a durable wearing course, offering a safe contact zone between vehicle tyres and the road surface. The function of the bituminous binder in a seal is essentially to provide adhesion between the various aggregate particles and between the entire seal and the existing road surface. In addition, it should have sufficient cohesive strength to resist brittle fracture. It is important to follow manufacturer's recommendations on storage of emulsions before use; occasional gentle agitation or circulation is necessary to prevent build-up of sedimented bitumen particles.

General reviews of bituminous surfaces for LMICs are contained in TRL (2000), SABITA (2011) and Roughton International (2012), and World Bank (2020), while TRL (2003) focuses on labour-based approaches.

As well as being used as a surfacing binder, bitumen has other uses in LVRR flexible pavements:

1. As a *prime coat*, which is a spray application of a suitable bituminous binder onto a granular base layer prior to the application of overlying bituminous materials or layers. The function of a prime coat is to:

Table 14.1 Typical LVRR Bitumen Options

Bitumen type	Description	Description	Uses and limitations
Hot bitumen	Penetration grades.	Penetration-grade bitumen commonly 60/70; 80/100; 150/200: usually produced at the refinery.	80/100 penetration is commonly used for thin seals. Penetration-grade bitumen can be used at higher road surface temperatures than cut-back. High initial adhesion. Can be used on higher gradients (12%) than cut-back or emulsion. High temperatures of placement (130°C–170°C) brings Quality Control and safety issues.
	Cut-back: medium curing (MC) 30-to MC 3000.	Made by blending appropriate amounts of kerosene, diesel or a blend of kerosene and diesel. Higher the % solvent the lower the viscosity. MC 3000 grade cut-back is normally blended using 12%–17% of cutter.	Cut-back grades of bitumen are most appropriate at the lower road temperatures. High initial adhesion. Usually, an MC 3000 grade used for surface dressings. MC30-70 commonly used for prime coat. Usual alignment gradient limit of 8%–10%. High temperatures of placement (130°C–170°C) implies Quality Control and safety issues.
Bitumen emulsion	Cationic or anionic; can be slowing setting or fast setting.	Cationic more common than anionic.	Rapid setting cationic bitumen emulsion with a normal bitumen content of 65%–75% is recommended for most surface dressing work. Anionic has adhesion problems with high silica aggregates (e.g. granite, quartzite). Gradient limit 6%–8% without special precautions. Lower spraying temperatures (75°C) make it easier for labour-based operations and local maintenance. Less risk of early bleeding but low initial adhesion. A 1:1 dilution of emulsion with water commonly used as a tack coat. Slow setting also used for cold-pre-mix.

- Provide adhesion between a granular layer and a bituminous layer.
- Inhibit ingress of water while not hampering the evaporation of water in the layer being primed.
- Limit absorption of a sprayed binder application by the base.
- Bind finer particles of the upper zone of the layer being primed.

2. As a *tack coat*, which is a sprayed bituminous binder applied to an existing bituminous layer to promote adhesion between it and a new overlying bituminous layer. Its main purpose is to prevent undue movement of the newly placed layer during its rolling-in, and to provide a bond between the new layer and the substrate.

Bitumen may also be used as modifier or stabiliser to sandy materials for use as base or sub-base, as described in Chapter 8. Common types of bituminous surface treatment suitable for LVRRs are shown in Figure 14.3 and summarised in the following sections (see also Cook et al. 2013; World Bank 2020).

Sand Seal

This seal consists of a spray of bituminous or bituminous emulsion binder followed by the application of a coarse, clean sand or crusher dust as a fine aggregate. This surfacing can be used on low volume, low axle-load roads, especially in drier regions, but can also be used for maintenance resealing or for temporary by-passes or diversions. For new construction, two layers are usually specified. Single Sand Seals are generally not a recommended option, although a single Sand Seal may be used in conjunction with other options, for example in a single chip seal and Sand Seal combination. There is a recommended extended curing period (typically 8–12 weeks) between the first and second Sand Seal applications to ensure complete loss of volatiles from the first seal and thus minimise the risk of bleeding.

Slurry Seal

A Slurry Seal consists of a homogeneous pre-mixed combination of fine aggregate, stable mix-grade emulsion or a modified emulsion, water and filler (cement or lime). The production of slurry can be undertaken in simple concrete mixers, and laid by hand or more sophisticated purpose-designed machines which mix and spread the slurry.

Slurry Seals can be used for treating various defects on an existing road surface carrying relatively low traffic. The following are typical uses:

- Arrest loss of chippings.
- Restore surface texture.
- Reduce unevenness because of minor bumps or ruts.
- Rectify minor surface cracking.

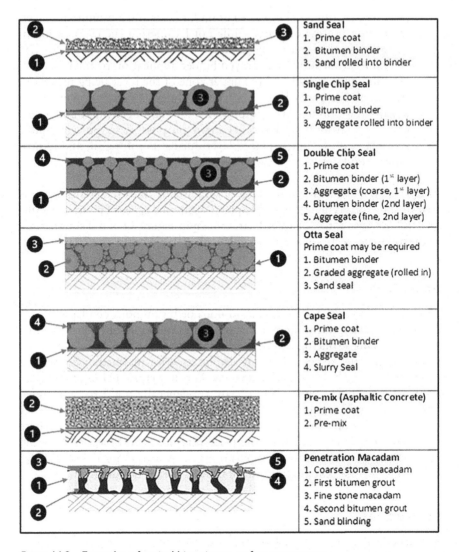

Figure 14.3   Examples of typical bituminous surface treatments.

- As a grout seal following a new single chip seal.
- In multiple layers directly on the base of low-traffic roads.
- A component of a Cape Seal.

Chip seal

Chip Seals consist of a spray of bituminous binder (either hot bitumen or bitumen emulsion) followed by the application of a layer of aggregate (stone chippings). A second layer of bitumen and chippings may follow as double treatment.These are commonly referred to Single Bituminous Surface Treatment (SBST) or Double Bituminous Surface Treatment (DBST). The binder acts as a waterproofing seal preventing entry of surface water into the road structure, while the chippings protect this film from damage by traffic. DBST has a proven performance record in most climatic environments and well suited to LVRR situations with typical initial life of 8–12 years, if well-constructed and maintained. Chip Seals are the commonest the option used in rural road design. Chip Seals can be used as a labour-based option, particularly if bitumen emulsion is used as the binder. They can be used for a number of purposes, including:

- In new construction (normally DBST).
- Temporary by-passes or diversions (normally SBST).
- Maintenance resealing (normally SBST).
- First layer of a Cape Seal or in combination with a Sand Seal.

Otta seal

An Otta Seal is a sprayed bituminous surfacing comprising an application of a heated, relatively soft, low-viscosity binder followed by a mixture of graded aggregates ranging from natural gravel to crushed rock, with or without a sand cover seal. This type of seal contrasts with the single-sized crushed aggregate relatively harder (high viscosity) binders used in Chip Seals. Otta seals can be single or double (two layers). The technique also differs from chip seals in that a heavy pneumatic tyred roller is used to knead the aggregate into the binder, working it up between the aggregate particles.

Otta Seals can be used for a variety of purposes, including:

- New construction (single or double Otta Seals with/without Sand Seal).
- Temporary seal (normally single Otta Seal – diversions, haul roads, temporary accesses, etc.).
- Maintenance reseals (normally single Otta Seal).

Cape seal

A Cape Seal is a multiple bituminous surface treatment consisting of an application of a single Chip Seal followed by a single or double application of bitumen Slurry Seal. Usually, a 13 mm aggregate first seal is combined with a single slurry application or a coarser 19 mm first seal with a double slurry application.

The technique is durable (typical initial life 8–14 years) and suitable for construction by small contractors or community groups, as it does not require expensive equipment. It is considered to be climate resilient and appropriate for moderate to high rural traffic.

Penetration macadam

Penetration Macadam is a well-understood sealing option that has been used for decades in many LMICs and consists of layers of successively finer broken, or crushed rock, interspersed with applications of heated bitumen to infill voids and seal the surface. In a typical procedure, an initial layer of 40–60 mm aggregate is keyed-in and rolled onto the underlying base. A first penetration of bitumen is sprayed onto the initial aggregate layer (commonly 5–6 kg/m²), and immediately afterwards a second stone application is made by hand onto the grouted aggregate, using 10–20 mm chippings. A second penetration of bitumen is then (commonly 3–5 kg/m²) sprayed onto the cover layer of chippings. Sometimes a final sand blinding layer is applied.

The effect is to achieve a matrix of keyed stones grouted and sealed with bitumen to a depth of about 60–80 mm. It is laid as a surfacing on a previously prepared stone road base. Penetration Macadam provides a structural contribution as well providing a seal. It is normally used with Dry-Bound (DBM) or Water-Bound Macadam (WBM) structural layers (200–300 mm total thickness). This design option is not normally recommended above 300,000 esa. The high bitumen application rates (commonly 8–11 kg/m²) make this technique particularly sensitive to bitumen costs and increasing CO_2 concerns.

Cold pre-mix asphalt

Cold mixes normally consist of a mixture of graded crushed aggregate and a stable, slow-breaking emulsion (see Figure 14.4). For LVRRs these can be mixed by hand or in a concrete mixer on site. After mixing, the material is spread on a primed base and rolled. The thickness could vary, depending on nominal size aggregate 20–40 mm.

It is very suitable for labour-based construction, requires very simple construction plant and reduces the potential hazard of working with hot bitumen. It makes a contribution to pavement structural strength, has good climate resilience, is less sensitive to poor maintenance capacity and has a longer design life than most thinner surfacing options. Its disadvantages are its higher cost, less flexibility and sensitivity to poor base support, compared to thin seals.

Hot mix bitumen macadam

This type of surface is more usually applied to higher traffic volume routes. It normally requires high capital, specialist equipment for mixing and laying, a high level of Quality Control, and can involve substantial mobilisation and demobilisation arrangements and costs. It is therefore not usually appropriate for LVRR use.

Figure 14.4   Cold asphalt pre-mix – cored pavement, Vietnam.

SELECTING OF SEAL OPTIONS

Factors impacting bituminous seal performance

The performance of a bituminous surfacing treatment can vary widely in relation to a number of factors, such as:

Climate: Very high temperatures cause rapid binder hardening through accelerated loss of volatiles, while low temperatures can lead to brittleness of the binder leading to cracking or aggregate loss resulting in reduced surfacing life. Future climatic increases in temperature need to be considered in bitumen-type selection.

Pavement strength: Insufficient underlying pavement stiffness will lead to fatigue cracking and reduced surfacing life.

Base materials: Unsatisfactory base performance and absorption of binder into porous base materials will lead to reduced surfacing life.

Binder durability: The lower the durability of the binder, the higher the rate of its hardening, and the shorter the surfacing life.

Design and construction of surfacing: Improper design and poor construction techniques (e.g. inadequate prime, uneven rate of binder/

Figure 14.5   DBST damaged by high axle loaded vehicles on climbing corner.

aggregate application or 'dirty' aggregates) will lead to reduced surfacing life.

Traffic: The higher the volume of heavy traffic, the shorter the surfacing life. The impact of heavier traffic may be particularly evident on tight curves or hairpin bends, especially those on gradients more than about 6%, as illustrated by the DBST less than 1 year old in Figure 14.5. Junctions, steep sections and laybys can be susceptible to deterioration from lubricant dropping from slow or stationary vehicles.

Stone polishing: The faster the polishing of the stone and loss of skid resistance, the earlier the requirement for resurfacing.

Maintenance: A poor maintenance regime that doesn't include sealing of cracks in the surfacing will allow further crack deterioration and eventual loss of integrity for the whole surface.

The various factors affecting the choice of surface treatments are summarised in relation to the operational requirements in Table 14.2, which is based on a review of work reported in Intech-TRL (2006), SABITA (2011) and Roughton (2012).

Key issues in the design of bituminous surfacing

Following the selection of the bituminous sealing option, a number of key issues require definition at the design stage:

- Binder selection and spray rates.
- Aggregate size, grading and spreading rates.
- Rolling and post-construction procedures.

Table 14.2 Appropriate Surfacing Selection

Parameter	Sub-division	Type of surfacing						
		SS	SIS	SCS	SOS	DOS	CS	AC
Service life required (years)	Short (<5)	A	A	A	B	B	B	X
	Medium (5–10)	X	X	X	A	B	A	B
	Long (>10)	X	X	X	B	A	B	A
Traffic level (AADT)	Light (<100)	A	A	A	B	B	B	B
	Medium (100–300)	X	X	X	A	B	A	A
	Heavy (>300)	X	X	X	A	A	A	A
High axle loads	Low (no trucks)	A	A	A	B	B	B	B
	Medium (trucks)	X	X	X	A	B	A	A
	High (three-axle trucks)	X	X	X	P	A	B	A
Gradient	Mild (<5%)	A	A	A	A	B	A	A
	Moderate (5%–12%)	X	X	X	A	B	A	A
	Steep (>12%)	X	X	X	A	A	A	A
Material quality	Poor	X	X	X	A	A	X	X
	Moderate	B	B	B	A	A	B	B
	Good	A	A	A	B	P	A	A
Quality of base	Poor	X	X	X	B	A	B	B
	Moderate	B	B	B	A	A	A	A
	Good	A	A	A	A	A	A	A
Labour-based approaches[a]		A	A	A	X	X	A	A
Contractor experience.	Low	B	B	X	A	A	X	X
	Moderate	A	B	P	A	A	B	B
	High	A	A	A	A	A	A	A
Likelihood of good maintenance	Low	X	X	X	A	A	X	A
	Moderate	X	X	X	A	A	A	A
	High	A	A	A	A	A	A	A

A = most suitable, B = possible, not optimal, X = not recommended.

SS, Sand Seal; SIS, Slurry Seal; SC, single chip seal; DCS, double chip seal; SOS, single otta seal; DO, double otta seal; CS, cape seal; AC, cold pre-mix.

[a] Otta seal requires significant amounts of heavy tyred rolling to bring up the bitumen during construction.

The following notes provide general guidance on the above issues, while detailed guidance is included in ORN 3 (TRL 2000) with specific guidance tables on spray rates in TRL (2003). In addition, some LMIC rural road manuals provide detailed seal specifications within their specific environment.

Table 14.1 provides general descriptions of the bitumen options for LVRR pavements, with detailed selection being a function of factors such as range of road temperature, type of seal and traffic. For chip seals the amount of bitumen sprayed on the road must be adequate to bind the chippings to the road surface, by coating the bottom of the chippings and being squeezed up between them. If too much bitumen is applied, it will rise up to the top of the chippings when the road is trafficked and cause the surfacing to 'bleed', leading to surface deterioration as well as reducing skid resistance.

A key element in the quantity bitumen to be used is the average thickness of the single layer of chippings. When the chippings are rolled or trafficked, they move and tend to rotate until the vertical thickness of the chippings is their smallest dimension (The Least Dimension). The Average Least Dimension (ALD) is the average thickness of the chippings when laid flat, the vertical thickness being the 'least' or the smallest dimension of each chipping. A representative sample of approximately 200 chippings is taken and the least dimension of each is measured manually. The average value is then taken as the ALD. ALD can also be linked to the nominal aggregate size by considering its flakiness. Table 14.3 provides some typical ranges.

ORN 3 (TRL 2000) provides formulae for initial estimation of spray rates binder and spread rates for chip seals based on ALD.

For Binder rate:

$$Rb = 0.625 + (F \times 0.023) + [0.0375 + (F \times 0.0011)] ALD$$

where F=Sum of the site weighting factors as shown in Table 14.4.
ALD=The average least dimension of the chippings (mm).
Rb=Basic rate of spread of bitumen (kg/m²).
For aggregate rate

$$Ra = 1.364 \times ALD$$

Where ALD=The average least dimension of the chippings (mm).
Ra spreading rate in kg/m²

An additional 10% can be allowed for whip off during initial trafficking. Storage and handling losses for source estimations should also be allowed for when stockpiling chippings.

Table 14.3 ALD as a Function of Flakiness

Aggregate nominal size (mm)	Flakiness index ALD (mm)			
	10%	20%	40%	60%
19	14.0	12.6	10.9	9.5
14	11.0	10.2	8.7	7.7
9	8.3	6.5	5.5	5.0
6	5.0	4.5	4.0	N/A

Table 14.4 Binder Weighting Factor Adjustments for Chip Seals

Site factor	Weighting factors
Traffic (ADT/lane)	
<50	+3
50–250	+1
250–500	0
>500 (Not LVRR)	−1 to −3
Surface condition	
Untreated base	+6
Existing very lean bituminous	+4
Existing lean bituminous	0
Rich bituminous	−1 to −3
Climate	
Wet, cold	+2
Tropical (wet, hot)	+1
Temperate	0
Sem-arid (hot, dry)	−1
Arid (very hot, dry)	−2
Chippings	
Round, dusty	+2
Cubical	0
Flaky	−2

Table 14.5 summarises some typical spray rates for binders and Table 14.6 some typical rates for aggregates for specific surfaces. These are general figures that should be adjusted for local conditions.

The rolling of a surface dressing plays an important part in ensuring the quality of the seal by assisting in the initial orientation of the chippings. Traditionally, steel-wheeled rollers have been used, but these tend to crush weaker and poorly shaped aggregates, and if used they should not exceed 8 tonnes in weight and should only be used on chippings which are strong enough. In general, pneumatic tyred rollers are preferred, if available, because the tyres have a kneading action that tends to manoeuvre the chippings into a tight mosaic without splitting them.

Aftercare is also an essential part of the surfacing process and consists of removing excess chippings within 24–48 hours of the construction of a dressing. Some of the excess chippings will have been thrown clear by passing vehicles, but some loose chippings will remain on the surface. They can be removed by brooming, although care must be taken with brooming to avoid damage to the new dressing.

Table 14.5 Typical Binder Spray Rates

Surface	Spray rate	Bitumen
Prime	0.3–0.8 kg/m²	Penetration grade/cut-back
	1.1 kg/m²	Emulsion
Penetration Macadam	Layer 1: 3.0–5.0 kg/m²	Penetration grade
	Layer 2: 8.0–11 kg/m²	
Sand Seal	1.1–1.3 l/m²	Penetration grade
	1.4–1.6 l/m²	Emulsion
Chip seal	14 mm: 1.2–1.3 l/m²	MC3000
	10 mm: 1.1–1.2 l/m²	MC3000
	6 mm: 1.0–1.1 l/m²	MC3000
	14 mm: 1.9–2.0 l/m²	Emulsion (60% bitumen)
	10 mm: 1.6–1.8 l/m²	Emulsion (60% bitumen)
	6 mm: 1.4–1.6 l/m²	Emulsion (60% bitumen)
Otta Seal (Single)	2.0 l/m²	MC3000
	2.0 l/m²	Emulsion

Table 14.6 Aggregate Spreading Rates

Surface treatment	Spread rates (kg/m²)
Sand Seal	13–19
Single chip seal, 14 mm agg. (First layer of DBST)	12–13
Second layer in DBST, 6 mm agg.	4.0
Otta seal (single)	15–20

STRUCTURAL DESIGN OF LOW VOLUME BITUMINOUS PAVEMENTS

General approach

Chapter 13 summarised the options for the structural design of pavements and indicated that the pragmatic application of nationally-relevant pavement design catalogues or charts is the most common approach. Many LMICs now have developed LVRR design manuals or guides that include pavement design charts, which in detail are based around traffic and subgrade strength, the latter most commonly represented by CBR. Although they have many similarities and frequently have their origin in variations on ORN 31 (TRL 1993), the individual catalogue or charts have been assembled bearing in mind specific local Road Environments.

Design charts for bituminous-sealed options

Table 14.7 lists some typical LVRR design catalogues now in use, with their associated key Road Environment criteria. Note that in some countries with a relatively dry climate, a relaxation in specification items is possible; the definition of 'dry climate' in this case is defined as a Weinert 'N' value>4 (Weinert 1974, see Chapter 6).

Most design charts do not cater for very weak subgrades (CBR<2%–3%) and other problems soils. In these cases additional measures should be taken to replace, improve or overlay the weak subgrade, as outlined in Chapter 10.

The following tables and figures illustrate some typical charts for the used for the selection of structural pavement layers and materials. These tables are presented as illustrative examples only. They should not be directly applied without reference to their Road Environment envelope and the original assumptions.

Tables 14.8 and 14.9 present typical examples of wet environment design chart using granular materials and its dry condition variation. In these tables 'G80, G65', etc. refer to granular materials with a soaked CBR of 80%, 65%, etc.; Chapter 8 provides further material description. These particular examples are based on the Ethiopian Low Volume Road Design Manual (ERA 2016).

In a similar style, Table 14.10 presents a typical design for chart for using water-bound macadam (WBM) structural layers in conjunction with a bitumen seal (AfCAP 2012).

Recent research funded by UKDFID (RECAP 2019) has re-evaluated or back-analysed the performance of a wide range of existing sealed and unsealed LVRRs. This has indicated that many, well-constructed, sealed LVRRs are performing better than their designs anticipated. This can be related to work previously undertaken on sealed roads in Vietnam (Intech-TRL 2006) and on very low-traffic low-axle-load designs for Laos (SEACAP 2008). The consequence from this work is that some reductions in layer thickness and materials specifications could be possible in the appropriate circumstances. In an example from Laos, Table 14.11 presents proposed structural thicknesses for very low traffic and with axle loads below 4.5 tonne. This approach has also emphasised the use of imported lower quality (and cheaper) capping layer materials as a means of reducing the base/sub-base thickness.

Table 14.12 makes some sample comparisons to illustrate the effects of taking into account differing axle loads and different rainfall regimes, using typical low-traffic scenarios:

- Working subgrade CBR: 6%.
- Traffic requirement: 100,000 and 1,000,000 esa.

This comparison confirms that by controlling axle loads on low volume roads, there can be a significant reduction in material quality in the structural layer materials. This is an important issue when aiming to use local materials use as much as possible. Some savings are also evident when assuming a 'dry' environment or opting for a sealed shoulder construction as an equivalent to a 'dry' condition.

Table 14.7 Typical Catalogue/Chart Approaches to This Bituminous-Sealed LVRR Pavements

Country location	Catalogue description	Key environment factors	Comment
International ORN 31-(TRL 1993) – updated 2023	Catalogue of charts based on CBR (2%–>30%) and traffic (0.3–3.0 mesa) for thin seals and up to 30 mesa for other options.	No specific environmental factors other than moisture condition included in CBR assessment.	A template for many country specific design charts. Not explicitly designed for LVRR and hence lacks detailed differentiation for lower traffic levels.
South Sudan AfCAP 2013 (Similar for Ethiopia, Malawi, Tanzania, Liberia)	Based on CBR (3%–30%) versus traffic up to 1.0 mesa Alternative design process included for some countries using DCP direct penetration method in appropriate circumstances included.	Climate variation based on Weinert 'N' value. Southern/ East African weather patterns.	Recommends use of DCP as a means of assessing CBR. DCP-DN approach included for upgrading gravel to paved road within the Southern Africa environment (See Chapter 13).
Afghanistan (RECAP 2019)	Based on CBR (3%–30%) versus traffic (0.01–1.0 mesa).	Climate very dry to wet (75–1,200 mm/year). Flat to mountainous terrain.	Although a DCP-DN design alternative is included, the limited correlation with southern Africa should be noted.
Malaysia (JKR 2012)	Based on CBR (2%–24%) versus traffic (up to 0.5 modified mesa). Uses thicker sub-base rather than imported subgrade.	Wet climate. Includes terrain and lane number in modified esa figures.	Includes alternative options such as AC and as stabilised base up to 1.0 mesa.
Myanmar (RECAP 2020)	Based on CBR versus traffic 0.01–1.0 mesa	Climate taken as being overall wet on basis of Weinert N values even in locally defined Dry Zone.	Options are included for reducing materials CBR strength and layer thicknesses where axle loads are constrained.
Lao PDR (SEACAP 2008)	Based on CBR versus traffic (esa) with 4 tonne axle load limitation and up to 100,000 esa design life	Wet climate. Low-axle-load limitation.	Special case for low-axle-load tertiary roads.
Australia; (ARRB 1995;Austroads 2017)	Based on diagrams involving CBR (3%–30%) and traffic (1,000 to 5×10^5 esa) and overall pavement thickness with minimum 100 mm base thickness.	Allows very low traffic. Dry to wet environment. CBR testing assumed to model the working moisture environment.	Allows variation of layer composition and thickness depending on strength of available materials.

Table 14.8 Typical Thin Bituminous Pavement Design Chart for Structural Layers for a Wet Climate

		Structural layer thicknesses (mm) and associated material (G15 -G80)				
		Traffic range (T) layer thickness in mm				
Subgrade CBR	Layer	<0.01 mesa	0.01–0.1 mesa	0.1–0.3 mesa	03–0.5 mesa	0.5–1.0 mesa
<3%		Special measures required				
3%–4.9%	Base	150 G45	150 G65	150 G65	175 G80	200 G80
	Sub-base	150 G15	150 G30	150 G30	175 G30	175 G30
	Subgrade		130 G15	175 G15	175 G15	175 G15
5%–7.9%	Base	125 G45	150 G65	150 G65	175 G65	200 G80
	Sub-base	150 G15	100 G30	150 G30	150 G30	150 G30
	Subgrade		100 G15	125 G15	125 G15	150 G15
8%–14.9%	Base	200 G45	150 G65	150 G65	175 G65	200 G80
	Sub-base		120 G30	200 G30	200 G30	200 G30
15%–29%	Base	175 G45	125 G65	175 G65	175 G65	175 G80
	Sub-base		100 G30	125 G30	150 G30	150 G30
>30%	Base	150 G45	100 G65	175 G65	200 G65	200 G80

Table 14.9 Typical Thin Bituminous Pavement Design Chart for Structural Layers for a Dry Climate

		Structural layer thicknesses (mm) and associated material (G15 -G80)				
		Traffic range (T)				
Subgrade CBR	Layer	<0.01 mesa	0.01–0.1 mesa	0.1–0.3 mesa	03–0.5 mesa	0.5–1.0 mesa
<3%		Special measures required				
3%–4.9%	Base	150 G45	150 G65	150 G65	175 G80	175 G80
	Sub-base	150 G15	125 G30	150 G30	150 G30	175 G30
	Subgrade		100 G15	150 G15	150 G15	175 G15
5%–7.9%	Base	125 G45	150 G55	175 G65	175 G80	175 G80
	Sub-base	125 G15	175 G30	175 G30	200 G30	250 G30
8%–14.9%	Base	175 G45	150 G55	150 G55	175 G65	175 G80
	Sub-base		100 G30	150 G30	150 G30	175 G30
15%–29%	Base	150 G45	200 G55	125 G55	125 G65	150 G80
	Sub-base		125 G30	125 G30	125 G30	125 G30
>30%	Base	150 G45	175 G45	175 G55	175 G65	175 G80

Table 14.10 A Typical Structural Design Chart Using WBM

Subgrade CBR	Structural layer thicknesses (mm) and associated material (G15 -G30) Traffic range		
	<0.01 mesa	0.01–0.1 mesa	0.1–0.3 mesa
3%–4.9%	150 WBM	150 WBM	150 WBM
	150 G30	150 G30	175 G20
		150 G15	200 G15
5%–7.9%	150 WBM	150 WBM	150 WBM
	125 G30	125 G30	150 G30
		100 G15	150 G15
8%–14.9%	150 WBM	150 WBM	150 WBM
	100 G30	150 G30	200 G30
>15%	150 WBM	150 WBM	150 WBM
>30%	150 WBM	150 WBM	150 WBM

Table 14.11 Revised Design Chart for Roads with Very Low Traffic and Axles <4.5 tonnes

Subgrade class	Layer	Traffic (mesa): axle loads < 4.5 tonnes	
		<0.01	0.01–0.1
2%–4%	Base	100 G45	100 G45
	Sub-base	100 G25	125 G25
	Subgrade	200 G10	250 G10
5%–7%	Base	100 G45	100 G45
	Sub-base	100 G25	150 G25
	Subgrade	100 G10	175 G10
7%–11%	Base	100 G45	100 G45
	Sub-base	100 G25	150 G25
	Subgrade		100 G10
>11%	Base	100 G45	100 G45
	Sub-base	100 G25	150 G25

The slightly different Austroads (2017) approach applicable to LVRRs, as noted in Chapter 13, aims at identifying the total pavement thickness over a subgrade, which can be subdivided into required layers based on the materials available; always allowing for a minimum 100 mm base. A series of design curves are derived from the empirical formula:

$$T = 0.475 \times \left[219 - 211(\log CBR) + 58(\log CBR)2 \right] \times \log(14 \times esa)$$

where T = Layer thickness (mm)

Table 14.12 Chart Comparisons.

Scenario	Layer	Structural layer thicknesses and associated material. Foundation CBR 6%		
		Table 14.4 (Typical wet area)	*Table 14.5* (Typical dry area)	*Table 14.6* (Low axle Load wet environment)
100,000 esa	Base	150 mm G65	150 mm G55	100mm G45
	Sub-Base	100 mm G30	175 mm G30	150mm G25
	Subgrade	100 mm G15		175mm G101
	Total thickness	350 mm	325 mm	425 mm
1,000,000 esa	Base	200 mm G80	175 mm G80	
	Sub-Base	150 mm G30	250 mm G30	N/A
	Subgrade	150 mm G15		
	Total thickness	500mm	425mm	

where:

T = required structural thickness over the subgrade (or sub-base)

CBR = Appropriate CBR strength of underlying layer (subgrade or sub-base)

esa = design equivalent standard axle value

Using the Austroads formula, or chart, in the previous example would give a total cover thickness of 300 mm over a subgrade of CBR 6% with an esa of 100,000; similar to the dry option in Table 14.8. Using the minimum recommended base thickness of 100 mm (with G80 material), this would allow a sub-base of 200 mm with G30 material, with an option of using 100 mm of G30 and 100 mm of G15 instead.

REFERENCES

AfCAP.2012. South Sudan low volume roads design manual. UKAID-DFID for Ministry of Roads and Bridges, S Sudan. https://www.gov.uk/research-for-development-outputs/south-sudan-low-volume-roads-design-manual-volume-1-2-and-3.

ARRB. 1995. *Sealed Roads Manual: Guideline to Good Practice*. Australia: ARRB Transport Research Ltd.

Austroads. 2017. *Guide to Pavement Technology Part 2: Pavement Structural Design*. AGPT 02–17. Sydney, NSW: Austroads. https://austroads.com.au/publications/pavement/agpt02/media/AGPT02-17_Guide_to_Pavement_Technology_Part_2_Pavement_Structural_Design.pdf.

Cook, J. R., R. C. Petts and J. Rolt. 2013. Low volume rural road surfacing and pavements: A guide to good practice. Research report for AfCAP and UKAID-DFID. https://www.research4cap.org/index.php/resources/rural-access-library.

Ethiopian Roads Authority (ERA). 2016. *Manual for Low Volume Roads*. https://www.research4cap.org/index.php/resources/rural-access-library#.

Intech-TRL. 2006. SEACAP 1 rural road surfacing trials final report. UKAID-DFID report for MoT, Vietnam. https://www.gov.uk/research-for-development-outputs/seacap-1-rural-road-surfacing-research-rrsr-final-report-volume-1-main-report.

JKR Malyasia. 2012. Design Guide for Alternative Pavement Structures, Low Volume Roads. Road and Geotechnical Engineering Unit, Public Works Dept, Malaysia.

RECAP. 2020. Low volume road design manual: Section B design. UKAID-DFID for Department of Rural Road Development, Ministry of Construction. https://www.research4cap.org/ral/DRRD-2020-LVRRDesignManual-SectionB-Ch5t011-AsCAP-MYA2118A-200623-compressed.pdf.

Roughton International. 2012. Best practice manual for thin bituminous surfacings. AFCAP report no. ETH/076/A.

SABITA. 2011. *A Guide to the Selection of Bituminous Binders for Road Construction.* Southern African Bitumen Association. http://www.sabita.co.za/wp-content/uploads/2017/08/Download-for-Manual-30.pdf.

SEACAP. 2008. Low Volume Rural Road standards and specifications: Part II: Pavement options and technical specifications. SEACAP 3, UKAID-DFID for Ministry of Public Works and Transport. https://www.gov.uk/research-for-development-outputs/seacap-3-low-volume-rural-roads-pavement-options-and-technical-specifications.

Skempton, A. W., M. M. Chrimes, R. C. Cox, P. S. M. Cross-Rudkin, R. W. Rennison and E. C. Ruddock. 2015. *A Biographical Dictionary of Civil Engineers in Great Britain and Ireland: McAdam, P141.* ICE and Thomas Telford Limited.

TRL. 1993. *Overseas Road Note 31 A Guide to Structural Design of Bitumen Surfaced Roads in Tropical and Sub-tropical Countries.* 4th edition. UK: TRL Ltd for DFID. https://www.gov.uk/research-for-development-outputs/orn31-a-guide-to-the-structural-design-of-bitumen-surfaced-roads-in-tropical-and-sub-tropical-countries.

TRL. 2000. *Overseas Road Note 3. A Guide to Surface Dressing in Tropical and Sub-tropical Countries.* UK: TRL Ltd, for DFID. https://trl.co.uk/uploads/trl/documents/ORN3.pdf.

TRL. 2003. *Manual for the Labour-Based Construction of Bituminous Surfacings on Low-Volume Roads.* Report ref. R7470 for DFID. https://www.gov.uk/research-for-development-outputs/manual-for-the-labour-based-construction-of-bituminous-surfacings-on-low-volume-roads.

RECAP. 2019. Development of guidelines and specifications for low volume sealed roads through back analysis. Draft final report. ReCAP reference number: RAF2069A. https://www.gov.uk/research-for-development-outputs/development-of-guidelines-and-specifications-for-low-volume-sealed-roads-through-back-analysis-phase-3-final-report.

Weinart, H. H. 1974. A climatic index of weathering and its application in road construction. *Geotechnique* 24:475–488.

World Bank. 2020. *To Pave or Not to Pave. Developing a Framework for Systematic Decision Making in the Choice of Paving Technologies for Rural Roads.* Mobility and Transport Connectivity Series. https://openknowledge.worldbank.org/handle/10986/35163.

Chapter 15

Non-bituminous and concrete pavement options

INTRODUCTION

Background

There is a range of non-bituminous proven pavement options with varying characteristics that can be appropriate for specific Low Volume Rural Road (LVRR) circumstances. Despite being proven in various climatic and traffic circumstances, some of the surface options discussed in this chapter are not included or permitted in some national specifications. In these cases, it may not be possible to benefit from their use without local demonstration and research on their performance and official regulatory acceptance. Table 15.1 summarises the options considered in this chapter (Henning et al. 2005; Cook et al. 2013; World Bank 2020).

Construction of many of these surfaces are labour-based and local materials orientated. Therefore their application can support local-resource-based policies and strategies, and some also offer low carbon footprint solutions. Further consideration of the application of appropriate construction technology is included in Chapter 22.

Some of the unbound (non-concrete) options are ideally suited to stage construction, where the initial surface can be later overlaid by a bituminous or concrete surface, subject to availability of resources and/or increase in traffic. The various options can also be an integral part of Spot Improvement strategies.

Approaches to non-bituminous detailed pavement design

The design approaches to non-bituminous pavement vary from very simple, locally-derived procedures to more complex structurally-based solutions. As discussed in Chapter 13, the recommended LVRR approach to the design of these options is a pragmatic use of established design catalogues and charts. As noted in Chapter 14, they should be used in the knowledge that they were developed for specific Road Environments and may need adjustment for local conditions and materials.

DOI: 10.1201/9780429173271-15

Table 15.1 Non-Bituminous Pavement Options

Group	Option
Unpaved	Earth and natural surface (ENS)
	Gravel wearing course (GWC)
Paved – non-bituminous	Stone macadam
	Hand-Packed Stone
	Stone Setts/Dressed Stone/Cobble Stone
	Fired Clay Bricks
Concrete	Concrete slabs
	Geocells
	Concrete Blocks

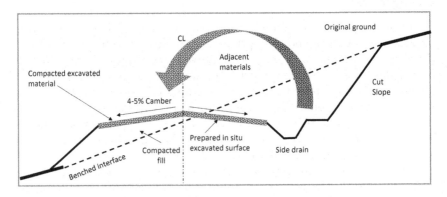

Figure 15.1 An ENS cross-section.

The advice in this chapter takes on board recent advances in pavement design developed through recent UKDFID, World Bank or ADB-funded research in Asia and Sub-Saharan Africa, much of which is now included in LMIC road design manuals or guides.

EARTH AND GRAVEL ROADS

Engineered natural surface (ENS)

Engineered Natural Surface (ENS) or earth roads form the major portion of classified and unclassified rural networks in many developing regions. An ENS uses natural material at the road location, either in situ or from within the right of way, to form a basic surface for traffic, as shown in Figure 15.1. Essential provisions for an ENS being Fit for Purpose are a good cross-sectional shape and an effective drainage system (Hindson 1983).

A cross-sectional camber of around 4% is recommended, up to 6% may be recommended where regular routine maintenance of shape cannot be guaranteed. However, this could compromise non-motorised traffic (NMT) safety, especially on plastic materials in wet (slippery) conditions.

Side drains may be trapezoidal, V or rectangular (lined) in shape. The edge of road/shoulder should normally be at least 30 cm above the bed or invert of the side drain, whatever the side drain cross-section shape, and the crown height of the pavement should be at least 50 cm above the bed or invert of the side drain or the adjacent ground level. If the alignment is liable to flood, then it should be raised to be above the design flood level (taking account of future climate). Special requirements may apply in expansive soil subgrades (Chapter 10).

Research indicates that in situ materials with an in-service CBR of a minimum of about 15% can provide a year-round running surface for light motor traffic (Rolt 2007). Alignment sections with steep gradients or weak or problematic soils can be improved in situ by upgrading to a higher standard surface under a Spot Improvement strategy (see Chapter 16).

An ENS can be constructed by equipment or labour-based methods. Using the labour-based option, LVRR earth road construction typically requires between 500 and 2,000 person days per km including a basic drainage system, with culverts (Intech 1992, 1993; ILO 1984, 2014). This surface option is very susceptible to deterioration by weather in combination with traffic, and adequate maintenance is essential. This can be achieved principally using motor graders, towed graders, or labour. Length worker labour routine maintenance can be achieved with an input typically of 50–150 person days per km per year (Intech 1992).

Natural gravel surfaced pavements

In their simplest form gravel roads comprise one or more layers of natural gravel placed directly on the existing shaped earth formation and compacted with an appropriate (4%–6%) surface camber (ARRB 2000; Intech-TRL 2007). The natural gravel material can be mechanically stabilised or blended with other material to improve the natural material properties. Alignments with steep gradients may need to be improved in situ by upgrading to higher standard surface under a Spot Improvement or EOD strategy.

A natural gravel or Gravel Wearing Course (GWC) surface is often considered as the usual upgrade option for ENS roads where improvement is justified. However, particular care should be taken in considering this option. Local environment factors may restrict the satisfactory use of natural gravel surface for sections of route that are affected by:

• Traffic AADT >200 vpd.
• Longitudinal gradient >6%–10% (depending on rainfall intensity).
• Annual rainfall >2,000 mm.

- Excessive haul distances for initial and maintenance (re-) gravelling.
- Available gravel material does not meet specifications (see Chapter 8).
- Dust emissions in settlements or adjacent to high-value crops.
- Seasonal flooding.

As annual gravel loss rates or costs may be excessive in these cases, other surface options should be considered as part of the Phase I selection process (Cook and Petts 2005).

LMICs may well have their own procedures for determining gravel layer thickness, with some using simple default options of 150 or 200 mm. However, the following is a simple generic approach for gravel road layer thicknesses:

1. Base gravel layer thickness (D1): The minimum thickness necessary to avoid excessive compressive strain in the subgrade caused by traffic loading.
2. Gravel wearing course (GWC) layer thickness (D2): Essentially a sacrificial layer based on the extra thickness needed to compensate for the gravel loss due to traffic and weather during the period between initial placement and realistically predicted first re-gravelling maintenance operations (see Figure 15.2).

The total initial total gravel thickness (Dt) required is found by adding the above two thicknesses: Dt=D1+D2. The wearing course will suffer material losses due to traffic and weather, and should be regularly reshaped and replenished under the maintenance regime to ensure that the structural gravel layer retains at least the minimum design thickness. Table 15.2 presents a simple design approach for very low traffic gravel roads.

In practical terms, the layers will usually be of the same material and source and be laid in accordance with the material requirements discussed

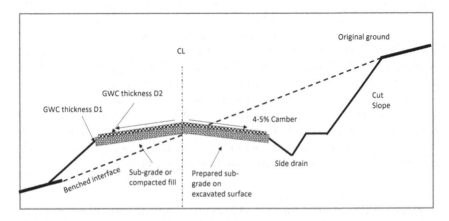

Figure 15.2 A typical gravel road cross-section.

Table 15.2 Minimum Thickness (DI) for
Simple Gravel Road Design

Subgrade strength	Layer thickness
CBR 3%–4%	200 mm
	100 mm
CBR 5%–7%	200 mm
CBR >8%	150 mm

Gravel: Soaked CBR >15%

Table 15.3 Typical Thicknesses (DI) from the Austroads Structural Layer
Approach

Subgrade strength	Total cover thickness over subgrade			
	10^3 esa (mm)	10^4 esa (mm)	10^5 esa (mm)	5×10^5 esa (mm)
CBR 3%	240	280	340	380
CBR 5%	160	200	240	260
CBR 7%	140	160	200	220
CBR 15%	100	110	120	160
CBR >30%	100	100	100	120

Austroads recommends a minimum base (CBR 50%) of 100 mm in their chart with a
sub-base of CBR 30% material. They recommend an overlying GWC of CBR 40%
material.
See Chapter 8 for other material requirements.

in Chapter 8. The thicknesses are based on a suitable gravel material with a soaked CBR of >15% (G15). Stronger G25 to G45 gravels may be available for use, and in this case some designs allow reductions in thickness. However, this approach should be used with caution, given that surface deterioration may be dominant over structural deterioration, and CBR is not a suitable indicator for this. Experience suggests that the expected maintenance regime will be a dominant decision factor.

Some road authorities prefer a more engineering-based approach whereby a gravel road can be structurally designed in line with the procedures outlined in Chapter 14 for bitumen sealed pavements. In this model, the pavement consists of a wearing course overlying structural layers (base and sub-base) that cover the in situ material and provides adequate structural protection for the road foundation.

This structural approach is not appropriate for most LMIC networks, where the simpler approach is usually adequate. However, in situations where there are significant numbers of commercial vehicles (around 20%), this approach may be justified.

Austroads (2009) describe the gravel option in detail, and Table 15.3 presents a gravel road design chart layout derived from their approach in which the layer thickness design has incorporated subgrade strength and traffic.

An advantage of this approach is that it could be more easily upgraded to a sealed condition with less additional design and construction.

In both the above approaches, the thicknesses are the minimum required for load carrying and subgrade protection. Assessment of the likelihood of routine maintenance reshaping and realistic probability of the re-gravelling financing and resourcing should be a vital input to the design process.

The additional sacrificial gravel wearing course (GWC) layer (D2) thickness will depend on expected material losses and will be influenced by a range of factors such as:

- Rainfall (annual quantity and intensity).
- Longitudinal gradient.
- Traffic (size, numbers and speed).
- Gravel type.
- Maintenance regime.

Some predictability of surface gravel loss is provided by TRL Laboratory Report LR 1111, where an estimate of the annual average gravel loss can be estimated from the following equation based on research in Kenya (TRL 1984):

$$GL = f.\left(4.2 + 0.092T + 3.50R^2 + 1.88V\right).T^2\Big/\left(T^2 + 50\right)$$

Where
GL = The annual gravel loss measured in mm/year
T = The total traffic volume in the first year in both the directions, measured in thousands of motor vehicles
R = The average annual rainfall measured in m

$$V = \text{the total}\left(\text{rise} + \text{fall}\right)\text{as a percentage of the length of the road}$$

f = 0.94–1.29 for lateritic gravels
   = 1.1–1.51 for quarzitic gravels
   = 0.7 – 0.96 for volcanic gravels (weathered lava or tuff)
   = 1.5 for coral gravels
   = 1.38 for sandstone gravels

These gravel loss estimates are only indicative, but research on a comprehensive sample of Road Environments in Vietnam supported the above figures on the basis of measured gravel losses, indicating that gravel loss of 20 mm/year could be taken as the limit of operational sustainability (Cook and Petts 2005).

Therefore, the wearing course of a new gravel road can be considered as requiring total thickness, DT, calculated from: DT=D1+N.GL (normally with a minimum of 150 mm)

Where:
- D1 is the minimum structural thickness.
- N is the period between construction and initial re-gravelling operation in years.
- GL is the annual gravel loss calculated above.

For planning purposes, labour-based roadworks experience indicates that labour outputs of 1.6–2.4 cubic metres of gravel excavation and adjacent stockpiling per person per day can be achieved (Intech 1992).

STONE ROADS

Waterbound/drybound macadam

Waterbound (WBM) or Drybound Macadam (DBM) may be used in unsealed pavement design as well as in sealed road designs. A WBM layer consists of a stone skeleton of single sized coarse aggregate. The considerable voids in the skeleton are filled by finer aggregate which is washed or 'slushed' into the coarse skeleton with water. DBM is a similar, however, instead of water and deadweight compaction being used, a small vibrating roller is used. to vibrate the fines into the voids. The structural designs for WBM and DBM are outlined in Chapter 14. WBM/DBM may be overlain by a GWC to provide a sacrificial smooth running surface, as for example in localities with abundant suitable stone but limited gravel resources, for example as used in Myanmar highland areas where there is an abundance of hard rock but a scarcity of gravel (RECAP 2020), as shown in Figure 15.3.

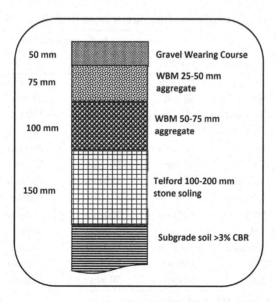

Figure 15.3 Local design for unsealed WBM from highland area in Myanmar.

Hand-packed stone

Placement of Hand-Packed Stone (HPS) is a proven labour-based technique (IFG 2008). The surfacing consists of a layer (typically 150–250 mm thick) of large broken stones pieces, laid by hand, and tightly packed together and wedged in place with smaller stone chips rammed by hand into the joints using hammers and steel rods. The remaining voids are filled with sand. HPS is normally placed on a thin Sand Bedding Layer (SBL) of sand or gravel. The layer can be compacted with a vibrating or heavy non-vibrating roller, if heavy commercial traffic (>4.5 tonne axle load) is anticipated. An edge restraint or kerb constructed of large or mortar-jointed stones improves durability and lateral stability.

A degree of interlock is assumed to have been achieved in the example of structural design shown in Table 15.4. There is an option that includes a mortared surface (Figure 15.4). However, although this potentially provides some water proofing, its resistance to cracking has been shown to be doubtful (OTB-LTEC 2009).

For planning purposes, the placing and packing of the stones by labour can achieve about 6.5 square metres per person-day. Placing and slushing/vibration, the fines can achieve 30 square metres per person per day (CIDB 2005). HPS can be later upgraded using a thin regulating layer of granular material and appropriate bituminous seal in a staged construction strategy.

Stone setts or cobble stones

Dressed Stone Block, Stone Sett and Cobble Stone surfacing are historically well-established labour-based techniques that have been adapted

Table 15.4 Thicknesses Designs for Hand-Packed Stone (HPS) Pavement

Subgrade strength	Layer	Traffic range (mesa)			
		<0.01	0.01–0.1	0.1–0.3	0.3–0.5
CBR 3%–4%	HPS	150 mm	200 mm	200 mm	250 mm
	SBL	50 mm	50 mm	50 mm	50 mm
	Sub-base	150 mm	150 mm	150 mm	150 mm
	Capping		150 mm	200 mm	200 mm
CBR 5%–7%	HPS	150 mm	200 mm	200 mm	250 mm
	SBL	50 mm	50 mm	50 mm	50 mm
	Sub-base	125 mm	125 mm	150 mm	150 mm
	Capping			150 mm	150 mm
CBR 8%–14%	HPS	150 mm	200 mm	200 mm	250 mm
	SBL	50 mm	50 mm	50 mm	30 mm
	Sub-base	100 mm	150 mm	200 mm	200 mm
CBR >15%	HPS	150 mm	200 mm	200 mm	250 mm
	SBL	50 mm	50 mm	50 mm	50 mm

HPS, hand-packed stone; SBL, sand bedding layer.

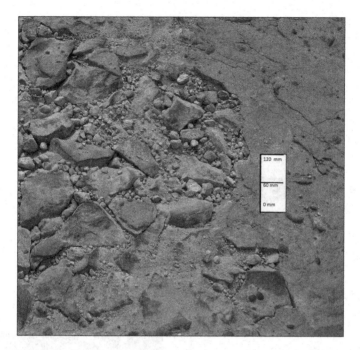

Figure 15.4 Degraded mortar surface on HPS pavement.

successfully as a robust option on LVRRs where there is a good local supply of suitable stone (Intech-TRL 2007). These options are suited to homogeneous rock types that have inherent orthogonal stress patterns (such as granite) that allow for easy break of the fresh rock into the required shapes by labour-based means (Figure 15.5).

The cobble option consists of a layer of roughly cubic (100 mm) stone blocks laid on a bed of sand or fine aggregate within mortared stone or concrete edge restraints (kerbs). The individual stones should have at least one face that is fairly smooth, to be the upper or surface face when placed. Each stone sett is adjusted with a small (mason's) hammer and then tapped into position to the level of the surrounding stones, as shown in Figure 15.6. Sand or fine aggregate is brushed into the spaces between the stones, and the layer is then compacted with a roller or plate compactor.

The Dressed Stone/Stone Sett technique is similar to the above; however, the individual stones are larger; normally of size 100–300 mm. They are cut from suitable hard rock and 'dressed' manually to rectangular shape with a smooth flat finish on at least one face using hammers and chisels. As above, the stones are laid on a bedding sand layer and tapped into final position with a hammer. Suitably graded sand is compacted into the inter-block joints.

Compacting with a heavy roller can improve durability. An edge restraint or kerb constructed, for example, of large or mortared stones is required

Figure 15.5 Rectangular granite blocks to be used on road surfacing.

Figure 15.6 Cobble stone pavement construction. (a) Laying stones within edge constraints. (b) Final cobble stone surface laid to camber.

for durability. Alternatively, sand–cement mortar joints and bedding can be used to improve durability and prevent water penetrating to moisture susceptible foundation layers and weakening them.

A typical design chart is shown as Table 15.5. There may be local variations in the specifications of material and the use of imported subgrade for different countries or environments (Roughton International 2008; Intech-TRL 2007).

For planning purposes, labour productivity could be estimated as chiselling cobbles/stone setts from quarried stones – typically 20–60 per person-day, depending on materials characteristics; Laying and sand jointing stone

Table 15.5 Atypical Thicknesses Design for Dressed Stone/Cobble/Stone Sett Surfacing

Subgrade strength	Layer	Traffic range (mesa)				
		<0.01	*0.01–0.1*	*0.1–0.3*	*0.3–0.5*	*0.5–1.0*
CBR 3%–4%	Stone	Cobble/ sett	Cobble/ sett	Cobble/ sett	Cobble/ sett	Cobble/ sett
	Sand bed	25 mm	25 mm	25 mm	25 mm	25 mm
	Base	150 mm	150 mm	150 mm	150 mm	150 mm
	Sub-base	100 mm	150 mm	175 mm	220 mm	250 mm
CBR 5%–7%	Stone	Cobble/ sett	Cobble/ sett	Cobble/ sett	Cobble/ sett	Cobble/ sett
	Sand bed	25 mm	25 mm	25 mm	25 mm	25 mm
	Base	150 mm	150 mm	150 mm	150 mm	150 mm
	Sub-base		125 mm	150 mm	200 mm	250 mm
CBR 8%–14%	Stone	Cobble/ sett	Cobble/ sett	Cobble/ sett	Cobble/ sett	Cobble/ sett
	Sand bed	25 mm	25 mm	25 mm	25 mm	25 mm
	Base	150 mm	150 mm	150 mm	150 mm	175 mm
	Sub-base			150 mm	200 mm	225 mm
CBR 15%–29%	Stone	Cobble/ sett	Cobble/ sett	Cobble/ sett	Cobble/ sett	Cobble/ sett
	Sand bed	25 mm	25 mm	25 mm	25 mm	25 mm
	Base	125 mm	150 mm	150 mm	150 mm	150 mm
	Sub-base			125 mm	125 mm	150 mm
CBR >30%	Stone	Cobble/ sett	Cobble/ sett	Cobble/ sett	Cobble/ sett	Cobble/ sett
	Sand bed	25 mm	25 mm	25 mm	25 mm	25 mm
	Base	100 mm	150 mm	150 mm	150 mm	175 mm

Typical granular materials: Base G50-55, Sub-base G25, (Chapter 8).
Capping layer (G15) could be used in CBR < 7% subgrade situations to reduce sub-base.

setts/cobblestones; typically, 6–8 square metres per person-day (GTZ 2009; Beusch 2013). An interesting variation (Irregular Cobble Stone Paving) has been developed and applied in China using untrimmed cobbles produced from the quarry crusher. The technique is described in Yun Nan RP&DRI (2005).

FIRED CLAY BRICK

Engineering quality bricks suitable for road surfacing can be produced by firing clay in large- or small-scale kilns using coal, wood, or some agricultural wastes, such as rice husk, as a fuel. The fired bricks are generally laid on edge to form a layer of typical 100 mm thickness on sand or sand–cement bedding layer (Howard Humphreys & Partners 1994; Intech-TRL

(a)   (b)

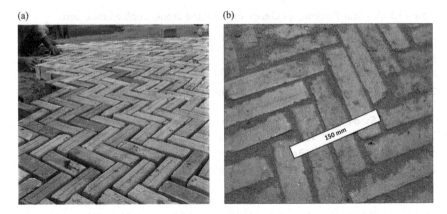

Figure 15.7 Fired clay brick construction. (a) Bricks laid to herring-bone pattern. (b) Bricks after sand brushed-in and compacted.

2006). Sand or sand–cement filling is used for the inter-brick joints. Kerbs or edge restraints are necessary, and can be provided by bedded and mortared fired bricks. The fired bricks are normally laid in a herring-bone or other approved pattern to enhance load spreading characteristics (as illustrated in Figure 15.7) The bricks are laid on a sand bedding layer or on a previously placed 'soling layer' of flat-laid bricks (LGED 2005). Joints between the bricks may be either in-filled with suitable sand or the bricks may be mortared in.

Unmortared brick paving is compacted with a plate compactor and the jointing sand is topped up if necessary. For mortar-bedded and jointed-fired clay brick paving, no compaction is required. When the mortar has set, the layer should be covered in sand or other moisture retaining material, and kept wet for a few days to aid curing (Cook et al. 2013). Typical thickness designs are as shown in Table 15.6.

Mortared brick pavements are a variation to be considered if waterproofing of the base or sub-base materials is required. However, their behaviour is different to that of sand-bedded pavements, and is more analogous to a rigid pavement or semi-rigid option. There is, however, little formal guidance on a mortared option, although empirical evidence indicates that inter-block cracking may occur. For this reason, the option is currently only recommended for the lightest traffic divisions (Dzung and Petts 2009).

CONCRETE PAVEMENTS

General

This section provides general guidance on the design of the following concrete pavement types that are commonly used in LMIC rural roads, namely:

Table 15.6 Sample Thicknesses Designs for Brick or Block Surfacing

Subgrade strength	Layer	LVI traffic range (mesa)			
		<0.01	0.01–0.1	0.1–0.3	0.3–0.5
CBR 3%–4%		Block/brick	Block/brick	Block/brick	Block/brick
	Sand bed	25 mm	25 mm	25 mm	25 mm
	Base	150 mm	150 mm	150 mm	150 mm
	Sub-base	100 mm	150 mm	150 mm	175 mm
	Capping		150 mm	175 mm	200 mm
CBR 5%–7%		Block/brick	Block/brick	Block/brick	Block/brick
	Sand bed	25 mm	25 mm	25 mm	25 mm
	Base	150 mm	150 mm	150 mm	150 mm
	Sub-base	100 mm	125 mm	150 mm	150 mm
	Capping				150 mm
CBR 8%–14%		Block/brick	Block/brick	Block/brick	Block/brick
	Sand bed	25 mm	25 mm	25 mm	25 mm
	Base	150 mm	150 mm	150 mm	150 mm
	Sub-base		150 mm	200 mm	200 mm
CBR 15%–29%		Block/brick	Block/brick	Block/brick	Block/brick
	Sand bed	25 mm	25 mm	25 mm	25 mm
	Base	125 mm	125 mm	150 mm	150 mm
	Sub-base		125 mm	125 mm	125 mm
CBR >30%		Block/brick	Block/brick	Block/brick	Block/brick
	Sand bed	25 mm	25 mm	25 mm	25 mm
	Base	125 mm	150 mm	150 mm	150 mm

Typical granular materials: Base G50-55, Sub-base G25, Capping G12-15 (Chapter 8).

Concrete slabs (reinforced and non-reinforced).
Concrete block paving.
Geocell concrete.

Additional options such as continuously laid Reinforced Concrete, Roller-Compacted Concrete (RCC)and Ultra-Thin Reinforced Concrete (UTRC) are not common options in the rural Road Environment.

Most concrete surfacing is expensive in terms of initial costs in comparison with unsealed or bituminous sealed options. However, typically maintenance requirements are low, which can be beneficial in terms of whole life costing and demands on maintenance capacity. In addition, there are some environments, such as low lying floodable coastal areas or very steep curved alignment where some form of concrete pavement may be an engineering requirement.

World Bank (2019) provides details for use of concrete pavements as a climate resilient option, including design and construction information on jointed concrete slab and Geocell concrete and RCC concrete.

Concrete slabs

Non-reinforced or steel-reinforced cement concrete slabs are a well-established form of rigid pavement, designed to spread the applied load due to traffic through a slab effect. The concrete slabs are normally cast within formwork onto sand bedding layer or preferably a plastic membrane overlying a previously prepared and compacted sub-base, as illustrated in Figure 15.8. For many LVRR situations there is no requirement for reinforcement, provided the slabs are well constructed and the sub-base provides an adequate, even, support. The function of steel reinforcement in concrete slabs is to inhibit post-placement cracking and prevent 'hogging' and 'curling' as much as adding compressive strength.

Contraction joints of 10 mm width between slabs relieve tensile stresses. Additionally, expansion joints are normally required at 250 m intervals, with all joints to be filled and sealed with a mixture of sand and bitumen, with a reservoir of bitumen provided at the top of each joint. All joints should be provided with load transfer steel dowels (Figure 15.9). At expansion joints the dowel bar should be anchored into the concrete at one end and the other end coated with bitumen and fitted into a PVC sleeve. The PVC tube is to be omitted at contraction joints.

Figure 15.8 Casting of concrete slabs.

Figure 15.9 Load transfer dowels between slabs.

Reinforcement, if included, should be mild steel mesh or round bars complying with local standards for concrete reinforcement; but for LVRRs this is commonly 6 mm diameter rods at 200 mm centres. The reinforcing steel mesh should be placed at the top one-third of the concrete slab with minimum concrete cover on the steel reinforcement at any point of 40–50 mm. If the mesh is formed of reinforcing rods, they should be secured with steel binding wire.

TRL (2019) in a design review document (RECAP 2019) noted that the design process was 'not complex' for LVRRs, and the standard designs generally involved 100–200 mm thick slabs placed on a minimum 100 mm subbase of CBR 30% with a subgrade >3% CBR. Typical thickness designs for unreinforced slabs from current LVRR manuals and based to some extent on trial research in Vietnam and Laos (Intech-TRL 2006; Roughton International 2007; OTB-TRL 2009) are given in Table 15.7.

On single lane LVRRs, the concrete slabs should normally be constructed at full carriageway width (3–4 m). Two-lane rural roads, with total width greater than 5 m, will require half carriageway width slabs. In some other circumstances (e.g. to allow traffic to continue to flow in narrow constrained alignments) construction of half-width concrete slabs can be also used, as shown in Figure 15.10. Longitudinal joints between slabs should have a load transferring dowel and sealing arrangement similar to that of transverse contraction joints.

Table 15.7 Example of Thicknesses Designs: Non-reinforced Concrete Pavement

Subgrade strength	Layer	Traffic range (mesas)				
		<0.01 (mm)	0.01–0.1 (mm)	0.1–0.3 (mm)	0.3–0.5 (mm)	0.5–1.0 (mm)
CBR 3%–4%	Slab	160	170	175	180	190
	Sand bedding	25	25	25	25	25
	Sub-base	150	150	150	150	150
CBR 5%–7%	Slab	150	160	165	170	180
	Sand bedding	25	25	25	25	25
	Sub-base	125	125	125	125	125
CBR 8%–29%	Slab	150	150	160	170	180
	Sand bedding	25	25	25	25	25
	Sub-base	100	100	100	100	100
CBR >30%	Slab	150	150	160	170	180
	Sand bedding	25	25	25	25	25

For concrete cube strength of 30 MPa at 28 days. Sub-base granular materials 25%–30% CBR. Sand bedding layer may be replaced by impervious membrane.

Figure 15.10 Concrete pavement with half-width slabs.

Construction and immediate post-construction procedures have a key impact on the performance of concrete slab pavements (Rolt and Cook 2009). Key issues are:

- Even well compacted non-plastic sub-base.
- Good control on concrete mix, including water content.
- Adequate curing of concrete slabs – using damp sand or wet sacking – for 7 days.
- Compacted shoulders to resist erosion and under-cutting of slabs.
- Maintenance of bitumen–sand joint infill.

Where a section of concrete slabs adjoins an unpaved surface section, additional load spreading is necessary under that first concrete slab due to the impact loading of heavy vehicle wheels, as they traverse onto the slab.

Concrete block paving

Concrete Block paving is a system of individual blocks arranged to form a continuous hard-wearing surface pavement. The blocks are typically laid in a herring-bone pattern on sand bedding layer with a fine-to-coarse sand brushed into the inter-block joints. They are confined by edge constraints that are placed before or after the blocks are laid, as illustrated in Figure 15.11. A more waterproof option may be considered using a single or double Sand Seal over the blocks, although, in practice, this has proved to be prone to cracking and erosion without regular maintenance. Concrete blocks, typically 200×100×90 mm thick, composed of concrete with a minimum 28-day cube strength of 25 MPa are recommended (Cook et al. 2013).

Concrete block paving may be used for LVRR surfacing and designed in a similar manner to fired clay brick incremental paving, as shown previously in Table 15.6 (Jones and Promprasith 1991; Cement and Concrete Assoc. of Australia 1997; Intech-TRL 2007).

Geocells

Rather than being considered for design purposes as a concrete or rigid pavement, this construction should be considered more like a flexible block paving surface where the blocks are cast in situ. Specialist suppliers will normally provide detailed design requirements as different geocell products may have important differences in the way that they impact on concrete slump requirements, block interlock, and construction speed and cost (Roughton 2008).

Although Geocells have internationally been constructed with 75, 100 and 150 mm thicknesses, the current guidance is to use 150 mm thick cells with 20–30 MPa 28 day strength concrete (Roughton International 2007, RECAP 2019). Figure 15.12 shows the condition of a geocell paving surface after 6 years of very light traffic.

Figure 15.11 Concrete blocks placed with sand brushed into joints.

Figure 15.12 Geocell pavement 6 years after construction.

OPTION COMPARISON

Table 15.8 Summarises the advantages and disadvantages of non-bituminous pavement options. Further detail on these options and their construction can be found in Intech-TRL (2007) and Cook et al. (2013).

Table 15.8 Summary of Principal Non-bituminous LVRR Pavement Options

Option reference	Principal advantages	Principal concerns
ENS Rolt (2007), Hindson (1983)	Very low-cost option. Construction by local contractors and communities. Constructed with labour and hand-tools. No imported materials. Easy to maintain. Can be used in a planned stage construction strategy.	Only appropriate for light traffic, not suitable for heavy vehicles. Requires regular maintenance to keep crossfall between 4% and 6%. May be impassable in wet weather Possible dust pollution. Rainwater erosion on gradients more than 6%.
Gravel wearing course TRL (1984), ARRB (2000), Cook and Petts (2005), Austroads (2009)	Proven performance in tropical and sub-tropical, gravel-rich environments. Suitable for light-to-medium traffic <200 motor vehicles per day (MVPD). Often lower initial cost than most other surfacing options, except ENS. Can be used as an intermediate surface in a stage construction' strategy.	Natural gravel may occur in limited natural deposits of variable quality. Often difficult to meet standard grading and plasticity specifications fully. Requires a sustained maintenance programme and regular re-gravelling. Traffic, climatic and longitudinal gradient (>6%–8%) constraints Possible dust pollution in dry weather. Not suitable for flooding situations.
WBM/DBM CIDB (2005), Cook et al. (2013)	Suitable for light-to-medium traffic. Straightforward well-proven construction technique. Local contractors able to undertake this procedure following initial guidance. Can use locally produced aggregate, does not require sophisticated crushing plant. Provides an appropriate base for bitumen or bitumen emulsion seals for later upgrade. DBM suitable for zones with lack of construction water.	Requires source of strong stone WBM unsuitable for moisture susceptible subgrades. Requires good site control on materials and site procedures. Susceptible to erosion and increased roughness in high-rainfall, steep gradient environments. DBM requires the use of both static and vibrating compaction plant. WBM requires good supply of water.

(continued)

Table 15.8 (Continued) Summary of Principal Non-bituminous LVRR Pavement Options

Option reference	Principal advantages	Principal concerns
HPS IFG (2008), SEACAP (2008)	Suitable for light-to-medium traffic. Does not require expensive equipment. Suitable for construction by communities or small contractors. Crushing equipment or heavy plant not required. Can be constructed at steep gradients. Low maintenance, easily repairable. Can be later upgraded in a stage construction strategy.	Requires strong, angular, stone to be available locally. Requires skill to achieve a reasonable finished surface. For heavy traffic use, heavy compaction equipment should be used. HPS surface, unsuitable for road-base susceptible to soaking. High surface roughness disadvantageous to bicycles and motor-bicycles. Mortared HPS susceptible to cracking
Block stone (cobble etc) GTZ (2009), Cook et al. (2013)	Suitable for light-to-heavy traffic. Does not require expensive equipment. Suitable for construction by small contractors or community groups. Can be constructed at steep gradient. Low maintenance, easily repairable. Can be later upgraded in a stage construction strategy.	Requires strong stone to be available locally. Cobble stones must have at least one fair face. Requires skill in laying for a smooth finished surface. Light vibrating equipment preferable. Unsuitable over moisture sensitive road-base. Medium surface roughness. Potential safety issue with polished stones in wet condition (particularly two-wheeled traffic)
Bricks or concrete blocks Howard Humphreys (1994), Jones and Promprasith (1991)	Proven performance in all climates. Suitable for urban application if mortar jointed/sealed. Social and economic benefits to the communities through local brick manufacture. Local labour employment both in construction and in ongoing maintenance. Good durability, load bearing and load spreading characteristics. Low-cost maintenance procedures.	Unsuitable over moisture sensitive road-base. Requires consistent production of good quality engineering bricks of >20–25 MPa crushing strength. Requires good control of construction using string lines within pre-constructed edge constraints (kerbs). Light vibrating equipment preferable.

(continued)

Table 15.8 (Continued) Summary of Principal Non-bituminous LVRR Pavement Options

Option reference	Principal advantages	Principal concerns
Concrete slabs Parry et al. (1993), ARRB (1995), Cook et al. (2013)	Suitable for all climates and rural or urban application. Robust option suitable for high-rainfall and flood-prone regions. Generally resistant to axle overloading if well constructed and founded. General concreting procedures understood by local small contractors. Minimal maintenance if properly constructed and cured. No requirement for expensive construction plant.	High initial construction cost in relation to most other options. Usually requires expansion and contraction joints with steel load transfer dowels. May be susceptible to shrinkage cracking unless well constructed and cured. Concrete must not be mixed or placed in ambient shade temperatures above 38 degrees centigrade. First and last slabs of a section subject to impact loading. Requires at least 7–14 days curing time following initial construction. Requires good sub-base with shoulders maintained against erosion
Geocells RECAP (2020), Roughton International Ltd (2008)	Suitable for rural or urban application in all climates and steep gradients. Good durability, load bearing and load spreading characteristics. Suitable for light-to-heavy traffic. Low-cost maintenance procedures, easily repairable. Does not require expensive equipment. Suitable for construction by small contractors or community groups.	May not be locally available. High initial cost. May require plasticiser for acceptable workability. Medium surface roughness. Requires good site Quality Control of preparation, mixing, placing and curing. Needs control of construction using string lines within pre-constructed edge constraints (kerbs).

REFERENCES

ARRB. 1995. *Sealed Roads Manual: Guideline to Good Practice*. Australia: ARRB Transport Research Ltd.

ARRB. 2000. *Unsealed Roads Manual: Guideline to Good Practice*. Australia: ARRB Transport Research Ltd.

AUSTROADS. 2009. *Guide to Pavement Technology Part 6: Unsealed Pavements*. AGPT 06-09. Sydney, NSW: Austroads. https://austroads.com.au/publications/pavement/agpt06.

Beusch, A. 2013. Concrete block and cobblestone pavements – Work implementation guideline, for Kenya Roads Board.

Cement and Concrete Assoc. of Australia. 1997. Guide to residential streets and paths. C&CAA T51. https://www.ccaa.com.au/documents/Library%20Documents/ CCAA%20Technical%20Publications/CCAA%20Guides/CCAAGUIDE2004- T51-ResStreets-TBR.pdf.

CIDB. 2005. Labour-based methods and technologies for employment intensive construction works – A CIDB guide to best practice, Construction Industry Development Board, RSA. https://www.cidb.org.za/wp-content/ uploads/2021/04/Labour-based-Methods-and-Technologies-for-Employment-Intensive-Construction-Works.pdf.

Cook, J. R. and R. C. Petts. 2005. Rural road gravel assessment programme. SEACAP 4, module 4, final report. DFID report for MoT, Vietnam. https:// www.gov.uk/research-for-development-outputs/seacap-4-rural-road-gravel-assessment-programme-final-report.

Cook, J. R., R. C. Petts and J. Rolt. 2013. Low volume rural road surfacing and pavements: A guide to good practice. Research report for AfCAP and UKAID-DFID. https://www.research4cap.org/index.php/resources/rural-access-library.

Dzung, B. T. and R. Petts. 2009. *Report on Rice Husk Fired Clay Brick Road Paving*. Vietnam: gTKP.

GTZ. 2009. Cobblestone sector guide for Ethiopian cities, creating jobs, strengthening economies, revitalising cities, Ethiopia.

Henning, T., P. Kadar and R. Bennet. 2005. Surfacing alternatives for unsealed roads. World Bank TRN-33, Washington. https://documents1.worldbank.org/curated/ en/473411468154454358/pdf/371920Surfacing0Alternatives01PUBLIC1.pdf.

Hindson, J. 1983. *Earth Roads: A Practical Guide to Earth Road Construction and Maintenance*. Intermediate Technology Publications, p. 123.

Howard Humphreys & Partners. 1994. Road Materials & Standards Study (RMSS) Bangladesh. Reports for the Ministry of Communications Roads & Highways Department.

IFG. 2008. Hand packed stone surface. gTKP information note. Global Transport Knowledge Partnership (gTKP) Information Focus Group. https://www.gtkp. com/assets/uploads/20100323-010819-6222-RR%20SURFACE%207b%20 HPS%20Guideline-a.pdf.

ILO. 1984. The rural access roads programme, appropriate technology in Kenya. https://www.ilo.org/Search5/search.do?searchWhat=ILO%2c+1984.+The+Ru ral+%22access+road%22+Programme%2c+Appropriate+Technology+in+Ke nya.

ILO. 2014. Labour based technology tender and pricing manual. ILO Timor Leste. https://www.ilo.org/wcmsp5/groups/public/---asia/---ro-bangkok/---ilo-jakarta/documents/publication/wcms_470657.pdf.

Intech Associates. 1992. Technical and road maintenance manuals, Ministry of Public Works, Minor Roads Programme.

Intech Associates. 1993. Roads 2000, A programme for labour and tractor based maintenance of the Classified Road network, pilot project, final report, for MPWH, Kenya.

Intech-TRL. 2006. SEACAP 1 Rural road surfacing trials Final Report. UKAID-DFID Report for MoT, Vietnam. https://www.gov.uk/research-for-development-out-puts/seacap-1-rural-road-surfacing-research-rrsr-final-report-volume-1-main-report.

Intech-TRL. 2007. Rural road surfacing research. RRST construction guidelines. Report for UKAID-DFID and Ministry of Transport, Vietnam. https://www.research-4cap.org/ral/IntechTRL-Vietnam-2007-RRST+Construction+Guidelines-SEACAP1-v070910.pdf.

Jones, T. E. and Y. Promprasith. 1991. Maintenance of unpaved roads in wet climates. 5th Int Conf. On Low Volume Roads, TRR 1291. https://onlinepubs.trb.org/Onlinepubs/trr/1991/1291vol1/1291-030.pdf.

LGED. 2005. *Road Design Standard, Rural Road*. Local Government Engineering Department (LGED). Government of the People's Republic of Bangladesh.

OTB-TRL. 2009. Performance monitoring of the NEC-ADB Package 1 trial and gravel roads. SEACAP 17.02 final report. https://www.research4cap.org/ral/OTB+LTEC-LaoPDR-2009-Phase2+Final+Report-SEACAP17-v090614.pdf.

Parry, J. D., N. C. Hewitt and T. E. Jones. 1993. Concrete pavement trials in Zimbabwe. Research Report 381. TRL Ltd for DFID. https://assets.publishing.service.gov.uk/media/57a08dcd40f0b64974001a40/RR381.pdf.

RECAP. 2020. Low volume road design manual: Section B design. UKAID-DFID for Department of Rural Road Development, Ministry of Construction., Government of Myanmar https://www.research4cap.org/ral/DRRD-2020-LVRRDesignManual-SectionB-Ch5t011-AsCAP-MYA2118A-200623-compressed.pdf.

Rolt, J. 2007. Behaviour of engineered natural surface roads. SEACAP 19 technical paper 2.1. DFID for MRD, MPWT, Cambodia. https://www.gov.uk/research-for-development-outputs/seacap-19-technical-paper-no-2-behaviour-of-engineered-natural-surfaced-roads.

Rolt, J. and J. R. Cook. 2009. Mid term pavement condition monitoring of the Rural Road Surfacing Research. Final report. Ministry of Transport Vietnam Southeast Asia Community Access Programme (SEACAP 27) for UKAID-DFID. https://assets.publishing.service.gov.uk/media/57a08b6640f0b652dd000c58/SEACAP27-FinalReport.pdf.

Roughton International. 2008. Local resource solutions to problematic rural road access in Lao PDR. SEACAP 17 rural access roads on route no. 3. Module 2 – Completion of construction report. https://www.research-4cap.org/ral/Roughton-LaoPDR-2008-Module2+Construction+Report-SEACAP17-v080918.pdf.

SEACAP. 2008. Low Volume Rural Road standards and specifications: Part II: Pavement options and technical specifications. SEACAP 3, UKAID-DFID for Ministry of Public Works and Transport. https://www.gov.uk/research-for-development-outputs/seacap-3-low-volume-rural-roads-pavement-options-and-technical-specifications.

TRL. 1984. *The Kenya Maintenance Study on Unpaved Roads Research on Deterioration*. UK: TRL LR 1111. http://transport-links.com/wp-content/uploads/2019/11/1_274_LR1111-Kenya-Maintenance-study-research-on-deterioration.pdf.

RECAP. 2019. Development of guidelines and specifications for low volume sealed roads through back analysis. Draft final report. ReCAP reference number: RAF2069A. https://www.gov.uk/research-for-development-outputs/development-of-guidelines-and-specifications-for-low-volume-sealed-roads-through-back-analysis-phase-3-final-report.

World Bank. 2019. Concrete Pavements for Climate Resilient Low-Volume Roads in Pacific Island Countries. Transport Global Practice World Bank Group, World Bank Group, Washington. https://openknowledge.worldbank.org/bitstream/handle/10986/32394/Concrete-Pavements-for-Climate-Resilient-Low-Volume-Roads-in-Pacific-Island-Countries.pdf?sequence=1&isAllowed=y.

World Bank. 2020. *To Pave or Not to Pave. Developing a Framework for Systematic Decision Making in the Choice of Paving Technologies for Rural Roads.* Mobility and Transport Connectivity Series. https://openknowledge.worldbank.org/handle/10986/35163.

Yun Nan Road Planning & Design Research Institute. 2005. Irregular cobble stone pavement specification. https://www.gtkp.com/assets/uploads/20091229-171644-4097-Technique%20of%20Cobblestone%20pavement-a-en.pdf.

Chapter 16

Spot improvement

INTRODUCTION

The development of a road network represents a large capital investment and there are frequently insufficient funds to undertake all the identified road rehabilitation or new construction works. This problem is especially acute for Low Volume Rural Road (LVRRs), as relevant road authorities frequently need to be able to prioritise the investments within often severely constrained budgets. Assessment and prioritisation of roads or road programmes, as a whole, are discussed in Chapter 5, but an additional approach is possible, that of Spot Improvement. In this option the proposed overall works are scaled down to address individual prioritised sections or assets within a road link. The use of Spot Improvement involves the appropriate upgrade of specifically identified road sections or drainage assets, and allows limited resources to be targeted at key spots along an alignment.

The appropriate application of Spot Improvement can allow the target of Fit-for-Purpose access to be achieved for a lower cost than a traditional whole link design approach. Importantly, this approach not only ensures that the problematic lengths are provided with a more robust pavement but also ensures that resources are not wasted on the existing good areas. Spot Improvement is particularly relevant when considering climate resilience improvements.

This chapter briefly outlines Spot Improvement principles and provides guidance on the application of the use of a Spot Improvement strategy to provide effective access in situations of constrained budget.

THE SPOT IMPROVEMENT CONCEPT

Environmentally optimised design

Spot Improvement is closely linked to the principle of Environmentally Optimised Design (EOD), which is design approach focussed on creating the most efficient solution for a given road or sections of road, taking account of an assessment of the Road Environment (as discussed in Chapter 6).

Some of the Road Environment impact factors vary not only from road to road within a network and but also from site to site along a road; therefore, good-practice road design options may vary along the length of a road. The principle of EOD can provide a variable design solutions for a road alignment, one of which may be to deal with individual critical areas on a road link by implementing a Spot Improvements strategy.

Spot Improvement principles

Spot Improvements are engineering-based solutions to specific and prioritised underlying engineering issues, or problems, in order to ensure a road or section of a road is capable of delivering a level of access compatible with its defined purpose. By using a Spot Improvement approach it is possible to balance low-cost surfacing solutions such as gravel or even engineered natural surfaces for low-risk areas with higher cost solutions for the higher risk areas. Individual spots may comprise amendment or construction of:

- Short lengths of minor horizontal or vertical alignment.
- Short lengths of carriageway.
- Drainage.
- Culverts, minor bridges or parts of major bridges.
- Earthworks.
- Safety hazard mitigation options.

From an asset management and contractual viewpoint, it is important to distinguish Spot Improvement applications from routine, periodic or emergency maintenance. Maintenance is aimed at continuity of road performance throughout the design life of a road by processes of upkeep, preservation and repair within the original design envelope. Spot Improvement is focussed on access improvement through addressing and improving underlying engineering issues. Figure 16.1 presents a summary illustration of the Spot Improvement concept.

APPLICATION

Programming

A Spot Improvement programme requires appropriate planning, investigation and design, if it is to deliver the required enhancement in access service level. Figure 16.2 outlines the stages involved in identifying, defining and implementing a Spot Improvement initiative. The process uses similar investigative and design tools to those described in other chapters, but with the addition of a greater degree of prioritisation based on risk to access as well as other issues.

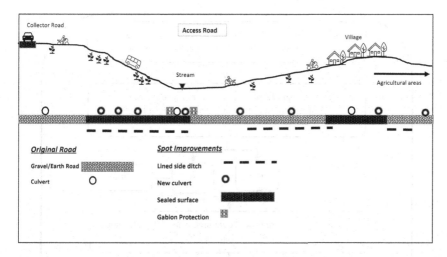

Figure 16.1 Typical Spot Improvements along an access road alignment.

Prioritisation

Spot Improvement is usually implemented within constrained budgets and frequently requires a rigorous assessment of proposed 'spots' in terms of their relative importance in providing safe sustainable access and socio-economic return on investment. Whilst the overall use of a Spot Improvement may be guided by policy or general planning decisions, the final selection of individual spot applications requires a pragmatic decision based on the assessment of defined priorities. Table 16.1 is an example of a priority list used to rank improvements to be implemented as budgets become available (SEACAP 2009).

Links to climate resilience strengthening

There are clear links between rural road climate resilience strengthening and Spot Improvement. The climate strengthening of existing roads in engineering terms is commonly considered by defining a list of options (Table 16.2). These are further discussed in the context of climate change and climate resilience in Chapter 17.

Implementation

The spot solutions should be designed and constructed to the appropriate standards and specifications so that if, with increased funding, the Spot Improvement solutions can be joined up under a phased construction

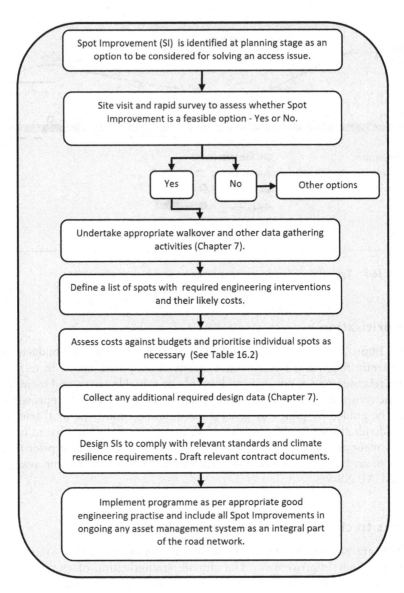

Figure 16.2   Spot Improvement stages.

approach, then the road should be able to deliver the same service as if it been fully rehabilitated as an EOD single project. Spot Improvements can also be included within annual routine maintenance or periodic maintenance packages as additional items, particularly as climate strengthening or disaster mitigation options.

Table 16.1 Typical Ranked Spot Selection Criteria

Priority	Criteria	Description
1	Current location is unsafe; high risk.	Safety concerns put road users or others at high risk of injury or death, for example, unstable roadside slope above a village/school or health centre.
2	Impassable access.	Existing location is completely blocking access or limiting it to pedestrian traffic only.
3	Impassable in wet season.	Motorised traffic unable to pass along the road in the wet season, although closures up to 24 hours after rainfall are accepted.
4	Unstable slope – low risk to life.	The slopes above or below the road are unstable and at risk of slipping and cutting access.
5	Condition likely to deteriorate.	Vehicles or climate impacts are likely to cause significant deterioration of the road within the next year.
6	Lack of shoulders.	Trafficked road with no shoulders/narrow shoulders poses risk to school children.
7	Health risk.	The health of road users and others is at risk, typically due to dust from a gravel or earth road.
8	Drainage in poor condition.	Drainage capacity or performance is reduced, and retained water is likely to damage the road.
9	Environmental concerns.	Current or future usage will cause environmental impacts along the road such as erosion of bare soil, disruption of a water course or contamination of a water supply.
10	Inhibited traffic.	Vehicles can only pass with difficulty or very slowly along the road due to its poor condition.
11	Geometry below standard.	The curvature, sight distance or gradient of the road do not meet the required standard; consequent impacts on road safety.
12	Narrow carriageway and shoulders.	Areas next to the road used as a bus stop or market with risk to pedestrians.
13	Pavement below standard.	The pavement, although passable, does not meet the required design standards.

There are some potential challenges to a Spot Improvement approach which must be acknowledged and addressed:

1. The Spot Improvement approach requires that spot sites must be identified and prioritised on a rational basis. It is vital that any underlying engineering or climatic issues are fully understood. This may not be an easy task and can require significant engineering judgements, as well as a knowledge of local conditions. Appropriate training and guidance will be required if this is to be undertaken by local transport staff.

Table 16.2 Typical Climate Resilience Spot Improvement Options

Ref.	Adaptation group	Comment
1	Pavement strengthening	Usually required for steep (>8%–10%) gradients on unsealed roads, within village areas and areas subject to erosive flood not mitigated by raised alignment.
2	Pavement and earthwork drainage	Lined drains likely required where gradients >6%. Additional side drains and associated turn-outs may also be recommended. Drainage required above slopes and on earthwork benches.
3	Cross drainage	Bridge, causeways, and culvert designs adjusted to take account of forecast increased river/stream flows and storm intensities. Additional culverts recommended where considered essential to improve overall road drainage as relief culverts, for example, on steep sections. Occasionally, existing fords, causeways or bridges might be replaced by climate resilient structures, such as vented fords/causeways or submergible bridges.
4	Alignment	Horizontal alignment may be shifted to avoid high climate vulnerable sections. Vertical alignment rising of earth embankments can be recommended where the alignment is too low and is being impacted by flooding and/or the weakening of the pavement by saturation.
5	Earthwork slope protection	Protection where erosion of exposed soil or rock slopes either in cut or embankment is identified as a significant risk.
6	River/stream bank protection	May be recommended where erosion of the alignment by rivers or streams is identified as a significant risk.

There may be resistance to the approach from local communities who may regard an apparently 'unfinished' road as a consequence of poor management, bad contracting practice, or corruption. Cooperation with local communities in the selection of Spot Improvement sites and on-going involvement of local stakeholders in the road rehabilitation programme will do much to allay these fears.

2. There has in the past been some confusion of Spot Improvement with periodic maintenance. Spot Improvements must be seen and designed as a fully engineered responses to defined requirements and not as a series of repairs within existing designs that may just perpetuate an underlying problem.

The application of Spot Improvement has been shown to be effective in cases where an un-engineered rural track or trail can be significantly improved in terms of rainy season access by the application of low-cost Spot Improvement. An important element in these cases is the use of a 'bottom-up' driven process to allow local communities to identify key access 'hotspot' interventions on access trails and improve community access. For example, in a poverty reduction project in Laos, hotspots were primarily associated with rainy season floods and rain-induced washouts or landslides. Spot Improvement of culverts, small bridges and trail drainage then allowed increased motorised transport to health centres and village markets. (World Bank 2016).

REFERENCES

SEACAP. 2009. Low volume rural road environmentally optimised design manual. SEACAP 3.02 manual for MPWT, Laos. https://assets.publishing.service.gov.uk/media/57a08bbb40f0b64974000d28/Seacap3-LVRR3.pdf.

World Bank. 2016. Lao Poverty Reduction Fund phase II impact evaluation: Final report. documents1.worldbank.org/curated/en/765151491464033254/pdf/114048-WP-Lao-Povery-Reduction-PUBLIC-PRF-small.pdf.

Chapter 17

Climate resilience

INTRODUCTION

Rural transport infrastructure is particularly vulnerable to climate threats and associated impacts, and it represents a significant challenge to the sustained eradication of poverty. The risks arising from these impacts are considerably increased when the likelihood of increasing climate threats from future climate change is taken into account. For rural roads, the climate strengthening costs for some assets can comprise a larger proportion of the construction costs of rural roads than for higher volume roads. It is therefore a major challenge, within available budgets, to ensure that the rural roads are made more resilient to current and future climatic threats (Cook 2022).

This chapter outlines climate change principles within the context of the threats to rural road networks. A framework is presented for assessing and prioritising climate risk, identifying engineering and non-engineering resilience options, followed by their construction, monitoring, and maintenance.

CLIMATE CHANGE

Background

Climate change is not a recent occurrence. The Earth, throughout its history, has undergone radical climate changes on a geological timescale, and within the last 5 million years alone climatic variation has influenced major steps in human evolution (Arnaud et al. 2011; Slezak 2015). Current climate change concerns are much shorter term and are focussed on current socioeconomic vulnerability to projected anthropogenic-driven climate threats.

There is now general agreement on the broad premise that climate change will continue to occur over the coming decades, and that it can involve issues such as a rise in sea level, changes in seasonal rainfall, increased temperature and increased major climatic events. Reports from the Intergovernmental Panel on Climate Change (IPCC) clearly show that land

DOI: 10.1201/9780429173271-17

surface air temperatures and sea surface temperatures have both increased in the last century with the maximum and the minimum temperatures increasing over land since the mid-20th century. Further increases in temperature, further rise in sea levels, and greater frequency and/or higher intensity of extreme weather events are forecast (IPCC 2014).

Global climate change projections indicate a combination of gradual changes to average climatic conditions and increasing frequency, severity and location of extreme weather or 'shock' events. Gradual changes related to climatic variables are those that are experienced over a period of time from months to years, decades or even centuries. Extreme events are typically those that occur suddenly, sometimes with limited warning, typically over a period of hours, days or weeks (World Bank 2017). These events include heavy and/or prolonged precipitation events, heatwaves, single very hot or cold days and prolonged periods of drought.

Global Climate Change Models (GCMs) are used to define alternative greenhouse gas emission levels known as Representative Concentration Pathways (RCPs). The projections from RCPs are then used as inputs to GCMs in generating climate change projections. In doing so, it is possible to model future change attuned to regional and national levels. However, it is a major challenge to further downscale climate change to rural road network level, and a significant amount of interpretation and judgement is required. Climate modellers use differing assumptions of the global climate system, resulting in differing climate projections for the same RCP scenario between models. The World Bank Climate Change Knowledge Portal (CCKP) provides general on-line guidance at a regional scale as well as presenting specific country data and advice on likely climate change in relation to a range of future scenarios (World Bank 2021).

Definitions

In recent years, a distinct climate change terminology has emerged that can be confusing and sometimes contradictory. The following definitions have been adopted for this book. They are derived from and, in some cases, slightly modified from UNISDR (2015) definitions.

Climatic impact: The effects that climate may have on people, their livelihoods, physical structures and environments.

Climate change: Changes, or likely changes, in the future state of the climate as compared to its current state that can be anticipated by scientific and statistical analysis.

Climate threats: In the context of this document, climate events that could be detrimental to the transport infrastructure, its operation and the people that use it.

Climate vulnerability: The extent to which engineering structures, populations or regions are susceptible to climate impacts.

Climate resilience: The ability of a structure or entity to resist climate-related factors. It may also be used in an adjectival sense (climate-resilient) to indicate the ability of structures, or the designs and procedures used in their construction or maintenance, to withstand climate impacts.

Climate adaptation: The process or processes, by which engineering or non-engineering measures are adopted and used to increase the resilience of engineering structures, populations or regions. To actual or expected climate and its effects.

Climate strengthening: The process or processes of making engineering structures climate resilient.

Disaster risk: Disaster risk is considered to be a function of hazard, exposure and vulnerability. It is normally expressed as a probability of loss of life, injury or destroyed or damaged assets that could occur to a system, to society or to a community in a specific period of time.

Disaster risk management: Disaster risk management is the application of disaster risk reduction policies, processes, and actions to prevent new risk, reduce existing disaster risk, and manage residual risk contributing to the strengthening of resilience.

Exposure: People, property, other assets, or systems exposed to hazards.

Hazard: A potentially damaging physical event, phenomenon or human activity that may cause the loss of life or injury, property damage, social and economic disruption, or environmental degradation.

Risk: The combination of the probability of a hazardous event and its consequences that result from interaction(s) between natural or man-made hazard(s), vulnerability, exposure, and capacity.

Risk assessment: An approach to determine the nature and extent of risk by analysing potential hazards and evaluating existing conditions of vulnerability that together could potentially harm exposed people, property, services, livelihoods, and the environment on which they depend.

Climate and rural roads

Climate change presents a strategic risk to transport infrastructure and rural road networks, in particular (ADB 2014; CSIR 2019a). In addition to gradual climate change within the normal design life of road infrastructure, it is reasonable to assume that their assets will be exposed to an increased frequency in shock weather events. The consequences of climate change in terms of climate threats to rural roads may be considered within a framework of cause and effect, as shown in Figure 17.1.

Investments in rural transport infrastructure are key to the sustainable economic growth of developing countries, but at the same time they may be highly vulnerable to projected climate change. Rural roads are built to provide rural access, and if this access is compromised by climate impacts, then this can have significant economic, social, and financial consequences and

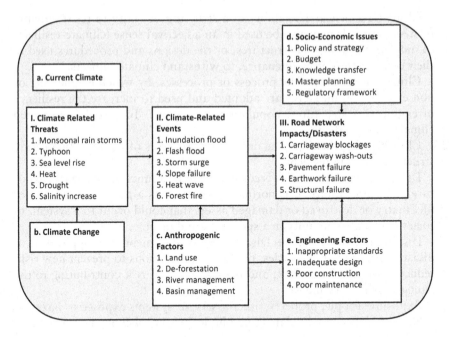

Figure 17.1   The climate change impact framework.

can threaten the hard-won gains achieved in the past few decades in terms of development and reductions in rural poverty (ADB 2014). Continued poverty reduction requires proactive efforts to address environmental sustainability, including mitigating the causes of climate change and supporting vulnerable communities in adapting to the unavoidable impacts of current and future climate.

There is already evidence of climate change having an impact on road infrastructure through increased frequency and severity of flooding events, increased landslide frequency, and damage to roads from excessive heat. The African continent is reported facing the potential of a $180 billion liability to repair and maintain roads damaged from temperature and precipitation changes directly related to projected climate change through 2100 (Chinowsky and Arndt. 2012; CSIR 2019a).

Although rural road networks are recognised as being vulnerable to current and future climate impacts, they are just one of several significant issues contributing to premature deterioration of road assets. The whole causal package of inter-related impacts needs to be understood, assessed and mitigated in a holistic manner. Figure 17.2 summarises the deterioration impact factors, in general, whilst Table 17.1 focuses on the specific engineering impacts associated with climate threats. Figure 17.3 illustrates some typical climate impacts on rural roads.

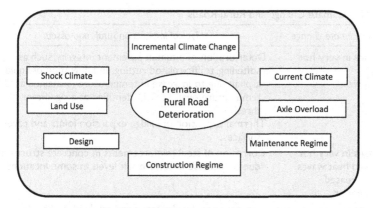

Figure 17.2 Road asset deterioration impact factors.

CLIMATE RESILIENCE STRATEGY

Climate resilience principles

A number of key principles can be identified that underpin effective and appropriate climate resilience. These are relevant to a range of scenarios: national planning, road network management, road projects, individual roads, and individual road assets.

Climate resilience assessments should be holistic and cross-sectoral. Strategies need to be integrated fully within cross-ministry government processes to be fully-effective and sustainable – from government policy down to on-the-ground application. A holistic approach should involve route corridors, land use, watersheds and a land systems approach as well as key non-engineering social, capacity building and regulatory issues.

Road infrastructure should already be designed to withstand current climate impacts to deliver the Fit-for-Purpose performance levels. In doing this, designers and engineers have typically relied on historical records of climate. However, using historical climate data alone is not a reliable predictor of future impacts. Most road infrastructure assets are now built to last from 10 years (light sealed pavements) to 50–100 years (bridges) and understanding how future changes in climate might affect this infrastructure is important to protecting long-term investments.

Adapting to climate change is a process, which should be built into a road authority's normal planning and risk management procedures. Successful forward planning – not just responding to emergency situations – will enable an authority to make investment decisions at the right time, making sure that the road infrastructure continues to provide the levels of service that its stakeholders and network users expect.

Table 17.1 Climate Change and Rural Roads

Potential climate change	Typical impacts on rural road assets
Increases in very hot days. Heat waves.	Deterioration of bitumen pavement integrity, such as softening, traffic-related rutting, and migration of liquid asphalt due to increase in temperature (sustained air temperature over 32°C is identified as a significant threshold). Thermal expansion of bridge expansion joints and paved surfaces.
Increases in very hot days and heat waves and decreased precipitation.	Corrosion of steel reinforcements in concrete structures due to increase in surface salt levels in some locations.
Increases in temperature in very cold areas.	Changes in road subsidence and weakening of bridge supports due to thawing of permafrost.
Later onset of seasonal freeze and earlier onset of seasonal thaw.	Deterioration of pavement due to increase in freeze–thaw conditions in some locations. Reduced pavement deterioration from less exposure to freezing, snow, and ice.
Sea level rise and storm surges.	Damage to roads, underground tunnels, and bridges due to inundation flooding and erosion in coastal areas. Damage to infrastructure from land subsidence and landslides. More frequent flooding of underground tunnels and low-lying infrastructure. Erosion of road base and bridge supports. Reduced clearance under bridges. Potential for salt intrusion to deteriorate concrete structures.
Increase in intense precipitation events.	Damage to roads, subterranean tunnels, and drainage systems due to inundation and flash flooding. Increase in scouring of roads, bridges, and support structures. Damage to road infrastructure due to landslides. Overloading of drainage systems. Deterioration of structural integrity of roads, bridges, and tunnels due to increase in soil moisture levels.
Increases in drought conditions for some regions.	Damage to infrastructure due to increased susceptibility to wildfires. Damage to infrastructure from erosion in areas deforested by wildfires.
Increase of storm intensity.	Damage to road infrastructure and increased probability of infrastructure failures. Increased threat to stability of bridge decks. Increased damage to signs, lighting fixtures, and signage supports.
Increase in wind speed.	Suspension bridges, signs, and tall structures at risk.

Figure 17.3   Typical climate impacts on rural roads. (a) Side drainage overwhelmed by localised rainstorm, central Tanzania. (b) Vented culvert approach washed away, central Laos. (c) Low-lying alignment impacted by rainy-season flood levels. South Sudan. (d) Temporary deviation bridge damaged by tropical storm, northern Laos.

Climate resilience performance levels

It is unrealistic to expect a developing country road network to be completely climate resilient, particularly with respect to shock climate impacts. Climate resilience network strategies must consider individual road functions, their socio-economic importance, their national importance and, where relevant, strategic disaster relief importance. Table 17.2 indicates an outline of the sort of approach that might be adopted. An actual framework must involve discussion with stakeholders, particularly local groups.

Alternatively, or in addition, it may be advantageous to set up defined levels of general road link climate resilience. Table 17.3 is a preliminary set of definitions that could be used as a basis requiring discussion and refinement. It is derived from the concept of the safety 'star system' for rating of road designs that has been successfully applied in several counties (https://www.irap.org/).

Table 17.2 Possible Performance Levels Based on Resilience to 'Shock Events'

	Impassability time for flood events: return period (RP)			
Road class	5 year RP	10 year RP	50 year RP	100 year RP
National highway	Nil	Nil	<2 hours	<12 hours
Provincial road	Nil	<6 hours	<24 hours	<2 days
Urban road	Nil	<6 hours	<24 hours	<2 days
District road	<2 hours	<12 hours	<2 days	<5 days
Village road	<12 hours	<2 days	<5 days	<7 days
Farm access	<24 hours	<5 days	<10 days	<10 days

Table 17.3 Possible Climate Resilience (CR) Levels for Roads

CR star level	Definitions
1	Road link is significantly deficient to current climate impacts to the extent that all-season access is consistently compromised. Largely not resilient to shock events.
2	Road link is largely resilient to current impacts in key areas in terms of drainage and surfacing. Adequate all-season access with some limited blockages. Requires additional adaptation. Largely not resilient to shock events.
3	Road link is resilient to current impacts in all areas but may still be vulnerable to future climate incremental in impacts. Key assets resilient to low-level shock events.
4	Road link is fully resilient for anticipated future incremental climate impacts within the design life. Key assets resilient to all except the highest level.
5	Road link is fully resilient for anticipated future incremental climate impacts. Key assets resilient to all shock events within design life. Other assets generally resilient to all events except very highest level.

The importance of policy and planning

There is a broad understanding of the need for climate strengthening but less agreement on how this is to be achieved and there is a need for a formal framework of policies, decrees and standards within which to work at national levels. Only through integration of climate resilience into policies and long-term strategic plans can real change to counter climate impact be embedded into general good engineering practice.

It is essential that the necessary climate resilience policy objectives are defined at the highest levels. Policy is required to underpin a focussed strategic framework for increasing the climate resilience of the road networks and ensuring they are Fit for Purpose now and in the future. Policy is required to give guidance and authority to this framework and the consequent

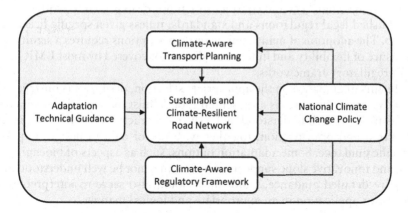

Figure 17.4 Policy as a driver of climate resilience.

allocation of necessary resources. The linkage of policy to application can then be driven by the specific issues contained in national policy and related ministry level decrees. Figure 17.4 illustrates the concept of policy as the initial key driver of climate change adaptation.

The United Nations Framework Convention on Climate Change (UNFCCC 2017) noted that *"To enable workable and effective adaptation measures, ministries and governments, as well as institutions and non-government organizations, must consider integrating climate change in their planning and budgeting in all levels of decision making"*.

Climate resilience adaptation plans at network level need to encompass the following key issues:

- Understanding of the implications of the climate resilience policy.
- Identification of network segments with high vulnerabilities.
- Identification of critical roads or assets within critical areas.
- Prioritisation of at-risk segments, roads, assets.
- Identification of the timeline of critical risks.
- Link budgets with the timeline of adaptation requirements.
- Identify non-engineering requirements.

In many LMICs, the development of sector-level planning is important, because it is often at this stage that criteria such as engineering designs, alignment, technology, and priority areas will be established, and crucially, budgets assigned. Incorporating climate adaptation considerations into these plans will increase the likelihood of meeting network objectives by assessing the budgeting requirements in a logical manner.

Regulatory frameworks of standards and specifications provide the necessary environment within which climate resilience can be effectively applied.

Road design engineers and contractors cannot be expected to work outside established legal regulations and standards, unless given specific licence to do so. The adoption of many climate resilience options requires a significant measure of flexibility and innovation that is not covered by most LMIC current regulatory frameworks.

Technical guidance on the appropriate selection, design and construction of adaptation measures is an essential tool. Whilst many of the adaptation measures may well understood by LMIC road practitioners, the criteria for selection and prioritisation, together with use of data in design, requires specific guidance. Some adaptation options, such as aspects of bioengineering and innovative slope support measures, may not be well understood and require detailed guidance. Guidelines should also serve to interpret standards for application in an appropriate and logical manner.

CLIMATE RESILIENCE APPLICATION

Climate resilience in the road life cycle

The vulnerability of a transport network is a function of the potential impact of climate change, based on location, its exposure and sensitivity to climate change, and its adaptive capacity or resilience to climate impact, both current and in the future. This can best be assessed and acted upon within the framework of the road Project Cycle. Table 17.4 lists network or road cycle project stages with the expected outcomes and required decisions.

Climate resilience application

"Adapting to climate change is a process, which should be built into a road authority's normal planning and risk management procedures" (PIARC 2015).

(PIARC 2015)

There is now an accepted general framework within which climate strengthening of infrastructure may be assessed and implemented. The details will vary, depending on location, scale, and whether an assessment is required at national, programme or individual road level (Carew-Reid et al. 2011; ADB 2014; CSIR 2019b–d). The principal steps are essentially:

1. Define the project aims and financial limitations.
2. Define the climate change threats within the current project environment.
3. Assess the vulnerability of assets within the region/programme/road corridor.
4. Define and, if necessary, prioritise adaptation options.
5. Implement the adaptation option.
6. Monitor and evaluate the performance of the options.

Table 17.4 Stages of Climate Resilience Activities

Level	Climate resilience (CR) input	Climate resilience (CR) procedure
General project planning.	CR base line summary for proposed initiative.	Broad impact types identified.
Pre-feasibility (Pre-FS).	Likely climate threats identified associated with each option and general budget levels. Climate threats/ impacts identified that could have major project implications. General levels of CR identified for individual projects.	Initial screening for climate threats, and assessment of project vulnerabilities.
Feasibility and preliminary engineering design (FS and PED).	Climate threats and impacts for road projects identified and prioritised. Range of relevant engineering and non-engineering adaptation options listed and outline-costed with preliminary BoQ.	Application of procedures for threat and vulnerability assessment. Identification of key data sources. Preliminary data collection.
Final Engineering Design (FED).	Adaptation options (engineering and non-engineering) detailed in project documentation. Engineering options designed in detail and quantities included in BoQ.	Detailed design of CR options linked to definition and prioritisation of risks. Engineering data collection and input. Drafting of detailed CR specifications.
Construction and Technical Audit.	Ensure non-engineering options are in place and being actioned. Supervision of engineering options.	Application of defined CR monitoring procedures.
As built review.	Review compliance of engineering options with CR specifications. Set up any required monitoring procedures. Review of efficiency of non-engineering options and ensure procedures in place to ensure their ongoing application, where appropriate.	Analysis of monitored data related to climate and climate event information.
Asset management planning.	CR adaptation options prioritised for maintenance planning. Ongoing monitoring of climate assumptions.	Monitoring of climate and climate threats within project area of influence. Possible set-up of weather and flood gauges where feasible and required.
Ongoing maintenance review.	Monitoring of efficiency of adaptation options and updating of specifications and procures considering performance – assumption studies.	Updating of CR engineering and non-engineering procedures.
Upgrade/ rehabilitation assessment.	Review of need for changes in CR approaches.	Review of CR assumptions in original designs and lessons to be learnt.

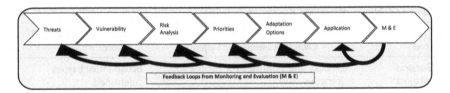

Figure 17.5   A risk assessment-based approach to climate resilience.

The World Road Association (PIARC 2015) in a review presented a frame-work to guide road authorities through identifying relevant assets and climatic variables for assessment, identifying and prioritising risks, and developing a robust adaptation response. Importantly, it is designed to be applicable at a range of scales (national, regional, local or asset specific). This was a further development of a risk-based approach outlined by ERA-NET (2010), and based on existing risk analysis and risk management tools (Figure 17.5).

Vulnerability assessment is recognised as being a crucial action to be conducted within the Project Cycle, preferably in the following key steps:

1. A preliminary climate risk screening to identify vulnerable projects or assets that may be at risk, undertaken at the planning stage to highlight any major concerns. This first step aims to provide an initial assessment of the level of sensitivity of the project to climate threats.
2. A Climate Risk Vulnerability Assessment (CRVA) is implemented at a Feasibility Stage to quantify climate change risks and identify costs of climate strengthening measures that may be required. This leads to the identification and screening of a list of possible adaptation options. A simple climate vulnerability inventory and strip-map procedure can be used to gather CVRA field data; an example is included within Appendix A – Standard Data Sheets.
3. At Final Engineering Design stage, a more detailed risk assessment, based on the Feasibility Stage CRVA, of individual roads or assets potentially at risk from climate change. This stage aims to detail the specific nature of the climate risks and leads to design of the resilience options.

Climate change and continual development (or deterioration) of infrastructure mean that the climate vulnerabilities, risks, and data sets are likely to all continue to change over time. Hence, a CVRA should be considered as continually evolving throughout the network asset life cycle. The CRVA should facilitate the planning and adoption of both engineering and non-engineering procedures. Engineering options are based around the design and construction of road assets, with an emphasis on the importance drainage and

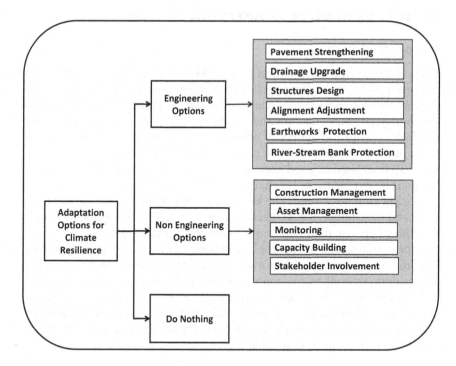

Figure 17.6 Strategic options for climate resilience.

drainage structures. Non-engineering options consist of a range of policy, management, communication, and capacity building improvements. Non-engineering options that are used to address adaptation for road infrastructure and asset management tend to be more strategic and organisational in their nature than engineering options.

Figure 17.6 outlines general engineering and non-engineering options available following CRVA and when considering the application climate resilience (ADB 2011).

It should be clearly appreciated that in most cases the 'do nothing' option is not equal to 'status quo'. Doing nothing to improve climate resilience will lead to decreasing levels of service, decrease in asset value and a risk to sustainable development. However, in some cases there is just not enough budget to deal with all impacted areas, roads and structures, or the consequences of climate change are too severe to justify comprehensive engineering adaptation. In these circumstances, the focus should shift to a planned programme of prioritisation. Apart from information-based decision on adaption priorities, this should also involve dialogue with affected communities, and contingency programmes may be necessary to minimise any adverse effects of prioritisation.

CLIMATE RESILIENCE ADAPTATION OPTIONS

Adaptation data

There is little experience within many LMICs in understanding the cost-effectiveness of the different climate resilience adaptation options. The reasons for this include the long-term nature of climate change, the need for appropriate baseline data and the difficulties in measuring climate vulnerability within the matrix of other sources of degradation. Fortunately, many of the practical aspects of climate strengthening of a road network or road can be related to good practice road engineering and can be classed as 'no regrets' interventions, in that they yield benefits even in the absence of significant climate change. The key engineering challenges lie not in having to amend the engineering, geotechnical or hydraulic design principles but in the climate change data required as input to the design process.

Information is key to the implementation of effective climate resilience. This includes current and historical climate information and road vulnerability data, as well as environmental, socio-economic, and socio-ecological information. The latter are recognised as important fields in making adaptation decisions. There is also a need to gather information about the impacts of adaptation projects that have already been implemented. Key data sets are summarised in Table 17.5.

Table 17.5 Key Climate Resilience (CR) Data Sets

Data sets	Description	Possible sources
Current and future climate.	Information on current climate and prediction on likely changes within the design life of the network assets.	World Bank Climate Change Knowledge Portal. Ministry relevant to climate/ weather data recording.
Network asset condition.	Location, condition, and hence potential vulnerability of road and other assets.	Relevant road asset management system.
Existing climate change impact data.	Information on currently identified climate risks and relevant historical events.	Research reports. Search World Bank, ADB, AfDB Foreign, UK Commonwealth & Development Office (FCDO) for on-line knowledge libraries.
CR option designs.	Information of possible engineering and non-engineering option, and their design and use.	Technical reports, within relevant ministries or MDB on-line libraries (as above).
Cost norms.	Information on adaptation cost options.	Ministries of public works/ transport contract records.

ADB have produced two knowledge products (ADB 2020a, b) that present and illustrate clear processes for incorporating allowances for climate change in engineering design. Attention is focussed on credible adjustments to extreme rainfall, with the 1-day maximum rainfall total (Rx1day) as the critical variable in formulas for hydrologic calculations of design water discharge, flood level, and river flow velocity, as input to bridge and culvert design.

There are three basic steps:

1. Obtain historic climate data ('base values').
2. Obtain and assess future climate data ('uplift' factor).
3. Adjust historic data with future change factors ('base' multiplied by 'uplift' factor).

Rx1day data can be extracted from the World Bank CCKP for different future time projection periods (e.g., 2025–2045 or 2045–2065), for different RCPs (e.g., 4.5 or 8.5) and for different storm return periods. The adjusted historic data can then be used within the relevant hydraulic procedures for design, as discussed in Chapter 11.

Engineering options

Pavement strengthening

Pavement climate strengthening is largely concerned with the following issues:

- Pavement type; selection of appropriate option.
- Cross section geometry.
- Construction materials.

A majority of LMIC rural roads are unsealed earth or gravel surfaced and potentially vulnerable to erosion (Cook and Petts 2005). Research has highlighted the unsustainability of unsealed granular roads in a seasonal-concentrated high rainfall environment at gradients of greater than 6%–8%. Options for providing a more resilient surfacing are contained in Chapters 14 and 15. In summary, these include:

- Bitumen sealing.
- Concrete paving.
- Cobble stones.
- Concrete blocks.
- Bricks.
- Hand-packed Stone.

Removing excess water either from the pavement layers or from its surface is key issue in combating the degradation impacts of climate. This can be achieved by:

1. Ensuring adequate cross-fall is maintained to allow surface water to flow-off unimpeded to side drainage.
2. Avoiding pavement 'canal' cross-sections, where permeable pavement layers are enclosed by impermeable clayey shoulder and subgrade; shoulders drainage is essential.
3. Ensuring internal pavement materials have adequate drainage capacity.

Natural materials selected for pavements at risk of being impacted by significant changes in rainfall pattern should be resistant to increased changes in moisture condition, either by decrease in strength or vulnerability to volume change (shrink/swell).

Drainage upgrade

Good drainage is an essential part of any road climate strengthening initiative, comprising

- Pavement drainage.
- Earthwork drainage.
- Cross drainage.

Drainage upgrade and improvement includes upgraded cross drainage (mainly culverts and bridges) and improved side drains that must be designed to cope with future rainfall; otherwise they risk being overwhelmed, as shown in Figure 17.3a, with consequent impacts on the pavement.

The number and location as well as size of culverts require particular attention within the climate resilience context. In steep terrain, relief culverts could be very closely spaced, although in very low volume access routes some limitations due to excessive cost must be considered.

Some culverts and small bridge approaches are vulnerable to storm events, as shown in Figure 17.3b. There is a need therefore to consider not only the design of structures but also to design appropriate protection against climate threat of the associated earthworks and abutments. The high cost of fully climate-protected river crossings may be mitigated by considering cheaper options that may inhibit access for limited periods but, at the same time, are inherently more resilient than traditional small bridges and approach earth works, as discussed in Chapter 12.

Alignment adjustment

Alignment adjustment in the context of climate strengthening is largely concerned with flood mitigation, although there can be requirements to reduce grade to counter drainage issues or select more suitable bridge locations. Horizontal realignment may be beneficial in steep terrain to avoid marginally stable slopes in residual soil profiles. As regards flood mitigation, the options are:

1. Horizontal realignment to avoid low-lying problem area, as illustrated in Figure 17.3c.
2. Raise the embankment. Increases in embankment height to ensure an adequate critical pavement crest level above defined flood level.

Earthwork protection

Increased rainfall and occurrence of climate shock events are acknowledged as having an impact on road-side slope stability and erosion. Saturation following heavy rain reduces pore suction, adds to the weight of the slope mass, and reduces both cohesion and the friction. It has been well recognised that the combination of a long period of rain followed by an additional rainstorm impacting on an already saturated slope is a worst-case condition for slope instability (Lumb 1975). Increased slope support measures, both hard engineering works and soft bio-engineering options, should be considered (Howell 1999, ICEM 2017). Drainage systems for earthwork protection structures need to be strengthen using combination of graded filter layers, cut-off drains, chute and cascade drains and face drains. Detail on earthwork protection and stabilisation are covered in Chapter 10.

Riverbank and coastal road protection

Riverbank and coast road protection is intended either to protect the alignment from erosion by nearby streams, rivers or the sea, as part of enhanced protection works associated with cross drainage structures. This protection includes installing gabions, block facing, bioengineering to strengthen slopes and improving face drainage against overtopping. In some cases, the development or maintenance of natural options such mangrove forests may be an option. For rural road networks the combination of engineering and bioengineering options can prove very cost-effective (ICEM 2017).

Erosion and road embankment failure is an important consideration for climate change strengthening in low-lying coastal areas, for example Bangladesh, or areas of Vietnam and Mozambique (De Souza et al. 2015).

Problem soils and rocks

Problem soils and rocks are a climate strengthening issue that cross-cut many engineering adaptation options. Their sensitivity to moisture change can increase their climate change vulnerability to a significant extent such that there may be a requirement to select more erosion resistant road materials or to overcome increased difficulties with, for example, problematic sub-grades (CSIR 2019d).

It is possible that some local construction materials that could have been assessed as marginal, but acceptable, need to be re-assessed in the light of increased exposure to moisture change or intense erosive rainfall. These problem materials include:

- Expansive soils/rocks.
- Erosive soils/rocks.
- Collapsible soils.
- Saline sensitive aggregates.
- Soft and organic soils.

Climate resilience non-engineering options

Construction management

Effective climate strengthening is reliant upon the adoption of good engineering practice during construction. There may be a requirement within the overall project procurement plan to identify specialist sub-contractors to undertake specific climate strengthening tasks – for example construction of gabion retaining walls or plant bioengineering options. No matter whether main contractors, specialist contractors, or even Local Village Groups are employed on the climate adaptations, it is essential that they are fully aware of the specifications and have appropriate experience. If there is an experience shortfall, then appropriate training and supervised construction programmes are essential and should be initiated.

The overall project Quality Control framework should be adjusted to accommodate the requirements of the climate adaptation elements. This may require initial additional specialist support and training for site inspectors.

Climate change will also have practical implications in terms of changes in the timing and length of the construction season for road construction and maintenance works, due to changes in rainy-season intensity and duration, changes in temperature, water shortages and accessibility constraints. Leaving construction works exposed to impacts from increased rainfall or storm events can have disastrous effects on incomplete drainage and earthworks, as shown in Figure 17.3d.

Asset management

Lack of effective maintenance is a major contributor to climate-related degradation of road network assets. Maintenance can be a significant proactive action to lessen climate impacts and should be an integral part of any climate strengthening programme. Related to this is the requirement for simple asset condition and climate monitoring procedures that should:

1. Identify areas of concern that require maintenance interventions prior to rainy seasons.
2. Collate data on actual climate impacts and the asset response so that assumed design input on climate strengthening may be reviewed and modified where necessary for future programmes.

Periodic maintenance of the assets also can provide opportunities to make incremental improvements to the climate resilience of the network. Some climate adaptation options such as grass and shrub slope protection may require maintenance activities outside normal practice and others such as maintenance of culverts and seepage pathways in retaining walls will require increased attention.

Monitoring

Monitoring and evaluation of existing climate strengthening projects, policies and programmes forms an important part of the adaptation process. Ultimately, successful adaptation will be measured by how well different measures contribute to effectively reducing vulnerability and building resilience. Lessons learned, good practices, as well as gaps and needs identified during the monitoring and evaluation of ongoing and completed projects, policies and programmes will inform future measures, creating an iterative and evolutionary adaptation process. Recording data to determine the level of resilience being achieved along with the associated costs can be part of longer-term target of establishing effective climate-maintenance models.

Capacity building

In terms of the resources, the number of climate adaptation experts is likely to be very limited in many LMICs. It is good practice that consideration be given to identifying training focussed specifically on climate resilience issues. Transferring knowledge on climate impact and climate resilience can be a key activity. In order to establish and implement climate adaptation successfully, capacity will need to be developed across all relevant stakeholders. This includes road and transport ministries, departments and agencies/authorities and will include a wide range of participants from central government agencies cascading all the way through to village groups.

Stakeholder involvement

Identification of adaptation options should involve inputs from a range of stakeholders from national down to local levels. Conducting these consultations provides useful input for the process of identifying and appraising the whole range of adaptation options. In this manner, the top-down initiatives driven by policy can effectively link with bottom-up demand-driven requirements.

The sub-regional or local road authorities and communities are key stakeholders for a local-level climate risk and vulnerability assessment for roads. The involvement of communities in data collection is crucial. It is also crucial that they understand factors that render roads in their community vulnerable to climate. The benefits of increased community awareness to road authorities include receiving early warning about emerging structural damage on the roads, reduction of climate impacts through modification of land use practices, and frequent clearing of debris and vegetation from culverts, bridges, etc. Local government representatives from the environmental, emergency and disaster management, agriculture, and social development departments are important stakeholders in terms of provision and uptake of information, ensuring that additional data gathered from local assessments is integrated into national spatial data repositories. They are also well positioned to implement change management recommendations on factors that are not intrinsic to the road infrastructure.

REFERENCES

ADB. 2014. *Climate Proofing ADB Investment in the Transport Sector: Initial Experience.* Manila. https://www.adb.org/sites/default/files/publication/152434/climate-proofing-adb-investment-transport.pdf.

ADB 2011. *Climate Risk Management in ADB Projects.* Asian Development Bank, Manila,Philippines,https://www.adb.org/publications/climate-riskmanagement-adb-projects

ADB. 2020a. *Climate Change Adjustments for Detailed Engineering Design of Roads. Experience from Viet Nam.* Wilby, R. for Asian Development Bank. http://dx.doi.org/10.22617/TIM200147-2.

ADB. 2020b. *Manual on Climate Change Adjustments for Detailed Engineering Design of Roads Using Examples from Viet Nam.* Wilby, R. for Asian Development Bank. Manual on Climate Change Adjustments for Detailed Engineering Design of Roads Using Examples from Viet Nam (adb.org).

Arnaud, E., G. Halverson and G. Shields-Zhou, Eds. 2011. The geological record of Neoproterozoic glaciations. *Geological Society, London,* 2 Memoirs 36. https://doi.org/10.1144/M36.0.

Carew-Reid, J., T. Ketelsen, A. Kingsborough and S. Porter. 2011. Climate change adaptation and mitigation (CAM) methodology brief. ICEM–International Centre for Environmental Management. Hanoi, Vietnam.

Chinowsky, P. and C. Arndt. 2012. Climate change and roads: a dynamic stressor-response model. *Review of Development Economics* 16: 448–462. https://doi.org/10.1111/j.1467-9361.2012.00673.x.

Cook, J. R. 2022. Reducing future climate impact on rural transport infrastructure in developing countries; the role of engineering geology. *Quarterly Journal of Engineering Geology and Hydrogeology.* https://doi.org/10.1144/qjegh2021-052?ref=pdf&rel=cite-as&jav=VoR.

Cook, J. R. and R. C. Petts. 2005. Rural road gravel assessment programme. Module 4: Final Report. Intech-TRL for DFID-SEACAP. https://www.research4cap.org/index.php/resources/rural-access-library.

CSIR. 2019a. Climate adaptation: Risk management and resilience optimisation for vulnerable road access in Africa. Climate adaptation handbook. ReCAP project GEN2014C for UKAID-DFID. https://www.research4cap.org/index.php/resources/rural-access-library.

CSIR. 2019b. Climate adaptation: Risk management and resilience optimisation for vulnerable road access in Africa. Change management guidelines. ReCAP project GEN2014C for UKAID-DFID. https://www.research4cap.org/index.php/resources/rural-access-library.

CSIR. 2019c. Climate adaptation: Risk management and resilience optimisation for vulnerable road access in Africa. Climate threats and vulnerability assessment. ReCAP project GEN2014C for UKAID-DFID. https://www.research4cap.org/index.php/resources/rural-access-library.

CSIR. 2019d. Climate adaptation: Risk management and resilience optimisation for vulnerable road access in Africa. Engineering adaptation guidelines. ReCAP project GEN2014C for UKAID-DFID. https://www.research4cap.org/index.php/resources/rural-access-library.

De Souza, K., E. Kituyi, B. Harvey, M. Leone, K. S. Subrammanyam and J. Ford. 2015. Vulnerability to climate change E.in three hot spots in Africa and Asia: Key issues for policy-relevant adaptation and resilience-building research. *Regional Environmental Change* 15. https://doi.org/10.1007/s10113-015-0755-8.

ERA-NET 2010. Risk management for roads in a changing climate. A guidebook to the RIMAROCC Method. Project No. TR80A 2008:72148. https://www.cedr.eu/download/other_public_files/research_programme/eranet_road/call_2008_climate_change/rimarocc/01_Rimarocc-Guidebook.pdf

Howell, J. 1999. Roadside bio-engineering: site handbook and reference manual. Department of Roads, Kathmandu. A comprehensive study of bio-engineering options in mountainous terrain Nepal.

ICEM. 2017. Promoting climate resilient rural Infrastructure in Northern Vietnam, Final Report. Prepared for Ministry of Agriculture and Rural Development and ADB. Hanoi. https://www.adb.org/sites/default/files/project-documents/41461/41461-042-tacr-en.pdfIPCC. 2014. *Climate Change 2014: Synthesis Report.* Contribution of Working Groups I, II and III to the Fifth assessment report of the intergovernmental panel on climate change IPCC, Geneva, Switzerland, 151 pp. https://www.ipcc.ch/report/ar5/syr/.

Lumb, P. 1975. Slope failures in Hong Kong. *Quarterly Journal of Engineering Geology & Hydrogeology* 8:31–65.

PIARC. 2015. *International Climate Change Adaptation Framework for Road Infrastructure.* Report 2015RO3EN, Paris. https://www.piarc.org/en/order-library/23517-en-International%20climate%20change%20adaptation%20framework%20for%20road%20infrastructure.htm.

Slezak, M. 2015. Key moments in human evolution were shaped by changing climate. New Scientist. https://www.newscientist.com/article/mg22730394-100-key-moments-in-human-evolution-were-shaped-by-changing-climate/#ixzz7C8aFIKw5

UNISDR. 2015. United Nations Office for Disaster Risk Reduction terminology on disaster risk. https://knowledge4policy.ec.europa.eu/organisation/unisdr-united-nations-office-disaster-risk-reduction_en.

United Nations Framework Convention on Climate Change (UNFCCC). 2017. Climate change impacts, vulnerabilities and adaptation in developing countries. https://unfccc.int/resource/docs/publications/impacts.pdf.

World Bank. 2017. *Integrating Climate Change into Asset Management.* Washington, DC. https://documents1.worldbank.org/curated/en/981831493278252684/pdf/114641-WP-ClimateAdaptationandAMSSFinal-PUBLIC.pdf.

World Bank. 2021. *The World Bank Climate Change Knowledge Portal.* https://climateknowledgeportal.worldbank.org/.

Chapter 18

Procurement and documentation

INTRODUCTION

When a road design is complete and is approved by the relevant authority, the process can proceed to the procurement of the defined works; the next stage in the Project Cycle, as shown in Figure 18.1.

The central aim of the procurement process is to obtain priced proposals from competent organisations or groups to carry out the defined work. If a competitive process is used, then offers from different tenderers are compared in a fair and transparent manner and the work is awarded to the bidder with the most favourable offer in terms of price and technical competence.

The preparation of appropriate, technically supported, contract documents allied to a clear and straightforward procurement process, relevant to the existing construction environment, is crucial to the effective delivery of road projects. In the rural sector it is also frequently an aim to ensure that the procurement process assists in developing local capacity so that the available pool of local contracting resources expands. In this context, contractual obligations in terms of utilising local labour, groups, or formal subcontractors may be inserted alongside encouragement for gender balance. Requirements in terms of training and complementary interventions may also be included (Mikkelsen and Waage 2004).

Although the principal focus in this chapter is on the procurement for road construction or rehabilitation, the general principles also apply to other works such as:

- Consultancy or Supervision Services.
- Ground Investigations.
- Maintenance.
- Specialist Works.

Contracting arrangements for maintenance works are dealt with further in Chapter 21.

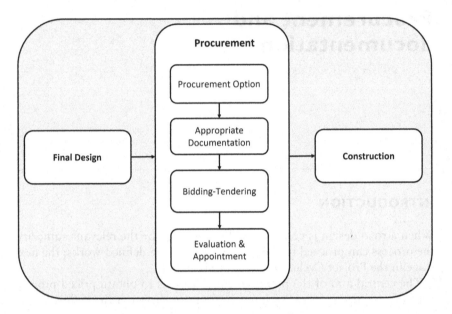

Figure 18.1　The procurement process.

PROCUREMENT STRATEGY

Contracting options

Several different contracting options exist that can be used for rural road-works. The most common options can be grouped as follows:

- In-house (Force Account) or Agency.
- Conventional Bill of Quantities (BoQ)-based Contract.
- Lump Sum or Negotiated Contract.
- Design and Build.
- Concessions.
- Performance-Based Contracts.

For most project implementation methods, with the exception of the in-house option, projects will be constructed under a formal contract agreement. Under a contract system, the employer enters into an agreement with a contractor, who is often chosen through competitive tendering from a number of bidders (SANRAL 2013a).

When the procurement process is open to competition from any company or consortium, irrespective of their country of origin, this procurement process is often referred to as 'International Competitive Bidding' (ICB). When competition is restricted to local firms, then the process is known as 'Local or National Competitive Bbidding' (LCB or NCB). In most cases, smaller individual rural road construction contracts will be on an NCB basis or on a Lump Sum, Negotiated or Design and Build approach, particularly when the location is remote.

In-house or agency

In-house implementation (also known as 'Force Account' or 'Agency') is a well-known method for rural roads, where all aspects of planning and execution of an infrastructure project are handled exclusively by a government department or designated road agency. Mikkelsen and Waage (2004) note that, previously, this approach has been common for all types of local rural road works. Today, it is still used in some regions for road maintenance, but its use for new construction and rehabilitation works is now very limited. An agency is an organization with delegated authority from the road owner for operation and management of a road network, or part of a network. Agencies have, in the past, operated under a framework agreement that sets out activities to be performed, but with considerable discretion to undertake work as they consider appropriate.

As the design, construction and supervision are handled by the Owner's organization, contractual disagreement with other parties does not occur, and the personnel involved are familiar with the requirements, policies and procedures related to the project. It is quite often possible to save time by adopting an in-house approach, because works can commence as soon as funding becomes available, without going through a sometimes lengthy bidding process.

A disadvantage is that internal cost control is often inadequate, and this can result in higher overall project costs than for other project execution methods, especially when there are unforeseen or complex ground conditions. There is also a tendency towards lax Quality Control. It can be difficult to hold anyone accountable for delays and poor quality of works, given that the same organization is both executing and supervising the works. Work is often implemented in a manner that makes use of available resources (work force, equipment, and materials), rather than applying rational and efficient work methods. There is also a perceived higher risk of corruption because of lax Quality Control.

Conventional contracts

Conventional contracts are suited to projects where the Client (Owner), Road Authority or their representative knows what is required, and can specify this in the tender and contract documentation, which includes a comprehensive Bill of Quantities. Conventional BoQ-based contracts are likely to be those most applied to rural road programmes. In this process, the contractor prices the tender with the knowledge of what is required, and the Client or Road Authority has resources in place to ensure that the defined works are completed satisfactorily within budget and time.

This form of contract is normally suited to road construction and rehabilitation, where all the parties have a good understanding of the necessary technologies, and where there has been sufficient design and contract documentation. All the items of work are properly quantified and specified in the design, schedule of quantities, and related specification and documentation. Variation orders can be issued during construction to cater for limited situations that were not catered for in the BoQ (e.g., unforeseen ground conditions).

Lump Sum or Negotiated Contracts are a potential variation on conventional contracts. Under a lump sum contract, a single price for all the works is agreed before the works begin. It is defined as a fixed price contract, where the contractors agree to execute the works for a stated total sum of money. Lump Sum contracts are generally appropriate where the project scope of works (and quantities) are well defined when the tenders are sought and significant changes to requirements are unlikely. This means that the contractor is able to accurately price the works they are being asked to carry out. They can also be used in situations where speed may be of the essence, for example, in disaster emergency repair work.

The main advantage to the client is that the overall price is known in advance. However, if the contractor feels that the scope of works provided by the client is not very clear, then the contractor may provide a higher price to cover any potential risks. Depending on the scope of works and project duration, the agreed payment system could be one payment when the works are completed or staged payments, based on percentage of work completed against an agreed implementation programme.

Design and build

Design and Build, or Engineer Procure and Construct (EPC) contracting generally shifts the responsibility for detailed design from the Owner, Road Authority or Consultant to the Contractor. They may also be referred to as 'turnkey' projects. The process involves a contractor or contracting group employing a designer to design the project to meet the intended purpose defined by the Client. The technical contract documents may only specify the intended purpose of the project and any related performance requirements, although some form of Feasibility Study may be included. The contracting team then design the asset to meet the performance requirements. This form of contract is unlikely to be suited to small rural contractors, although there are some specialist situations where it could be considered (with experienced contractors preferably), for example, at difficult bridge sites, or road alignments in steep mountainous terrain where ground conditions are highly unpredictable.

In Design and Build projects the client does not have to supply detailed control of quality, as the contractor has to deliver a product that is 'Fit for Purpose'. In roadworks where the underlying quality is difficult to discern immediately, the client needs to ensure that a sufficiently long defects liability period is in place to prove the quality of the works and their 'Fitness for Purpose' over the medium term. For this reason, such works normally have an extended guarantee period beyond the normal 12 months.

As it is unlikely that any variation orders will be issued during construction, almost all risks have to be identified and costed in by the contractor's team during the tender phase and allowed for in the price. Exceptions may be, for example, extreme climate events above a defined return period or other defined *force majeure* issues. One disadvantage of these contracts, therefore, is that the client pays for risks, whether or not they materialise.

Concessions

Concession contracts are known by a number of names such as BOT (Build Operate Transfer) or OPBC (Output and Performance -based Contracts). In BOT contracts, a contractor or consortium is allowed to manage the entire road facility and provide the required service levels for the duration of a concession period, which should include a defined post-construction period. In most cases, the concessionaire is paid for providing the required service levels through tolls or some form of shadow tolling. Different terminology may be used for different concession models. Other terms in common use include BOOT 'build, own, operate and transfer', BOO 'build, own and operate', BOOST 'build, own, operate, subsidise and transfer', BLT 'build, lease and transfer' and DBFO 'design, build, finance and operate'. All are types of concessions, and all require the provision of private capital to finance the works. It is important to note that this model is very rarely used for rural road projects in LMICs, although quite frequently applied in a more developed country with higher traffic volumes and easier access to affordable finance.

Performance-based contracting

In the OPBC model, the contractor is responsible for design and initial construction of the road, together with ongoing maintenance for a set period (normally 5–10 years). The contractor will be paid a monthly sum for the initial works and another monthly sum for maintenance the following years (Gericke et al. 2014). The advantage of the OPBC model is that the contractors have an incentive to undertake efficient planning and good quality works in the initial work on the road, since they will have to maintain the road for the contracted number of years. It also reduces the workload for the road authority, since final engineering design and maintenance responsibilities are outsourced to the contractor. The disadvantages are the need for experienced contractors and clients and a revised framework of contract and financial regulation. Although this model is being championed by funding agencies for larger road projects, it carries with it significant challenges unless all contracted parties are fully aware of their responsibilities and their levels of risk (Silva and Liautaud 2011), and the relevant regulatory framework is in place. In LMICs, there is increasing use of the OPBC option as a means of improving the delivery of appropriate maintenance. Issues of local contracting experience, capacity, healthy competition and access to finance will influence feasibility of this approach.

Supervision contracts

To assist in project administration and to supervise the work of the contractor, it is common for the Owner/Client to appoint a firm of consulting engineers; in some cases, the same firm that designed the project. Unless these are retained on long-term service contracts, the Supervision Engineers are usually procured through a competitive bidding process based around detailed Terms of Reference.

An important element of the advantage of road construction contracts is that responsibilities between the three principal parties (the Owner/Client, the Contractor and the Consulting/Supervision Engineer) are well understood and clear. These responsibilities will vary to some extent depending on the type of contracting process. For example, in the Design and Build Model the Contractor, or Contracting Group, may themselves employ a Consulting Engineering firm to undertake supervision duties or Technical Audits.

APPROPRIATE DOCUMENTATION

Final engineering design

When a design is complete, the responsible engineering organisation usually prepares contract documentation on behalf of the client. The contract documents are developed in a style and format specified by the client. A typical list of documents includes the following:

1. General Conditions of Contract for Construction Works.
2. Scope of Works.
3. Specifications for Road and Bridge Works.
4. Design Report.
5. Roadworks Drawings.

6. Structures Drawings.
7. Materials Report including Investigation and Utilization.
8. Bills of Quantity.
9. Requirements for Works Quality Plan and Construction Supervision Procedural Manual.
10. Environmental Impact Assessment and Related Social Safeguard Documents.

The use of simplified bidding documents should be considered for rural road projects rather than a full suite of standard international documents. This has been found to be a significant contributory factor to efficiency and cost savings (Cook et al. 2013). Some selected aspects that need particular attention when drawing up documentation for rural road works are:

1. The description of the works (The Scope of Works) needs to be clear and include all pertinent issues. For rural road projects using locally based contractors, it may be necessary for them to be in the local language. The supervising engineer will be expected to be able to display the same language capability so that no confusion arises.
2. The documents should specify an appropriate level of experience for the contractor and staff. This is particularly important in difficult Road Environments (for example, mountainous terrain) or with non-standard contracts.
3. The documents may need to specify the use of certain material sources for specific purposes (e.g. pavement layers). The designer should be reasonably sure that the materials from these sources are of adequate quantity and can meet the required specifications. Alternative sources may be indicated.
4. Seasonal rainfall can have a major influence on constructability. Where such risks could be significant, they should be highlighted in the documents. In addition, climate change impacts need careful consideration with regard to contract duration and the criteria for extension of time.
5. There may need to be a contractual control limiting the type and frequency of construction traffic running over completed or partially completed pavement layers, particularly so for low volume rural roads designed for low axle-load use.
6. In mountain or hill terrain, there is a need for a contractual clause prohibiting or controlling the 'blading' or dumping of excavated materials over the side without appropriate compaction, benching and drainage. Side-dumping is a frequent cause of below-the-road slope failure.

7. Traffic diversion often represents an important part of rural road-works, and this must be carefully considered and allowed for in the design and documentation.

8. Geotechnical reports and the information contained in the tender documents need to present a reasonable factual view of the expected conditions and include the assumptions used in the road or bridge design.

Standard documents

Where appropriate, contracts should be packaged to accommodate the capabilities and experience of the different categories of contractors. The principle of optimal use of local resources for rural road works should guide the contract documentation format and contract classification.

Widely recognised forms of international standard contracts are produced by the Fédération Internationale des Ingénieurs-Conseils (FIDIC). The best known of the FIDIC contracts are the Red Book (building and engineering works designed by the Employer) and the Yellow Book (M&E, building and engineering works designed by the Contractor). Organisations such as the World Bank, Asian Development Bank or the African Development Bank may specify their own standard documents for their funded projects, mainly based on FIDIC documents (ADB 2016; FIDIC 1992, 1999, 2005).

Alternatively, the New Engineering Contract NEC documentation was first published in 1993 as a suite for construction contracts intended to promote partnering and collaboration between the contractor and client. It was developed as a reaction to other more traditional forms of construction contract which have been portrayed by some as adversarial. The third edition, NEC3 was published in 2005. https://www.neccontract.com.

The NEC contracts are intended to:

• Stimulate good management.
• Be clear and simple, written in plain English, in the present tense and without legal terminology.
• Be useable in a wide variety of situations from minor works to major projects.

NEC3 has been criticised as requiring too much expertise to operate, focussing too much on project management and generating too much paperwork.

The preparation of contract documents will be part of the Final Engineering Design (FED). To assist in project administration and to supervise the work of the contractor, it is frequently the case that the owner or executing agency will appoint a firm of consulting engineers. Clear and understandable ToR for this appointment may also be required as part of the FED output.

Table 18.1 Typical Groups of Technical Specifications

Specifications	Description
Construction methodology	Define detailed methodologies by which elements of a road should be constructed and include such key issues as: • Pavement. • Earthworks and its support. • Drainage. • Small to medium structures. • Bioengineering.
Construction materials	Defines the acceptable limits (properties, strength, durability etc.) for the selection and use of construction materials, both natural and man-made, and will cover such items as: • Pavement/structure aggregate. • Rock (blocks, riprap, fill, etc.). • Fill materials. • Concrete. • Cement and additives. • Bitumen. • Wood (wooden bridges).
Quality Assurance and Quality Control	Defines the methods to be used in terms of supervising the quality of road elements and the use of specified equipment and testing procedures.
Maintenance activities	Defines the procedures to be used in undertaking the different types of maintenance, routine (mechanical and non-mechanical), periodic and emergency.
Laboratory testing	Defines the laboratory testing procedures to be used in collaboration with construction, materials and quality specifications. These can be based on, or derived from established international procedures, such as BS, AASHTO or ASTM.

Technical specifications

Whatever contracting mechanism is employed, appropriate and realistic technical specifications and construction drawings are an essential pre-requisite for successful construction. Ideally, specifications should be concise and capable of being clearly understood by the contractors and supervisors alike; see Table 18.1.

Technical specifications define and provide guidance for the contractor on the construction criteria for rural roads to meet their required level of service. They are inserted in contracts to ensure uniformity and Quality Control of the works, facilitate measurements, define acceptable tolerances and define units of payments. Specifications should be appropriate to the local engineering environment and are essential if the use of locally available, but potentially nonstandard, pavement construction materials are to be used.

There are two basic types of process specifications that can be used in contracts:

1. 'Procedural' (or 'method') specification, where the employer defines details of how the work is to be carried out.
2. 'Functional' (or 'end-product') specification, where the employer defines the result to be achieved by the contractor in terms of a functional or performance requirement.

Procedural specifications have been used traditionally for road works. These reflect the high degree of competence of road administrations and are relatively easy to specify and to measure. Functional specifications are used increasingly for road maintenance works contracts, where the amount of supervision otherwise required can present a problem for employer organisations (SANRAL 2013b).

Defining performance standards in functional terms, such as required road surface condition, means that supervision requirements are minimised, since it is only necessary to test the end result. Contractors can then determine the most appropriate way to meet the performance requirement that maximises the use of their own particular skills, equipment and use of materials. This approach also encourages contractor innovation. However, contractor inexperience is a limiting factor on the use of functional specifications for rural road construction or rehabilitation in many LMICs. Methods for determining pavement end-result performance are discussed in Chapter 19.

Some new procedural options are likely to be best controlled by a tightly overseen method specification approach. This is particularly true of operations where control testing may involve significant delays, e.g. concrete surfaces and lime or cement stabilisation.

THE BIDDING PROCESS

Prequalification

In some cases, where tenders involve complex and large-scale works, it is useful to pre-qualify contractors and to only allow contractors that can demonstrate adequate capacity to tender for the works. This reduces the number of tenders received that need to be evaluated and avoids unnecessary tender expenditure by contractors who may not have adequate capacity to carry out the works.

Interested contractors can respond by submitting information, such as:

- Past experience.
- Previous employers.

- Present labour force.
- Present plant and equipment.
- Current and known future commitments.
- Financial strength.

The number of contractors seeking prequalification can be very high. However, by a process of evaluating the information submitted, seeking further clarification and making other investigations, a short list can be compiled of the contractors who have the requisite qualifications for the job.

Bid submission

For ICB road projects, the main tender period is normally around 3 months. If, during this period, the tenderers are in doubt about any aspect of the project or the tender documents, clarification must be sought in writing from the engineer. Alternatively, the matter should be brought up at the tender meeting normally arranged by the engineer and the employer for all tenderers. Part of the tendering process includes a site inspection, during which all prospective tenderers are informed of all aspects of the project that may be relevant for the preparation of their offers.

The tenders must normally be delivered in sealed envelopes to the employer by the date and time stated in the instructions to tenderers. It is common practice that all tenders are opened immediately after the deadline in the presence of those tenderers who wish to attend.

Evaluation and appointment

The tender evaluation must be carried out to ensure fairness and transparency. It should follow a process that ensures that all the important aspects of the various offers received are properly evaluated. This typically involves an evaluation of all of the elements and issues within the offers received, and a preparation of a report on the findings. The owner or more likely an appointed Engineer will scrutinise the tenders received and prepare a tender evaluation report for the employer. This task comprises:

- Evaluating the work programme, construction methods, and proposed plant and equipment.
- Checking whether there are any unacceptable reservations or conditions.
- Evaluating alternative tenders, if any.
- Checking the amount of subcontracting and qualifications of subcontractors.
- Evaluating foreign currency requirements (if applicable).

- Checking for unbalanced tenders; for example, prices for early work items that are too high.
- Checking for arithmetical mistakes in the priced bill of quantities.
- Comparing tender sum with the Engineer's cost estimate.
- Reviewing the competency of tenderers.
- Undertaking interviews.
- Developmental aspect evaluation.
- Drafting conclusions and recommendations on selection.

Since a considerable period of time may have elapsed since prequalification, it is often advisable, before a final recommendation is made, to review the two or three best-placed tenderers, especially in terms of their financial strength and work commitments. If no particular problems are anticipated, acceptance of the lowest tender is almost always recommended. Exceptions to this could occur if the lowest tender is much less than other bids or the Engineer's estimate. In such a case, great caution should be exercised, as experience has shown that a loss-making contract is bound to be fraught with problems. Caution needs also to be exercised if the proposed contracting staff fail to meet specific experience requirements for complex construction works, for example large bridges or high cut-slope support.

After the employer has accepted the engineer's tender evaluation recommendation, and possibly after negotiation with the winning tenderer, a letter of acceptance is issued. This letter advises the contractor that the tender has been accepted and that an invitation will be issued to sign a contract with the employer. At the time of signing, the contractor is normally required to produce a bank guarantee, normally amounting to 5%–10% of the tender sum, as security for performance of obligations under the contract. Following the signing of the contract, the engineer will issue an order to commence work.

REFERENCES

ADB. 2016. *User's Guide to Procurement of Goods: Standard Bidding Document.* Philippines. www.adb.org/sites/default/files/institutional-document/32829/sbd-goods-users-guide.pdf.

Cook, J. R., R. C. Petts and J. Rolt. 2013. Low volume rural road surfacing and pavements: A guide to good practice. AFCAP report GEN/099, Crown Agents, UK. https://www.gov.uk/research-for-development-outputs/low-volume-rural-road-surfacing-and-pavements-a-guide-to-good-practice.

FIDIC. 1992. *Conditions Contract for Works of Civil Engineering Construction.* 4th edition reprinted with editorial amendments. Lausanne: Fédération Internationale des Ingénieurs- Conseils.

FIDIC. 1999. *Conditions of Contract for Construction.* Lausanne: Fédération Internationale des Ingénieurs-Conseils.

FIDIC. 2005. *Conditions of Contract for Works of Civil Engineering Construction.* MDB harmonised edition. Lausanne: Fédération Internationale des Ingénieurs-Conseils.

Gericke, B., T. Henning and I. Greenwood. 2014. A guide to delivering good asset management in the road sector through performance based contracting. World Bank TP-42B. https://openknowledge.worldbank.org/bitstream/handle/10986/18646/878270NWP0TP4200Box377314B00PUBLIC0.pdf?sequence=1&isAllowed=y.

Mikkelsen, T. and T. Waage. 2004. Chapter 16 Contracts and works procurement. In *Road Engineering for Development.* 2nd edition. Robinson R. and Thagesen B. Eds. Taylor and Francis. ISBN 0-203-30198-6.

SANRAL. 2013a. Chapter 11 Documentation and tendering. In *South African Pavement Engineering Manual.* South African National Road Agency Ltd.

SANRAL. 2013b. Chapter 12 Construction equipment and method guidelines. In *South African Pavement Engineering Manual.* South African National Road Agency Ltd.

Silva, M. M. and G. Liautaud. 2011. Performance-based road rehabilitation and maintenance contracts (CREMA) in Argentina. A review of fifteen years of experience (1996–2010). World Bank transport paper TP36. https://documents1.worldbank.org/curated/en/241151468219001625/pdf/815970NWP0TP0300Box379836B00PUBLIC0.pdf.

Chapter 19

Construction supervision and quality management

INTRODUCTION

Poor Quality Control of construction is frequently an issue when trying to achieve sustainable new-construction or rehabilitation of rural road networks in developing regions. Experience backed by recent research has clearly indicated that, irrespective of design quality, poor construction leads to poor road performance, and hence supervision and Quality Control are essential elements in the road cycle and must be given a high priority (Intech-TRL 2007a, b, Ministry of Rural Development India 2007a,b). Rural road projects are often small in scope and can be widely scattered in remote areas, and this contributes to the construction process in the rural road sector frequently not being well-controlled.

To ensure the quality of works, it is necessary to establish oversight and control over the contractor's workmanship and materials. At the same time, it is necessary that, while developing a suitable Quality Management System for both construction and maintenance work, local constraints are kept in mind. The methods of supervision and types of Quality Control tests and their frequency have also to be judiciously selected to be achievable under the prevailing conditions.

This chapter provides practical guidance on construction supervision and associated Quality Control processes including recommendations on Technical Audits with a specific focus on Low Volume Rural Roads (LVRRs). The comments and recommendations in the chapter are equally applicable to maintenance activities, particularly periodic maintenance; this is covered further in Chapter 21, Road Maintenance Procedures.

QUALITY MANAGEMENT

The framework

Rural road construction should be implemented within a clear Quality Management framework that comprises Quality Control undertaken by the Contractor and Quality Assurance) undertaken by the Client (Road Authority) or an appointed Consultant acting as the Engineer.

Quality Control is undertaken by the Contractor in line with a submitted Quality Plan. Quality Control is largely concerned with measuring quantities and confirming that specifications have been met satisfactorily throughout construction. This will involve the use of in situ and laboratory testing, where appropriate. An initial step in good practice for Quality Control is the development and use of a checklist for monitoring and inspecting the construction activities. Checklists must be specific to the project in question, but, in general, should deal with the following issues:

- Quality of as-delivered materials.
- Construction methodology.
- Asset geometry (e.g. pavement thickness, earthwork slope angle, bridge deck/supports).
- Built asset visual assessment.
- Sampling and laboratory testing (e.g. soils and materials, concrete cubes or cylinders).
- In situ testing.
- Health and safety issues.
- Environmental impact compliance.

It is important that the supervision organisation is already set up and functioning when construction work is started. Information on the Quality Plan and associated responsibilities must be defined. Preparations should include a clear organisation plan with lines of authority and delineation of responsibilities. The number of the staff required for Quality Control and Quality Assurance will depend on the size and complexity of the project.

Quality Control and the Quality Plan

To meet Quality Control responsibilities, the Quality Plan should be compiled and implemented by the contractor and should clearly establish the activities that are proposed, particularly in relation to their management and technical control. The Quality Plan should comprise a written document which is reviewed and approved by the Client or supervising Consultant following contract award and prior to construction mobilisation. This document must clearly demonstrate how the contractor will control the construction processes in order to meet the requirements set out in the contract and its technical specifications. The Quality Plan will include the sequence, frequency and quantity of tests, observations and measurements to be performed during construction, together with a description as to how they will be reported.

Quality Assurance

Quality Assurance (QA) is commonly undertaken by the Client or his appointed Engineer with the aim of checking whether or not the contractor is adhering to his Quality Plan and whether or not the road works are being undertaken to a satisfactory standard. It incorporates standard procedures and methodologies, and applies to all site activities aimed at significantly reducing or eliminating, non-conformance, preferably before it occurs. QA activities are determined before construction work begins and are performed throughout construction.

A number of key QA decisions need to be taken immediately prior to and during the construction phase; these include:

- Approval of construction plant.
- Acceptance of material.
- Acceptance of pavement layers.
- Acceptance of construction procedures.
- Modification to design.
- Variations in BoQ items.

SUPERVISION ACTIVITIES

Aims and responsibilities

Initially the responsibility rests with the contractor to produce work to the requirements of the contract specification and drawings, and to implement a Quality Control system by providing sufficient experienced personnel and equipment to ensure adequate control of his works. Under most contractual arrangements, the contractor is required to test and check materials, products and completed road assets for compliance with the specified requirements, then submit the results to the designated Engineer for approval.

In most contract models the Supervision Engineer or Supervision Consultant plays a key Quality Assurance role who, in addition to reviewing the contractor's Quality Control submissions, also undertakes site supervision activities.

Key supervision activities

Whatever the contractual model, there are a key number of supervision activities required to ensure the quality of the completed road works. These are summarised in Table 19.1, with some key issues illustrated in Figure 19.1.

In reviewing construction plant resources, it is acknowledged that small rural contractors may have limited plant resources; for example, they may rely heavily on standard 8–10 tonne, three-wheel, static rollers for

Table 19.1 Key Supervision Activities

Construction activity	Key supervision issues for rural roads
Equipment	Check on availability and condition of construction equipment either specified or otherwise essential for satisfactory construction of the works. Common areas of concern: compaction plant; bitumen heating plant for chip seals (see Figure 19.1a and b); quarry processing equipment; haulage trucks.
Pavement surfacing	Choice of equipment, choice of material type, visual assessment and measurement of application rates of bitumen and aggregate are of greatest importance for sealed roads. For concrete roads the slump test (Figure 19.1c) is an important back-up to visual assessment (including assessing the curing process), allied to sampling for laboratory testing of concrete strength (cubes, or cylinders).
Pavement structural layers	Layer thickness and compaction (Figure 19.1d). Method specifications: appropriate choice of method and equipment, visual assessment and proof rolling, in combination with regular testing for 'calibration'. End product specification: Visual assessment, in situ testing. Use of DCP for compaction control is advantageous.
Bridge works	Concrete strength. Foundation conditions. Visual assessment, slump test and concrete cube/cylinder strength are of greatest importance.
Sub-grade and earthwork	End product specification: condition, thickness, in situ density. Visual assessment and laboratory tests. Method specifications: visual control, proof rolling and appropriate choice of method and equipment is normally sufficient for site control of workmanship.
Drainage	Visual assessment/measurement to check it meets design requirements. Laboratory tests on materials in advance and after construction combined with indicative tests and/or observations during construction are essential.
Earthwork slopes	Visual assessment of construction method and final geometry. Link to drainage supervision.
Environmental impact	Visual assessment of compliance with environmental requirements as laid out in the environmental management plan. Common problems include the side-casting of uncompacted spoil, non-re-instatement of borrow areas and pollution of water courses.
Health and safety	Visual assessment of health and safety hazards including issues around potentially dangerous equipment, lack of safety equipment and unsafe construction practices, including the safe provision of road diversions for local communities.

(a)

(b)

(c)

(d)

Figure 19.1 Some typical supervision issues. (a) Appropriate application of bitumen by use of heater and lance for chip seal. (b) Unacceptable uncontrolled and unsafe bitumen heating. (c) Check on acceptable mix design by measuring slump in concrete slump test. (d) Inspection pit dug to measure suspect base and sub-base layer thicknesses.

compaction, which may place challenges in the way of the effective construction of some pavement options.

Visual inspection is a key supervision tool for the detection of any deviation from the specified requirements. It is, for example, an essential element of pavement layer approval, particularly in the immediate identification of oversized particles in lower pavement layers or gravel wearing course. Physical measurements of thickness, widths and crossfall are also an essential element of this assessment. Visual inspection is supplemented by simple in situ checking of procedures, for example, temperature of bitumen and spray rates, or concrete slump. Specialist site inspection will be required for some construction operations such as bridge sites or for bioengineering protection of earthwork slopes. Standard inspection sheets are a useful aid to this process.

Date and time-stamped photographs are an important part of quality assurance, particularly if local (non-professional) community or NGO staff are involved in supervision (Intech-TRL 2007b).

Laboratory and in situ testing

Laboratory and in situ testing are important elements of rural road construction and maintenance supervision. Typical tests or actions associated with site inspection are summarised in Table 19.2.

Special testing may be required for specific pavement options, for example, cement or lime content in stabilised materials, crushing strength of bricks or the compressive strength of stone blocks.

The quantity or spacing of quality testing will be a function of the type and size of the asset. Table 19.3 summarises some typical examples from rural road construction projects.

Existing central or regional approved laboratories may be used by the contractor for testing. Alternatively, the contractor may set up and maintain a field laboratory for routine tests for Quality Control required to be conducted on a day-to-day basis. On larger rural road programmes, the Supervision Engineer may conduct QA tests on selected samples in

Table 19.2 Supervision Testing

Procedure	Description
The DCP test	May be used as a rapid control on quality as construction proceeds. Penetration rates may be used directly as quality check on the CBR of already constructed layers. The DCP test may be undertaken in conjunction with in situ density testing and moisture content testing for correlation purposes (Intech-TRL 2007b).
Sand replacement density test: ASTM D1556.	This is a common requirement in specifications. It may be replaced in some cases for Quality Control purpose by the DCP test, but only after satisfactory correlations have been established for the specific constituent materials.
Concrete slump test: ASTM C143.	This is an essential on-site test for supervisors to use as a general control of the concrete mix being produced. Addition of excess water in the concrete mix is a common malpractice (high slump). Concrete samples should be taken from the mixer at the specified intervals for slump tests as well as concrete cube testing.
Tray test	Tests for bitumen and chipping spray rates are an essential element in the control of thin bituminous surfacings for either machine- or labour-based operations (Roughton 2012).
Schmidt rebound hammer: ASTM D5873.	This is a non-destructive index test used to indicate the strength of placed concrete. A very useful tool to give an indication of poor-quality concrete in bridges and culverts.

Table 19.3 Typical Supervision Test Spacing

Construction	Activity	Typical frequency
Embankment	Soil index and grading; MDD, OMC and CBR. In situ density/set of 3 DCP.	One set per 5,000 m³ with a minimum of two analysis per cut or borrow area, or at every change of material. Minimum one test per 500 m of length.
Imported sub-grade	Soil index and grading; MDD, OMC and CBR. In situ density/set of 3 DCPs.	One set per 4,000 m³ with a minimum of two analyses per cut or borrow area, or at every change of material. Minimum one test per 250 m of length.
Granular sub-base-base	Soil index and grading; MDD, OMC and CBR. In situ density/set of 3 DCP.	One set per 4,000 m³ with a minimum of two analysis per cut or borrow area, or at every change of material. Minimum one test per 200 m of length.
Surfacing aggregate	Water absorption, Los Angeles Abrasion (LAA), grading. Bitumen-stone adhesion.	One set per 1.0 km, more frequently if material character changes. One set per source of stone/bitumen.
Concrete/ clay bricks or blocks	Density, compressive strength, water absorption.	Two sets of five per material source (more frequently if material character changes) and one per km of road.
Stone macadam	Uni-axial compressive strength, LAA, sodium sulphate soundness, grading/max size.	One set per 500 m with a minimum of two analysis per quarry area or at every change of material.
Stone, cobbles, setts	Uni-axial compressive strength, LAA, sodium sulphate soundness. Size dimensions	Two sets of five per material source (more frequently if material character changes) and one per km of road.
Block/brick joint sands	Grading, sand equivalent.	One per material source (more frequently if material character changes) and one per km of road.
Concrete pavement	Aggregate properties, such as LAA, water absorption, particle size and shape, sodium sulphate soundness.	One set per 1 km or every 1,000 m³ or change of quarry (more frequently, if material character changes).
	Concrete cube strength.	One set of three cubes to be crushed at 7 days and one set of three cubes to be crushed at 28 days for mix design per materials source. One set of three cubes to be crushed at 7 days and one set of three cubes to be crushed at 28 days per 500 m of pavement
	Concrete slump test.	One test per concrete batch; minimum one test per work shift.

Note: If a visual inspection shows up a suspected weak spot, additional appropriate testing should be done.

a separate laboratory. All laboratories should be nationally licensed and accredited for the tests required by the project. Laboratories that are certified to ISO 9000 are NOT necessarily accredited. ISO 9000 relates only to Quality Management Systems, and there is no specific requirement to evaluate the technical competence of a laboratory.

Portable field test kits have been developed that are suitable for testing of rural roads and provide the simple equipment for basic control tests (Giummarra 2009).

Specifications should include requirements for aftercare, such as curing of concrete or stabilisation layers, or remedial work on minor defects such as aggregate loss or bleeding of bitumen seals. 'Aftercare' issues are an integral part of the construction process, and it is important that supervisors ensure that these requirements are adhered to.

Approving material use

Because of the focus on the use of local, sometimes marginal construction materials, supervision approval is central to the issue of rural road quality and in this context Quality Management should be normally undertaken in two distinct phases:

1. General approval of source materials.
2. Approval of materials as delivered to site.

The final approval of construction materials must be on the basis of the materials as-delivered on site. It is not unusual for delivered materials to have significantly different characteristics from those approved at source during planning and design stages (Roughton International 2000).

Stockpiling forms an important part of effective materials' Quality Control by allowing appropriate selection, testing and approval of materials well in advance of their use in construction. Combining stockpiling with mixing also allows opportunities for blending materials and providing more consistent quality.

The Observational Method

When prediction of geotechnical behaviour is difficult, for example, in mountainous terrain, it can be appropriate to apply an approach known as 'the Observational Method' (OM), in which the design is reviewed and, if required, amended during construction (Peck 1969). The procedure is based around the concept of adjusting an original base design in the light of the actual ground conditions observed or monitored during the construction process.

An effective application of the OM puts additional pressure on site supervision and requires a pre-planned monitoring strategy, comprising an appropriate choice of variables to monitor the set-up of a reliable observational system and the choice of criteria to evaluate the monitoring results. It also needs a real-time analysis of the observations and the planning of alternative construction strategies to be adopted, depending on the results of the data analysis. Consequent design changes can be implemented either through variation orders or an agreed contractual approval process.

This OM approach can be applied to rural road projects with significant cut and fill sections but with limited investigation budgets. Supervision, or monitoring, and assessment of as-exposed ground conditions are central to the effective application of the OM. Key supervision issues are likely to be:

- Identify exposed soil–rock profiles.
- Define soil–rock boundaries and layer thicknesses, and check if they differ from design assumptions.
- Identify any signs of failure – including up-slope and down-slope of excavations.
- Undertake detailed discontinuity scan surveys where bedrock exposed.
- Focus on high-risk slopes or slopes where movement has been noted.
- Measurement of ground water levels from piezometers at different levels.

TECHNICAL AUDITS

Objective

The use of a formal Technical Audit allows road authorities to identify whether the parties involved in the contract are giving the client value-for-money. Technical Audits should be over and above the standard quality assurance procedures. They should be undertaken by a person, or organisation, independent from the contractor or supervising engineer.

Procedures for Technical Audits

Project Technical Audits may be carried out in a number of phases, although small projects may just have one Final Audit only. Phases are normally:

1. An initial audit: usually shortly after construction starts or when work is 10%–20% complete.
2. An intermediate Audit when construction is approximately 50% complete.
3. A final Audit when construction is complete.

Relevant advice on the various actions for each Technical Audit phase are summarised in Table 19.4 (MWTC, Botswana 2001).

Table 19.4 Technical Audit Actions Relevant to Pavement or Surfacing

Phase	Actions
Initial audit	• Review the contractor's project management procedures. • Assess capacity of site staff. • Assess the quality of any work already completed. • Review the quality and appropriateness of the plant and equipment. • Assess operator skills. • Review methods of working. • Check materials and water supply. • Review site organisation and site management. • Review quality and detail of the construction programme. • Check site safety.
Interim audit	• Review the initial audit and the subsequent actions. • Review of construction records and minutes of meetings. • Check both the completed work and work in progress. • Assess construction methods. • Review progress against the programme. • Check the current costs against the budget price. • Check measurement records. • Check materials on site. • Check that all payments to the contractor have been made. • Check any variation orders still appropriate.
Final audit	• Check contractor's construction/completion report. • Assess performance of the road to date. • Assessment of construction records. • Check materials as per specification. • Confirm (or not) that construction has been as per specifications. • Assess overall construction quality. • Check drainage construction as per design. • Construction of pavement layers and shape.

As-built survey

An as-built survey may be either part of the Technical Audit process or as a component of the supervision process. Either way, this is an important action leading to the collation of the as-built records that form the base level of knowledge for the future operational management, maintenance and potential eventual upgrade of the road.

Quality Control and assurance records will form a key part of the as-built knowledge base and may broadly be divided into the following four categories:

1. Historical records, that is, work programmes and monitoring data, weather data, resident engineer's diary, and daily inspection records.
2. Quality records, that is, test results, survey control, etc.

3. Quantity records, that is, measurements for payment, monthly statements, payment certificates and variation orders.
4. 'As-built' drawings and descriptions of all completed parts of the project.

On projects where a significant amount of climate strengthening has taken place, it is recommended that consideration be given to a longer-term monitoring of the effectiveness of the resilience measures in terms of performance against known climatic events. These will provide valuable feed-back for future lessons to be applied to climate change design issues.

REFERENCES

Giummarra, G. 2009. Development of a road base test kit. Australian Road Research Board. *Road and Transport Research* 18(1):5–92.

Intech-TRL. 2007a. Rural road surfacing research. RRST construction guidelines. Report for UKAID-DFID and Ministry of Transport, Vietnam. https://www.research-4cap.org/ral/IntechTRL-Vietnam-2007-RRST+Construction+Guidelines-SEACAP1-v070910.pdf.

Intech-TRL. 2007b. Rural road surfacing research. RRST pavement & surfacing condition monitoring. Report for UKAID-DFID and Ministry of Transport, Vietnam. https://www.research4cap.org/ral/IntechTRL-Vietnam-2007-RRST+Monitoring+Guidelines-SEACAP1-v071002.pdf.

Ministry of Rural Development, India. 2007a. *Quality Assurance Handbook for Rural Roads, Volume-I Quality Management System and Quality Control Requirements*. National Rural Roads Development Agency, Government of India. https://www.india.gov.in/download-e-book-ministry-rural-development.

Ministry of Rural Development, India. 2007b. *Quality Assurance Handbook for Rural Roads, Volume-II Equipment and Test Procedures*. National Rural Roads Development Agency, Government of India. https://www.india.gov.in/download-e-book-ministry-rural-development.

MWTC. 2001. *Technical Auditing of Road Projects. Technical Guideline no. 7*. Ministry of Works, Transport & Communications, Botswana.

Peck, R. B. 1969. Advantages and limitations of the observational method in applied soil mechanics. Ninth Rankine Lecture. *Geotechnique* 19(2):171–187.

Roughton International. 2000. Guidelines on borrow pit management for low-cost Roads. DFID KaR project report (Ref. R6852). https://assets.publishing.service.gov.uk/media/57a08d7e40f0b649740018ba/R68524.pdf.

Roughton International. 2012. Best practice manual for thin bituminous surfacings. AFCAP report no. ETH/076/A.

Chapter 20

Road maintenance strategy

INTRODUCTION

Road maintenance is fundamental to sustaining hard-won improvements to LMIC road networks. Despite this, it continues to be poorly delivered, particularly at rural road levels, as most LMICs road maintenance expenditures are generally well below the levels needed to keep the road network in a satisfactory long-term condition. This has led to a maintenance crisis in many countries where there is now a backlog of roads in poor condition (Geddes and Gongera 2016).

From the moment that a road is constructed or upgraded, it begins to deteriorate due to the impacts of weather and traffic as well anthropogenic activities such as inappropriate land-use change, poorly coordinated irrigation, and rubbish dumping. Maintenance is required to preserve road condition to be close to the as-constructed state (whether newly built or upgraded) and be fit for its designated purposes over its design life. Failure to carry out effective maintenance results in continued deterioration making access increasingly difficult and expensive for road users through increased trip-time and damage to vehicles. Roads will eventually become impassable for part, or all, of the year and require costly rehabilitation to make them passable again (PIARC 2013; Salomonsen and Diachok 2015).

In summary, road maintenance plays a key role in the sustainable rural access continuum, which runs from the planning of access provision through to its effective use for the transport of people and freight, as shown in Figure 20.1.

The overall asset management process comprises a number of issues that will have effects on the road network; including.

- Maintenance of network service.
- Socio-economic factors.
- Upgrade of assets within a network.
- Road user costs.
- Road safety.
- Environmental change or degradation.
- Road administration costs.

DOI: 10.1201/9780429173271-20

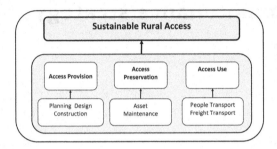

Figure 20.1 The provision of access continuum.

This chapter is primarily focussed on issues associated with rural road maintenance management, while at the same time acknowledging the wider context of road asset management (TRL 2003).

RURAL ROAD MAINTENANCE MANAGEMENT

Aims

In simple terms, maintenance management aims to mobilise the correct resources to the right place on the road network, to carry out suitable maintenance at the right time (Robinson 2004). Road infrastructure asset management should be based on the following principles:

- Systematic and strategic approaches over a long-term period (5–10 years).
- Full consideration of stakeholder needs.
- Optimal allocation of resources.
- Consideration of life-cycle costing.
- Meeting performance requirements in the most efficient way.
- Proactive management of risks.

Challenges

Multiple reasons are commonly put forward as being responsible for poor road maintenance; these include

1. Lack of funding: caused by the weak economic situation in many LMICs, compounded by funds being diverted to capital projects, or poorly utilised.
2. Shortage of qualified staff: With the skill base tending to be in the areas of design and construction, rather than road operation and maintenance.

3. Lack of equipment: Partly related to the lack of funds, but also to poor equipment management practices.

4. Deficient institutional arrangements: Many rural road administrations have too many responsibilities; emphasis is often put on works execution, while maintenance management is often neglected.

While the above issues may have a significant impact on maintenance delivery, the underlying reason for poor maintenance can most frequently be assigned to lack of a clear asset or maintenance management policy that has political support at all levels from central government down to village groups. Established policy, strategy and, by implication, budget support, are the fundamental drivers for effective asset management. A primary activity, therefore, should be the establishment of such a fundable policy and an associated strategy.

ROAD MAINTENANCE POLICY

Policy as driver of maintenance

For a rural road network to effectively serve the national economic and social demands, it is essential that an appropriate national transport sector policy exists for the investment in and the preservation and management of the roads (Petts et al. 2008). This will enhance political and public understanding of the importance and vital function of roads and encourage coherence and consistency leading to improved overall cost efficiency and performance of the transport network.

Policy framework

The policy framework should include aspects of:

- Classification or categorisation of the road infrastructure assets.
- The legal status and ownership of each road category.
- Network performance and service-level objectives.
- Financing arrangements for road improvements and maintenance.
- Establishing overall standards and performance levels.
- Planning and prioritising asset preservation in comparison with new construction works.
- Enhancing overall transport connectivity and strengthening the links within the road hierarchy.
- Interfacing with inland water transport and other relevant transport modes.
- Highlighting social, gender, and vulnerable group issues.
- Environmental, climate resilience, and sustainability issues.
- Road safety and health access issues.

Associated strategy

The associated strategy should take account of the following issues:

1. Efficiently implementing maintenance, rehabilitation and construction works.
2. Quantification, condition assessment, and prioritisation of the assets.
3. Allocation of responsibilities at all stakeholder levels for managing the assets.
4. Application of appropriate technology and optimal use of available local resources.
5. Creating local efficient implementation systems appropriate in the local operational environment.
6. Monitoring standards and specifications.
7. Establishing and maintaining realistic costing systems and databases for planning and management decision making, taking into account the specific features of local operational environment.
8. Enhancing overall transport connectivity and strengthening the links within the road hierarchy.
9. Sector human resource and sector capacity development, financing and resourcing.
10. Inclusion of social, gender and vulnerable group issues.
11. Road use and traffic restriction (including axle load control) issues.
12. Monitoring of performance of public assets and investments.

(Petts et al. 2008; Salter et al. 2020).

THE MAINTENANCE ENVIRONMENT

Key factors

The concept of Maintenance Environments sitting within the overall Road Environment (see Chapter 6) can be applied to the challenges of local road networks to provide a logical basis for developing and applying relevant maintenance models, as illustrated in Figure 20.2.

Key network, resource and implementation factors that should guide selection and influence performance of maintenance models are summarised in Table 20.1.

Climate change and maintenance

Effective maintenance is a core requirement for the sustainable climate resilience of rural road network, with a particular focus on those activities relating to drainage and the control of surface water. Maintenance

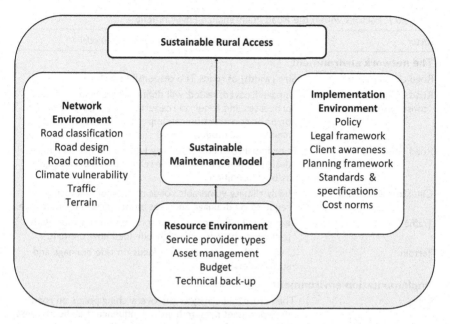

Figure 20.2 The road maintenance environment.

strategies and plans should include a specific focus on items that limit the impact from climate change.

Key actions with respect to maintenance and climate resilience may be summarised in the following points:

1. Incorporate climate change considerations into maintenance management goals and policies.
2. Assemble an inventory of all critical road assets that are susceptible to climate change impacts with an emphasis not only on pavements but also on earthworks, drainage, culverts and bridges.
3. Identify high-risk areas and highly vulnerable assets.
4. Use 'triggering' measures to identify when an asset or asset category has reached some critical level.
5. Monitor road asset conditions alongside actual weather conditions (e.g., temperature, precipitation, winds) to determine if climate change effects performance.
6. Incorporate climate change considerations into short- to long-range maintenance plans as part of Climate Change Action Plans (See Chapter 17).
7. Incorporate climate change into maintenance procedure guidelines.
8. Establish appropriate climate change mitigation strategies and agency responsibilities.

Table 20.1 Factors within the Road Maintenance Environment

Factor	*Implication for maintenance model*
The network environment	
Road classification	Size (width) of roads. The responsible authority.
Road design and construction	Unpaved, paved, sealed; will define maintenance processes and required resources. Inappropriate design or poor construction can impose severe strain on a maintenance model.
Road condition	Poor initial road condition may imply an unmaintainable designation and the need for rehabilitation back to an as-built condition.
Climate vulnerability	Highly climate vulnerable roads often implies a focus required on drainage and erosion protection maintenance.
Traffic	Traffic level influences rate of surface deterioration, with a particular challenge in a high axle load environment.
Terrain	Linked with climate implies a focus on side drainage and surface deterioration.
Implementation environment	
Policy	The nature and clarity of the overarching policy on road network asset management will influence the effectiveness of the procedures and their application.
Client awareness	Experience within the Client Authority in the selected mode of maintenance is essential.
Legal framework	The in-place legal instruments can either enhance or limit the adoption of a maintenance model.
Service provider capacity	Appropriate service provider experience is essential.
Planning framework	An effective medium to long-term planning framework enhances sustainability in maintenance programmes.
Standards/ specifications	Appropriate standards and specifications are required to define the activities and their service-level delivery.
Cost norms	Need to be aligned with the maintenance mode and targets.
Resource environment	
Available service providers	Availability of different groups will govern models, contractors, small enterprises, commune groups, special groups.
Asset management framework	An effective asset management framework enhances the options for model selection and their subsequent effectiveness.
Budgets	Appropriate sustainable budgets are a fundamental requirement, and limitations will constrain the extent of the activities. Budget preparation is essential for securing political, management and financial support.
Technical support and QA	The level of technical support and the QA arrangements will impact of the quality of the maintenance delivery. Poor support will limit the options for local maintenance groups.

Figure 20.3 Differing Maintenance Environments. (a) Cobble paved road in steep terrain (Myanmar). (b) Gravel road on low embankment in flat terrain (Laos). (c) Concrete slab road in flat coastal terrain (Vietnam). (d) Bitumen chip seal road in rolling hill terrain (Tanzania).

9. Include appropriate climate change strategies into programme implementation.
10. Monitor the maintenance management system to ensure that it is effectively responding to climate change.
11. Information management systems should include recording of climatic and impact data for planning purposes.

Flexibility of response

The flexibility required in terms of human and construction plant resources required for maintenance models can be gauged by considering contrasting Maintenance Environments, as illustrated in Figure 20.3a–d and summarised in Table 20.2.

Table 20.2 Example of Typical Variability in Maintenance Requirements

Maintenance Environment	General maintenance issues	Figure
Mountain road with an unsealed cobblestone pavement.	Possibility of landslip clearances; large amounts of side drainage and culverts to clear, roadside tree vegetation trimming. Individual pavement, occasional cobble and matrix replacement. Frequent culverts. Warning signage.	20.3a
Unsealed gravel pavement on minor embankment.	Regular (yearly) pavement grading/shaping and periodic re-gravelling (every 4–8 years depending on environment). Limited side drainage, minor embankment repairs, limited vegetation clearance and occasional culverts.	20.3b
Coastal flood area with concrete slab pavement.	Limited pavement and side drainage maintenance. Erosion protection and flood relief culverts maintenance. Limited vegetation clearance. Flood warning signage and pavement marking.	20.3c
Bitumen sealed hill-road with some slope support measures.	Pavement pothole and edge repair. Specialist care of the support measures, including bioengineering options. Roadside vegetation trimming, culvert cleaning; warning signage and pavement marking.	20.3d

The key lesson is that there is a matrix of road environmental and road design criteria that should be taken into account when planning a maintenance programme. This lesson is further developed in detail when considering the appropriate maintenance contract models and resources in Chapter 21.

PRINCIPAL ACTIONS

Maintenance cycle and budgeting framework

Maintenance management activities can be carried out by following steps in a maintenance cycle, which in turns fits within the overall road cycle. The maintenance management cycle is illustrated in Figure 20.4 and the overall road cycle is as shown in Chapter 2, Figure 2.1. The aim is to prepare work programmes that can be carried over the coming year for the budget that is likely to be available.

To be fully effective and sustainable, the rolling yearly cycle itself should be set within an overall Medium Term Expenditure Framework (MTEF) that is an integral part of the overall transport sector budgeting tool for translating government development plans into affordable and prioritised medium-term (3–5 years) expenditure plans.

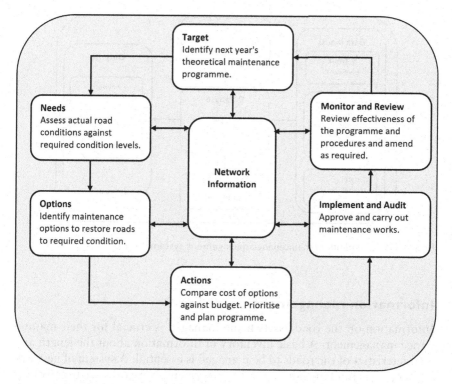

Figure 20.4 The yearly maintenance cycle.

Maintenance management system: Basic elements

The main objectives of a Road Maintenance Management System are to:

1. Provide a systematic and structured means of developing annual work programmes, resource requirements and budgets based on optimum economic standards, set within an MTEF.
2. Ensure an equitable distribution of funds over the country and enable priorities for allocations to be determined in a rational way when available funds are inadequate.
3. Authorise and schedule work.
4. Provide a system for monitoring the efficiency and effectiveness of maintenance works.

While the detail of such a system will be dependent of country or regionally specific factors, the key elements of a maintenance management system are summarised in Figure 20.5.

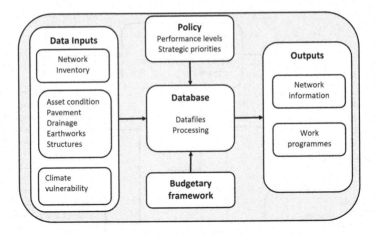

Figure 20.5   Elements of maintenance management system.

Information management

Information on the road assets being managed is crucial for their maintenance management. A basic inventory of information about the length and characteristics of the roads to be managed is essential. A system of network referencing is fundamental to road management. Networks are usually broken down into a series of links and sections, each defined by a unique label. The level of detail recorded in the inventory may depend on the importance of the road in the hierarchy. Generally, the inventory should be as simple as possible and not overloaded with unnecessary information. A rural road network inventory typically contains information such as the following:

1. Road sections: The length of each section in the network.
2. Cross-section: Width of the carriageway and shoulders, and whether side ditches are present.
3. Pavement: Type, thickness and, if possible, the age of the pavement.
4. Traffic level: Summary information of the volume of the various groups of motorised and non-motorised traffic using the road.
5. Alignment: Chainage of key points such as road junction, culverts, bridges, step sections and sharp curves.
6. Structures; Location, and size of major bridges, culverts and retaining walls.
7. Furniture: Information on road signs, guard rails, and other features.
8. Soils: Basic information about the soil type along the road (clay, sand, rock, etc.).
9. Materials: Chainage and offset of identified deposits of borrow pits, quarries.

10. Land use: Town, village, woods, cultivation, etc.
11. Climate: Basic data on current and likely future climate criteria in general summary format.

Condition data analysis

Maintenance requirement data may exist or be recovered in a combination of numerical, visual and descriptive formats, which poses a challenge to systematic interpretation for maintenance prioritisation. It is very useful, therefore, to transform the data sets into numeric code that can be easily manipulated and assessed.

The following sections outline as an example a typical numerical approach to assessing road condition. The system was developed for the monitoring and condition assessment of wide range of pavement types in S E Asia (Rolt and Cook 2007). Pavement condition assessment can be undertaken on sections of pavement using a range of physical criteria. Depending on the detail required, these sections may be 5–200 m in length. For each pavement group, a number of different criteria were identified to represent road performance (Table 20.3). These factors could be assessed using a numeric classification, and then be analysed to ascertain relative pavement

Table 20.3 Key Performance Indicators

Pavement group	Key indicative condition factors
Unsealed (Earth or GWC)	Erosion Ruts Potholes
Thin Bitumen Seals	Crack extent Ruts Potholes
Concrete slabs	Joint condition Crack extent Surface condition Potholes
Concrete geocells	Structural crack extent Cell condition Joint condition
Brick/blocks	Block condition Joint condition Ruts Potholes
Hand Packed Stone	Block condition Joint condition Depressions Potholes

deterioration and maintenance intervention level. Typical field data collection forms and codes are included in Appendix A: Standard Forms.

A Road Condition Deterioration Index (RCDI) was defined that could be calculated by combining the deterioration of the key factors as percentage of total deterioration (i.e. all numeric codes at their maximum defect values). The RCDI is then a percentage of the maximum deterioration possible for the key factors in a selected length.

The RCDI can be monitored, and the deterioration of pavements can be plotted versus time to provide information on when and where maintenance interventions are required. However, the RCDI only gives a measure of defect occurrence within a link but does not easily indicate whether this is an isolated badly deterioration issue or a widespread minor deterioration problem. To aid with this interpretation an associated Defect Extent Index (DEI) was developed, which is simply a measure of the percentage of the road affected by any deterioration, done by simply noting how many of the visual assessment blocks have a key defect. The combination of RCDI and DEI allows a rapid assessment for maintenance of deterioration seriousness and extent, for example,

- A high RCDI and high DEI indicates a widespread serious defect problem.
- High RCDI but low DEI indicates isolated serious defects.
- Low RCDI and high DEI indicates a minor widespread defect.

As a general guide it was assessed that deterioration should ideally be kept within 20% by maintenance and that a deterioration greater the 50% would require at least some rehabilitation. Typical output plots are shown in Figures 20.6 and 20.7. Figure 20.6 shows relative deterioration for three road lengths with differing paving options in the same high rainfall (3,000 mm/year) environment with no significant maintenance. Figure 20.7 shows differing plots for one option (concrete slab), including one for a specific indicator (inter-slab joint condition).

The plots would indicate that:

1. The gravel surface required re-gravelling maintenance after 1 year and after 2 years was in an unmaintainable condition. This emphasises the importance of assessing which roads in a network are sustainable and maintainable as they are, or whether they require rehabilitation or upgrade, in order to avoid ineffective and costly maintenance.
2. The cobble stone option required some maintenance after 5 years, but was still in a maintainable condition, without which the deterioration will rapidly escalate.
3. The concrete slab showed no major requirement for maintenance after 10 years, but the inter-slab joints (Figure 20.8a) ideally require regular

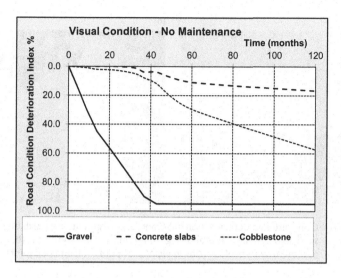

Figure 20.6 RCDI over time for different surfacing options.

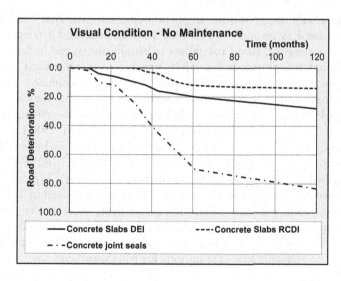

Figure 20.7 Different deterioration plots for one pavement option.

minor maintenance input every 2 years to prevent consequent general deterioration of the main slabs (Figure 20.8b).

4. So-called 'low-maintenance' paving options still require regular maintenance of apparently minor issues that, if not undertaken, could eventually lead to expensive failure.

(a)   (b)

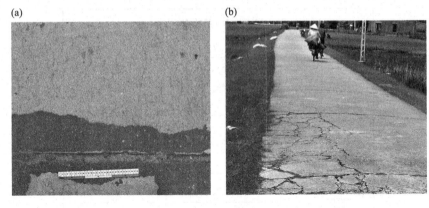

Figure 20.8   Concrete slab deterioration. (a) Concrete pavement, inter slab bituminous–sand joint eroded and allowing water ingress to sub-base. (b) Concrete pavement with broken slabs in vicinity of poorly maintained joints.

Low-cost roughness measurement

Pavement roughness is another important characteristic that can be monitored and used as an indicator of road performance and a triggering factor for maintenance. Road roughness is usually measured in terms of the International Roughness Index (IRI), which is a standardised roughness measurement, with recommended units: metres per kilometre (m/km or mm/m). It is a ratio of accumulated vertical motion of a vehicle or apparatus divided by the distance travelled during the test. Increasing IRI is an indication of deteriorating pavement surface (Sayers et al. 1986). However, this assessment does need to take account of the initial as-built IRI, given that some pavement options such as cobble surfacing (IRI 7–8) will of necessity be higher than a smooth bitumen seal (IRI 2–4). Typical relationships between IRI and travelling conditions are summarised in Table 20.4.

For main road network roughness assessment, there is the possibility of utilising automated profiling vehicles as part of a specialised pavement condition survey package. These are costly and are seldom applicable for LVRRs. The principle low-cost options for roughness assessment include:

- The Bump Integrator (BI)- either towed or in-vehicle
- The MERLIN (Machine for Evaluating Roughness using Low-cost INstrumentation).
- Visual assessment.

Robinson (2008) describes the Bump Integrator (BI) as a simple device for measuring roughness widely used throughout the developing world. BIs measure the sum of relative displacements between the axle and body

Table 20.4 Roughness-Road Condition Relationships

IRI (mm/m)	Typical road conditions
1.5 → 2.5	Ride comfortable at >120 km/h. No depressions or potholes. Typical high asphalt 1.4–2.3; high-quality bituminous surface treatment 2–3. Recently bladed surface of fine gravel or soil surface with excellent longitudinal and transverse profile.
3.5 → 4.5	Ride comfortable up to 100–120 km/h. Some surface defects with occasional depressions, or potholes. Bituminous surface showing some fretting, patches.
7.5 → 9.0	Ride comfortable up to 70–80 km/h but some sharp movements and some wheel bounce. Frequent shallow moderate depressions or shallow potholes. Moderate corrugations on unsealed surfaces. Areas of significant bituminous surfacing deterioration.
11.5 → 13.0	Ride comfortable at 50–60 km/h. Frequent moderate transverse depressions or occasional deep depressions or potholes. Strong corrugations on unsealed surfaces. Significant bituminous surfacing disintegration.
16.0 → 17.5	Ride comfortable at 30–40 km/h. Frequent deep transverse depressions and/or potholes or occasional very deep with other shallow depressions. Bituminous surface disintegrated.

of a running vehicle when installed in an ordinary passenger car, pick-up or within a proprietary small towed trailer. Details on their operation are summarised by Kumar et al. (2018).

Calibration of the BI and the vehicle are necessary, if the results are to be utilised for assessing maintenance needs. This enables values obtained from a response-type instrument to be converted to units of the International Roughness Index. A summary of a BI set-up and procedures is contained with TRL Overseas Road Note 18 (1999).

The MERLIN device works by transferring an expression of surface roughness onto a standard recording chart by means of the calibrated movement of a central foot and lever. The principal components of the MERLIN are as shown in Figure 20.9. Guidance on the operation of the MERLIN is detailed in Cundil (1996), and standard MERLIN chart is included with Appendix A. The MERLIN may be used to calibrate BI surveys. An example of the use of MERLIN to monitor performance of differing surfaces over time since construction is presented in Figure 20.10, using data from UKDFID-World Bank funded surfacing trials in Vietnam (World Bank 2014).

A visual assessment approach is sometime employed by Road Authorities to by-pass the challenges of budget and human resources required for the above options. In this approach a walkover or drive-over survey is used to assess general road condition in terms of potholes, surface erosion, corrugations, etc. and the assign an IRI figure to road section based on definition, such as summarised in Tables 20.5. This approach has significant inherent risks in terms of its susceptibility to operator error and bias, and the dangers of over-analysing subjective numbers.

(a)   (b)

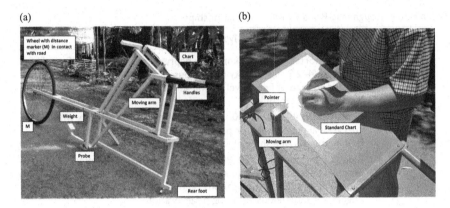

Figure 20.9   The MERLIN apparatus. (a) The overall apparatus. (b) The recording chart.

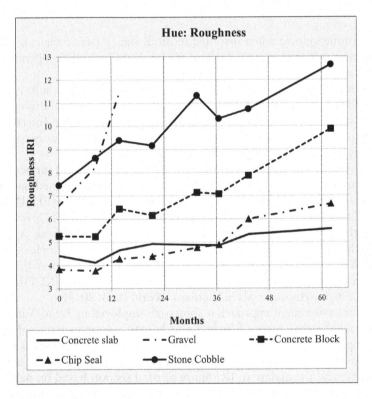

Figure 20.10   Monitoring variation of IRI roughness with time.

Table 20.5 Relationships between IRI and General Road Condition

Condition description	World Bank visual IRI for unpaved roads (mm/m)	World Bank visual IRI for paved roads (mm/m)
Very good	<6.0	<4.0
Good	6 0–11.0	4.0–7.0
Fair	11.0–15.0	7.0–8.0
Poor	15.0–19.0	8.0–11.0
Very poor	19.0–24.0	>11.0

High-tech approaches to rural road condition data collection

There are options to apply high-tech solutions to overcoming the challenges of acquiring road condition data for rural road maintenance management in LMICs. Workman (2018) undertook a review of potential high-tech options for this purpose and found the following grouped to have potential for use with low volume rural roads:
    Satellite imagery-related technologies:

*   Unmanned Aerial Vehicles (UAVs).
*   Smartphone applications, including apps to measure IRI.
*   Internet based solutions, such as: OpenStreetMap or Google Earth.

An associated guideline document (TRL 2017) contains specific guidance on a wide range of options within the above groups. In general, the benefits of using a remote or high-tech approaches to gathering network data over a traditional ground survey approaches were noted as being that they use less resources in terms of manpower and logistics, can be implemented more quickly over a very wide area, and reduce drastically the need to physically visit the roads being surveyed. The principal disadvantages could be summarised in terms of cost, limitations on the levels of detail and accuracy, and issues around the apparent differences in road condition between dry and wet seasons.

In summary, there are a number of high-tech solutions that have definite cost effective uses at the present time or increasing knowledge of rural road networks, for example UAVs and free-to-use mapping systems. While at the same time it should be noted that new technologies are being developed all the time and existing technologies are becoming more feasible economically, as competition grows and they become available to a wider market.

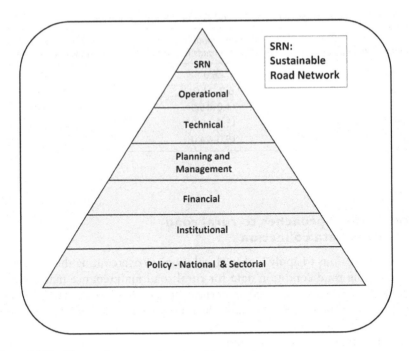

Figure 20.11 The road preservation pyramid.

MAINTENANCE MANAGEMENT PERFORMANCE

The road preservation pyramid

It has been recognised that the development and performance of maintenance management systems are dependent on a combination of inter-dependent external factors, internal institutional arrangements and technical capacity. The original concept of a maintenance sustainability pyramid has been developed over a number of years into its current form, as shown in Figure 20.11 (Brooks et al. 1989; Robinson 2008; RECAP 2019).

Good maintenance management practice needs to address all the building blocks in the pyramid to achieve affordable and sustainable preservation of the network roads. As an aid to this, an evaluation process has been developed that incorporates a structured methodology so that meaningful advances in road asset management can be monitored and improvement in rural road network performance realised (CDS 2019).

Key characteristics of the pyramid bocks can be summarised as follows:

Policy: This building block requires the existence of a management policy that is relevant to the rural transport sector, supported by senior decision makers and adopted at the highest level in government. Stakeholder engagement by the relevant authority requires informed consultation and open communications in order to understand stakeholder needs and expectations.

Institutional arrangements: This block is concerned with a range of issues contributing to the performance of the agency, including whether the relevant road authority:

- Has a corporate vision and mission statement which considers stakeholder needs and expectations?
- Has defined the basic levels of service for roads?
- Has organisational structure that identifies roles, responsibilities and competencies of key staff?
- Has a sustainable workforce structure and adequate human resources to staff the structure?
- Provides training opportunities for staff.

Funding: This block requires the existence of committed, adequate and sustainable funding for road maintenance. This includes whether an annual valuation is carried out of road infrastructure assets, a costing framework and a budgeting and programming processes is in place for prioritised maintenance plans.

Management and planning: This block requires the existence of appropriate maintenance management within an overall asset management capacity that contains network inventory definition, network condition data. This can facilitate the preparation of prioritised annual, medium and long-term maintenance plans.

Technical: This requires the existence of:

- An adequate road referencing system and inventory.
- A system for systematic and documented data collection for all principal road assets.
- Annual visual condition assessment surveys.
- Road link utilisation estimates and forecasts for criticality prioritisation.

Operational: This block is concerned with the efficiency of road authority operations including scheduling of maintenance, assessing procurement models, procurement of service providers and monitoring technical compliance of works undertaken.

Monitoring and audit

Recent UKDFID-funded research highlighted the value of the maintenance pyramid as basis for the assessment of road authority competence in managing maintenance (CDS 2019, 2020). Part of this approach has been to

Table 20.6 Levels of Maintenance Management

Stage	Level	Description of maintenance management level
0	Absent– very poor	No systems are in place, and the extent of the road assets or their condition is largely unknown. Maintenance planning for the network does not follow any prescribed criteria.
1	Limited– poor	There is an awareness of the potential benefits of adopting maintenance management approaches. However, they have not been adopted by the organisation and whatever is implemented is still largely through the effort of individuals using their own initiative.
2	Proficient– fair	The approaches are relatively well embedded in the authority, with a well-structured road inventory system. Road condition is regularly monitored, and the data are used to provide probable indications of amounts of funding required for maintenance works.
3	Advanced– good	The management system is well developed, and data are collected and analysed according to pre-set calendars. The results of the system are an influence on the implemented programmes and work procedures.
4	Mature– very good	All management policies and procedures are in place and are subject to continuous review and improvement. The results generated by the system dictate prioritised interventions on the network for all maintenance works.

develop a road administration specification that can be used as an assessment, or self-assessment, tool by rural road administrations. For a road authority, each performance block can be compared against a benchmark to target improvements in asset management either centrally or at district/province level. By understanding and working on the specification components, the road authority and sector stakeholders can monitor the improvement in asset management performance over time. Road authority maturity levels in developing maintenance can be characterised broadly as shown in Table 20.6.

REFERENCES

Brooks, D. M., R. Robinson, K. O'Sullivan. 1989. Priorities in improving road maintenance overseas: a checklist for project assessment. *Proceedings of the Institution of Civil Engineers, Part 1* 86:1129–1141.

Civil Design Solutions (CDS). 2019. Economic growth through effective road asset management (GEM). CDS final report for ReCAP-DFID Project No. GEN2018A. https://assets.publishing.service.gov.uk/media/5f9d91d1e90e07042c1e10b8/Geddesetal-CDS-2019-EffectiveGrowthThruRoadAssetManagement-FinalReport-AfCAP-GEN2018C-190829-compressed.pdf.

Civil Design Solutions (CDS). 2020. *Economic Growth through Effective Road Asset Management (GEM), Rural Road Asset Management Practitioners Guideline, SEACAP Project No. 10636A GEN2018A.* London: Recap for DFID. https://www.gov.uk/research-for-development-outputs/rural-road-asset-management-practitioners-guideline.

Cundil, M. A. 1996. *The Merlin Road Roughness Measuring Machine: User Guide. TRL Report No. 229.* The UK Overseas Development Administration (ODA). TRL. http://transport-links.com/wp-content/uploads/2019/11/1_268_TRL229_-_The_Merlin_Road_Roughness_Machine_-_User_Guide.pdf.

Geddes, R. and K. Gongera. 2016. Economic growth through effective road asset management. In *Conference on Transport and Road Research*, Mombasa, Kenya, 22 pp. https://www.gov.uk/research-for-development-outputs/economic-growth-through-effective-road-asset-management.

Kumar, V., L. S. Chowhan and R. S. Kumar. 2018. Pavement evaluation using bump integrator and MERLIN. *Indian J. Sci. Res. V.* 17(2). ISSN: 2250-0138(Online).

Petts, R. C., J. R. Cook and D. S. Salter. 2008. Key management issues for low volume rural roads in developing countries. Paper for INCOTALS 2008: South Asia Moves Forward. Colombo. 28 July.

PIARC. 2013. Best practices for the sustainable maintenance of rural roads in developing countries. Technical Committee A.4, World Road Association. Paris. https://www.piarc.org/en/order-library/ 19078-en-Best%20practices%20for%20the%20sustainable%20maintenance%20of%20rural%20roads%20in%20developing%20countries.

RECAP. 2019. Economic growth through effective road asset management (GEM), Final Report, Project No. 10636A GEN2018A, London: Recap for DFID. https://www.research4cap.org/index.php/resources/rural-access-library.

Robinson R. 2004. Maintenance management. Chapter 19. In *Road Engineering for Development,* Second edition, Robinson R. and Thagesen B., Eds. Taylor and Francis. ISBN 0-203-30198-6.

Rolt, J. and J. R. Cook. 2009. Mid term pavement condition monitoring of the Rural Road Surfacing Research. Final report. Ministry of Transport Vietnam Southeast Asia Community Access Programme (SEACAP 27) for UKAID-DFID. https://assets.publishing.service.gov.uk/media/57a08b6640f0b652dd000c58/SEACAP27-FinalReport.pdf.

Salomonsen, A. and M. Diachok. 2015. Operations and maintenance of rural Infrastructure in community-driven development and community-based projects. Lessons learned and case studies of good practice. World Bank Social, Urban, Rural & Resilience Global Practice. https://openknowledge.worldbank.org/handle/10986/22954.

Salter, D., R. C. Petts, J. R., Cook and S. Manibhandu. 2020. Key management issues for low volume rural roads in ADB Developing Member Countries. Asian Development Bank, Manila. Publication stock no. ARM200211. DOI: 10.22617/ARM200211.

Sayers, M. W., T. D. Gillespie and W. D. O. Paterson. 1986. Guidelines for conducting and calibrating road roughness measurements. World Bank technical paper no. 46, World Bank, Washington, D.C. https://documents1.worldbank.org/curated/en/851131468160775725/pdf/multi-page.pdf.

TRL. 1999. *Overseas Road Note 18. A Guide to Pavement Evaluation and Maintenance of Bitumen-Surfaced Roads in Tropical and Sub-tropical Countries.* Crowthorne: Transport Research Laboratory. https://www.gov.uk/

research-for-development-outputs/a-guide-to-the-pavement-evaluation-and-maintenance-of-bitumen-surfaced-roads-in-tropical-and-sub-tropical-countries-overseas-road-note-18.

TRL. 2003. *Overseas Road Note 20. Management of Rural Road Networks.* TRL for DFID. https://www.gov.uk/research-for-development-outputs/management-of-rural-road-networks-overseas-road-note-20.

TRL. 2017. *Guideline on the Use of High Tech Solutions for Road Network Inventory and Condition Analysis in Africa.* ReCAP report for UKAID-DFID. https://www.research4cap.org/index.php/resources/rural-access-library.

Workman, R. 2018. *The use of appropriate high-tech solutions for road network and condition analysis, with a focus on satellite imagery, TRL Ltd ReCAP for DFID.* https://www.gov.uk/research-for-development-outputs/the-use-of-appropriate-high-tech-solutions-for-road-network-and-condition-analysis.

World Bank. 2014. Improving Vietnam's sustainability. Key priorities for 2014 and beyond; Rural road pavement and surfacing design options. World Bank technical paper 91962, https://openknowledge.worldbank.org/handle/10986/20799.

Chapter 21

Road maintenance procedures

INTRODUCTION

Recognised good practice in road maintenance has now been collated and is well documented in national and international sector standard guidelines or included within road design manuals. Importantly, many of the national maintenance manuals are available in local languages.

The following is listing of some typical maintenance procedure manuals:

ADB (2011a): Community-Based Routine Maintenance of Roads by Women's Groups Manual for Maintenance Groups.

ADB (2011b): Performance-Based Routine Maintenance of Rural Roads by Maintenance Groups. Manual for Maintenance Groups.

Ethiopian Roads Authority (2016): Manual for Low Volume Roads. Part G Road Maintenance.

Larcher et al. (2010): Small Structures for Rural Roads Volume 2 Detailed Design, Construction and Maintenance.

PIARC (WRA) (1994–2006): International Road Maintenance Handbooks (4 Volumes).

TRL (1985): Overseas Road Note 2- maintenance techniques for district engineers.

TRL (1988): Overseas Road Note 7- Volume 1, A guide to bridge inspection and data systems for district engineers, Volume 2, Bridge Inspectors Handbook.

TRL (1999): Overseas Road Note 18. A guide to pavement evaluation and maintenance of bitumen-surfaced roads in tropical and sub-tropical countries. Crowthorne: Transport Research Laboratory.

TRL (2003): Overseas Road Note 20. Management of Rural Road Networks.

Key elements of this good practice are emphasised in this chapter in the context of the LMIC rural Road Environment. As distinct from purely technical issues, the procurement, contractual and implementation aspects of rural road maintenance have undergone some significant changes in the last

DOI: 10.1201/9780429173271-21

two decades, with a move away from in-house to 'force-account' activities to options involving contractors, small enterprises and local groups (ADB 2018). There has also been innovative development in terms of various forms of 'performance-based' contracting models for rural road maintenance. This chapter reviews and comments on the application of options for the implementation of road asset maintenance.

KEY RURAL ROAD MAINTENANCE ISSUES

Maintenance aims

The essence of road maintenance is to ensure that road networks and their constituent roads perform their required function for the duration of their design life. In most cases the principal focus is on the carriageway, and this can be summarised in three basic aims:

1. Keep the road surface in good condition (for example, the repair of ruts and potholes).
2. Maintain the road surface cross-sectional shape to shed water to the side of the road.
3. Maintain the drainage system to safely lead water away from, or across, the road.

Additionally, many roads will also require essential maintenance on earthworks and structures such as bridges and culverts or causeways.

Classification

There is a pragmatic primary division of rural roads within LMIC networks into maintainable and unmaintainable roads, where:

1. Maintainable: A road, or a section of road, in a good or fair, condition and serves the needs of the road users with only minor defects and inconvenience, which can be rectified through standard maintenance activities.
2. Unmaintainable: A road, or a section of road, is in an unmaintainable, or poor, condition if it falls significantly below its fitness-for-purpose performance level and does not adequately serve the needs of the road users because of major defects, which require rehabilitation actions rather than standard maintenance procedures.

Maintenance implementation activities themselves are usually grouped as outlined below.

Routine Maintenance (R): Maintenance activities that are normally required at least annually and are often specified on a repeated cycle. Routine maintenance does not have a high technical content and is basically a logistical challenge. It does not require significant engineering input (unlike construction, rehabilitation, or periodic maintenance). It can be managed as a simple operation at a local level, without the need for expensive survey and reporting systems This class of maintenance is sometime further divided into Routine 1 (R1) and Routine 2 (R2) activities, where:

1. R1 comprises unskilled activities, such as cutting grass and clearing drains, carried out mainly off-carriageway with no equipment other than appropriate hand tools, and
2. R2 comprises activities requiring some skill and limited equipment to undertake, for example, such activities as pothole filling, minor re-shaping and minor crack sealing.

Periodic Maintenance (P): Maintenance activities that are normally required to be undertaken on a 3–8 year cycle depending on the road type and its environment. These are activities, such as resealing, re-gravelling, that usually require the planned use of some equipment and significant technical skills, normally suitable for small and medium contractors.

Emergency Maintenance (E): Emergency, or disaster, maintenance may be required at a road location where extraordinary climate or seismic events, resulting in structural or earthwork failures, have significantly impacted the road and its fitness for purpose or raised major issues of safety.

These maintenance activities should be seen distinct from rehabilitation, which is normally specified in response to major road defects that are necessary to return a road to a maintainable condition.

Sequence of key actions

Figure 21.1 lays out the general sequence of practical actions in the delivery of maintenance interventions.

Key aspects of this figure are summarised below:

1. Input information: Questions that need to be addressed when planning maintenance interventions may be summarised in terms of what roads are being prioritised, what is the current condition and what is the target service condition. A combined deterioration index, such as that described in Chapter 20, is very useful for reporting the condition

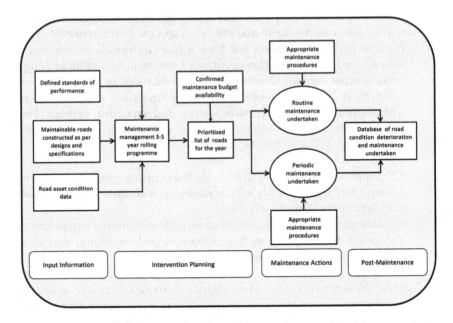

Figure 21.1 Key road maintenance activities.

of roads for prioritisation. It is important that the distress indicators of each road asset are reported, as different types of distress will require different treatments and resources.

2. Input information is then used in a planning process that should, ideally, involve a rolling 3–5 year programme.
3. Actual maintenance actions should be guided by country-specific guides or manuals that clearly define equipment and procedures.
4. Post maintenance activity should involve updating of a maintenance management database in terms of action, costs and effectiveness. This then feeds into the next round of maintenance planning to improve cost-effectiveness.

SPECIFIC MAINTENANCE ACTIONS

Carriageway and roadside

It is useful to consider the requirements for maintenance in relation to specific asset defects and required actions. Table 21.1 presents a summary of common on-carriageway defects and maintenance actions. Table 21.2 presents the same for off-carriageway assets. These defects and actions are also illustrated in Figures 21.2–21.9.

Table 21.1 On-Carriageway Defects and Actions

Defects/issues	Impact(s)	Maintenance action	Activity type
General: all pavement types			
Minor debris on pavement.	Safety issue.	Remove debris.	R1, E
Debris blocking carriageway.	Access impeded or cut. Safety issue. Figure 21.2a.	Remove debris. Check for damage to pavement and repair as appropriate (see below).	E
Carriageway washout.	Access cut. Safety issue.	Repair earthworks and pavement as appropriate (see below).	E
Vegetation on verges.	Safety issue. Impeding pavement drainage. Figure 21.2b.	Regular cutting and trimming.	R1
Road sign dirty.	Safety issue.	Regular clean.	R1
Road signs missing/broken.	Safety issue.	Repair/replace as required.	R2, P
Earth or gravel road			
Potholes in surface.	Degraded access, safety. Promotes further deterioration. Figure 21.3a.	Spot earth or gravel infill and compaction.	R1, R2
Shallow rutting on pavement.	Drainage impeded. Further deterioration.	Light grading, dragging, spot filling.	R2
Lack of camber.	Impedes drainage, increasing deterioration.	Light grading, dragging.	R2
Corrugations.	Degraded access, further deterioration, Figure 21.3b.	Light grading.	R2
Severe rutting/ loss of shape.	Severely degraded access and increasing deterioration, Figure 21.3c.	Earth/gravel infill and heavy grading/ compaction.	P
Loss of gravel.	Degraded access and increasing further deterioration.	Re-gravel, compaction, grade. Figure 21.3d.	P
Shoulder erosion.	Impeding pavement drainage.	Re-gravel, compaction, grade.	R2, P
Bituminous sealed pavement			
Potholes in pavement.	Degraded access, safety. Water ingress increasing deterioration, Figure 21.4a.	Spot base infill and re-seal. Base repair and resurface if severe.	R2, P P
Minor surface cracking.	Water ingress increasing deterioration.	Local sealing if minor.	R2

(continued)

Table 21.1 (Continued) On-Carriageway Defects and Actions

Defects/issues	Impact(s)	Maintenance action	Activity type
Severe rutting and/or cracking and potholing.	Degraded access, risk of loss of access, Figure 21.4b.	Repair/reshape base, local re-seal.	P
Edge fretting.	Further degradation, Figure 21.4c.	Trim and infill base edge and local re-seal.	R2, P
Severe cracking and/or loss of surfacing.	Degrading access leading to access failure. Figure 21.4d.	Re-surfacing/overlay; base repair as required.	P
Shoulder erosion.	Pavement erosion, impacts drainage and potential further deterioration.	Infill shoulder material and compact.	R1, R2
Bitumen bleeding.	Leads to surface deterioration.	Coarse sand spread and brush.	R1
Concrete slab			
Cracking. Figure 21.5a.	Water ingress weakened sub-base increasing deterioration.	Crack sealing.	R2
Severe cracking, block failure.	Degraded access, further deterioration, Figure 21.5b.	Replacement of part or all of block.	P
Inter-slab joint deterioration.	Water ingress increasing deterioration, Figure 21.5c.	Resealing with sand–bitumen.	R1, R2
Shoulder erosion.	Undercutting of slab, leading to slab crack and failure, Figure 21.5d.	Infill shoulder material and compact.	R1, R2
Clay brick or concrete block			
Brick/block broken or loose.	Degraded access, leading to further failure, Figure 21.6a.	Remove and replace block or brick and joint infill or mortar infill.	R1
Inter-block/brick infill.	Degraded access, leading to further failure.	Replace inter-block material and tamp down.	R1
Edge constraint damage.	Loosening of bricks/blocks and further failure, Figure 21.6b.	Repair or replace blocks/bricks or other edge restraint.	R1
Shoulder erosion.	Undercutting of pavement, leading to edge failure.	Infill shoulder material and compact.	R1
Cobble or hand-packed stone			
Broken stone.	Degraded access, leading to further failure.	Remove and replace stone and infill or mortar.	R1
Inter-stone infill erosion.	Degraded access, leading to further failure, Figure 21.7a.	Replace inter-stone material and tamp down.	R1
Edge constraint damage.	Loosening of stone and further failure.	Repair or replace edge restraint.	R1
Shoulder erosion.	Undercutting of pavement and failure. Figure 21.7b.	Infill shoulder material and compact.	R1

Table 21.2 Common Off-Carriageway Maintenance Requirements

Defects/issues	Impact(s)	Maintenance action	Activity type
Drainage			
Debris and vegetation in side-drain.	Blocked drainage, weakening of pavement, flooding, Figure 21.8.	Clean out debris and vegetation.	R I
Eroded side drains.	Impaired drainage, local flooding.	Trim drain, insert scour-checks. Local lining of drains.	R I
Blocked/missing mitre drains.	Impaired pavement drainage, pavement erosion. Weakening of pavement layer materials.	Clean out. Insert drains if missing.	R I
Missing side drains.	Impaired pavement drainage, weakening of pavement layer materials.	Insert either lined or unlined depending on gradient and material.	R I, R2, P
Earthworks			
Minor slope face erosion.	Further slope degradation and potential failure, drain blockage.	Reshape and trim face, bioengineering. Additional face drainage.	R I P
Major slope face erosion.	Slope degradation and potential failure. Access blockage. Figure 21.9a.	Reshape and trim face – check support requirement and improve if required.	P, E
Bioengineering degradation.	Slope erosion and potential failure.	Regular trimming and husbandry.	R I
Concrete/masonry retaining walls.	Slope failure and access loss.	Inspection and repair.	R I
Gabion walls...	Slope failure and access loss.	Inspection and repair. Figure 21.9b Improve if required.	R I P
Specialist support measures – netting, rock bolts, anchors, etc.	Slope failure and loss of access. Safety issue.	Inspection and specialist advice or input as required.	P, E

The following are some additional notes with reference to the listed maintenance activities.

1. Shape and drain: The maintenance of road cross-sectional shape and the clearing of drains are fundamental maintenance priorities for sustainable access on unsealed earth or gravel roads. Shape maintenance

Figure 21.2 Examples of general carriageway maintenance. (a) Debris blocking carriageway. Removal either labour-based and/or machine-based. (b) Vegetation on
verge; RI activity that is commonly community based.

Figure 21.3 Examples of unsealed carriageway maintenance. (a) Potholes leading
to further deterioration. Limited pothole repair RI; if widespread R2.
(b) Corrugations in gravel surfacing. Dragging or light grading required. (c)
Severe rutting, loss of shape and eventual loss of access. Periodic maintenance
required. (d) Spot re-gravelling to mitigate gravel loss. Usually, a periodic activity.

(a)

(b)

(c)

(d)

Figure 21.4 Examples of bitumen-sealed carriageway maintenance. (a) Scattered potholes. R2 maintenance can be done by local community if bitumen emulsion used. (b) Base failure. Periodic maintenance, but may need rehabilitation to replace and compact base and re-seal. (c) Edge fretting of seal. Trim, reshape shoulder and locally replace seal. R2 activity. (d) Widespread seal failure, some base deterioration. Scarify, reshape base and re-seal. Periodic maintenance activity.

may be carried out by means of dragging, as well as by light machine or towed grading. Dragging is usually carried out by towing a specially made drag behind an agricultural tractor. Figure 21.10 shows two different types of drag, one made from old truck or tractor tyres cut in half around their circumference and the other made from a steel rail. By adjusting the length of the towing cables on the steel rail some minor reshaping of the surface may be achieved.

2. The use of bitumen emulsion in low volume road seals allows R2-types activities such as crack sealing, pothole repair and surface patching to be safely undertaken by local community groups. Similar work undertaken using hot bitumen poses significant health and safety issues for these largely unskilled groups.

3. Maintenance of carriageway shoulders should be an essential aspect of carriageway maintenance. Erosion of shoulders has important consequences with regard to drainage of surface water, and the undercutting and failure of pavement edges.

Figure 21.5 Examples of concrete pavement maintenance. (a) Isolated slab crack. Requires R2 maintenance. Sealing with sand–bitumen mix. (b) Inter-connected cracks and broken slab. Requires break out followed by partial slab and possible sub-base replacement. (c) Eroded inter-slab joint erosion. Requires R2 sealing with sand–bitumen mix. (d) Eroded shoulder. Build up and compaction of shoulder required, possible R1 action.

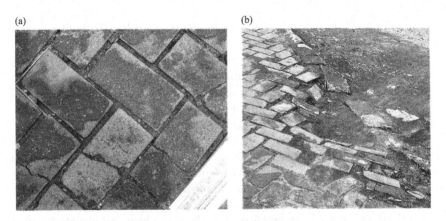

Figure 21.6 Examples of brick or block maintenance. (a) Broken blocks. R2 maintenance; block replacement and joint infill. (b) Damaged edge restraint to block pavement, replacement required. Routine maintenance.

(a)   (b)

Figure 21.7 Examples of stone block maintenance. (a) Eroded infill between stone setts. RI procedure to replace infill and brush in. (b) Eroded shoulder undercutting pavement. Build up and compaction of shoulder RI activity.

Figure 21.8 Debris in side-ditch.

(a)   (b)

Figure 21.9 Earthworks requiring maintenance. (a) Slope erosion requiring periodic maintenance. (b) Gabion wall requiring repair: periodic or emergency maintenance.

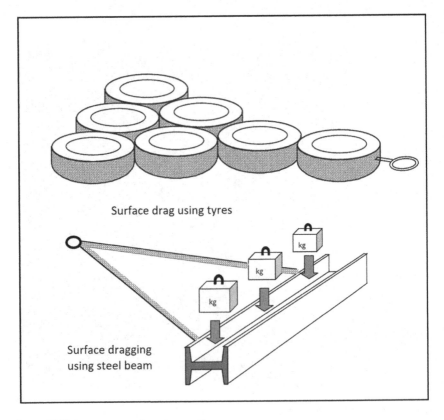

Surface drag using tyres

Surface dragging using steel beam

Figure 21.10 Low-cost surface drag options.

4. An important advantage of using brick, block and cobble surfacing for lower volume roads is that the majority of R2 carriageway maintenance can be undertaken effectively using local community groups.
5. Closely spaced cracks may be filled with a bitumen slurry rather than by local hot-bitumen sealing. A slurry is produced by mixing coarse sand with bitumen emulsion (in a ratio of approximately 3:1 by volume). After sweeping clean the slurry is spread out in a thin layer with a wooden board.

Timing

Rural road maintenance can be considered as a continuous series of actions, and the timing of the interventions depends on the purpose of each activity. A typical schedule of maintenance activities is indicated in Table 21.3.

Structures

The combination of replacement costs, long design life and access criticality requires that maintenance of cross-drainage structures such as bridges, causeways and major culverts should be a priority asset management focus. Maintenance of structures is specifically addressed by Larcher et al(2010)

Table 21.3 Timing of Typical Maintenance Interventions

Maintenance	Typical activities	Resources	Time period
Routine	Routine 1 (off-carriageway) Ditch, culvert cleaning Vegetation trimming Minor slip clearing Unsealed shoulder repair	Unskilled trained labour. Labour-based hand tools.	Minimum once per year, before rainy season. Ideally, twice per year, before and after rainy season.
	Routine 2 (carriageway) Pothole repair Minor re-shaping Minor slip clearing Minor surface repair	Unskilled and semi-skilled labour. Mainly labour-based tools, possible small hand compaction equipment.	Minimum once, before rainy season. Ideally, twice per year, before and after rainy season.
Periodic	Re-gravel and shape for unsealed roads. Bitumen overlay -sealed roads	Semi-skilled and skilled labour – support by unskilled labour.	Intervals: 2–5 years unsealed road. 5–8 years sealed road
	Slab repair for concrete roads Culvert repairs Bridge inspection and repair. Major earthworks repairs	Mechanical equipment.	6–12 years concrete road. Depending on Road Environment and road performance.

Table 21.4 Common Structural Maintenance Requirements

Defects/issues	Impact(s)	Maintenance action	Activity group
Culverts			
Aperture partially or fully blocked.	Structure overtopped. Pavement erosion and flood. Figure 21.10a.	Clean out debris.	R1
Inlet/outlet erosion.	Increased structure damage. Pavement erosion and flood.	Repair inlet or outlet platforms. Additional gabions, walls, bioengineering.	R2, P P
Wingwall damage.	Increased structure damage. Embankment and pavement erosion and flood. Figure 21.10b.	Repair wingwalls.	R2, P
Culvert washout.	Pavement damaged, access degraded.	Replace culvert with enhanced protection including bioengineering.	E, P
Bridges/vented culverts or causeways			
Foundation or abutment erosion.	Further degradation and possible bridge failure, Figure 21.11a.	Minor repairs. Structural repair.	R1 R2, P
Channels blocked.	Structure overtopped. Pavement erosion and flood.	Clean out debris.	R1
Damaged or missing signs or guideposts.	Safety issue for vehicles, cyclists and pedestrians.	Repair and/or replace.	R2, P
Degradation of superstructure or deck.	Weakening of bridge. Access degradation or blockage, Figure 21.11b.	Bridge survey and repair. Decking repair.	P R2

and in TRL ORN 7 (1988). Table 21.4 presents a summary of rural road structure defects and actions. These defects and actions are also illustrated in Figures 21.11 and 21.12.

In addition to the structures themselves, there is an accepted need to focus maintenance activities on potential river or stream erosion upstream and downstream, as well as adjacent abutments and piers. In addition, there also are clear safety concerns to be addressed by routine and periodic maintenance activities; these include:

- Bridge guard rails.
- Pedestrian walkway decks.

Figure 21.11 Required culvert maintenance. (a) Blocked culvert. R1 maintenance. (b) Culvert wingwall and apron erosion. R2 repair and maintenance.

Figure 21.12 Some typical bridge maintenance. (a) Bridge abutment eroded. R2 maintenance and repair. (b) Bridge deck requiring maintenance. R2 maintenance.

- Repair of depth guidance markers on causeways or fords.
- Cleaning or repair of warning signs on axle loads.
- Cleaning or repair of warning signs on safe crossing flood depths for causeways or fords.

PROCUREMENT AND IMPLEMENTATION

Contracting options

Many of the basic principles of procurement and documentation described in Chapter 19, apply to maintenance as well as construction and rehabilitation. However, a matrix of issues concerning contract models and implementation entities, groups or individuals, should be considered in more detail for maintenance actions. Options within this matrix include:

- In-house ('Force Account' or 'direct labour').
- Conventional quantity-based contracting.
- Performance-based contracting.
- Community or village groups.
- Large contractors.
- Small local contractor or SME.
- Contracted local groups.
- Single contracts (Length-persons).

Force Account

In many LMICs, the Force Account approaches to maintenance used to predominate. In this approach local road authorities plan and implement road maintenance works using their own staff and equipment utilising government assigned budgets. However, over recent decades, private sector approaches have been introduced by many road authorities, often with external encouragement or pressure. The commonly experienced Force Account issues are summarised below.

Advantages of Force Account Roadworks Implementation:

- Direct response to needs (operational/emergency).
- Rapid mobilisation when funds are available.
- Retain skills and experience, familiarity with the network.
- Direct control of personnel.
- Pride of 'ownership'.
- Security/continuity of employment, skills and career progression.
- Dealings/disputes with outside parties minimised.
- Provides benchmark for contractor performance/costs.
- Flexibility to target socio-economic groups.

Disadvantages of Force Account Roadworks Implementation:

- Remuneration usually inadequate to motivate sufficiently.
- Poor incentives, poor discipline.
- Slow equipment procurement.
- Slow, bureaucratic procedures, performance not actively encouraged.
- Erratic supply of funds.
- Poor quality assurance, financial and performance audit lacking.
- Low efficiency and poor management/use of available resources.
- Poor cost-awareness.
- Political interference and easy-to-divert funds/resources.
- Little pressure to try new methods/technologies.

Private sector: Conventional quantity-based contracts

This is the most common contract type for maintenance in which quantities of each activity are listed in a Bill of Quantities (BoQ). For each activity, the contractor estimates a unit cost, which is then multiplied by the quantities on the BoQ to give a total price. At regular intervals, normally monthly, the actual quantities of completed work are measured and the contractor is paid the value of the quantities multiplied by the contract unit rate. This contract requires careful measurement and regular site supervision.

Advantages of private sector implementation:

- Central and local road authorities released from direct organisational responsibilities.
- Plant funding, procurement and management transferred to contractors.
- Manpower sourcing and management delegated to private sector.
- Flexibility to hire/fire and motivate personnel.
- Able to respond to changing sector circumstances.
- Market forces can bring competition, efficiency, high utilisation of assets, and lower costs.
- Possible to gain political support for well-defined activity.
- Better accountability possible.
- Easier to resist political interference once contracts let.
- Greater chance of innovation to reduce costs.

Disadvantages of private sector implementation:

- Possible duplication of supervision.
- Duplication of equipment between contractors unless an active hire market exists.
- Long lead times in registration, tendering, evaluation, award of contracts.
- Risks difficult to qualify and value (particularly for unpaved roads).
- Civil service redundancies, client authority restructuring.
- Government employees require retraining and restructuring for new roles.
- Staff have to be trained and retained, and these costs accommodated.
- Higher cost of borrowing for contractor and difficulty in obtaining credit.
- Changes in legislation, and procedures may be required.
- Overall costs usually higher.

A variation on the standard BoQ model is the Grouped Measurement Contract in which work is measured on a larger scale, and the BoQ has much fewer items. The items will depend on local requirements, but may be as follows:

- Preliminaries, as a single sum.
- Routine maintenance, per kilometre.
- Carriageway formation, per kilometre.
- Gravelling, per kilometre (the haul distance is provided in the tender documents).
- Structures, an itemised list of requirements.

This contract model removes much of the measurement burden of a normal measurement contract. This contract requires skilled and experienced inspectors to ensure that all necessary work is carried out under the composite activities.

A lump sum contract takes the process of grouping activities to the extreme. It is, in effect, a single activity measurement contract. The contractor is paid a single price on full completion of the required works, although advance payments and part payments for partial completion may be made. There is considerable risk for the contractor in this contract type. It is not common for rural road works.

In a cost-plus contract, the actual costs incurred by the contractor are repaid with an agreed percentage addition for profit. The contractor has little incentive to work efficiently, and the contract is generally not cost effective. In rural road networks, this contract type is normally only used for specialised items of work or for emergency maintenance works.

A term measurement contract can be used for routine activities that will be repeatedly required during the year. Expected quantities are listed on a BoQ, bids are assessed and contracts awarded as above, but the contract will also include information on when the activities should be carried out. The contract may also include provision of an emergency response capability.

Performance-based contracts

In Performance-Based Contract (PBCs) contractors submit a price for their planned activities to achieve and maintain roads to a required level of performance, usually defined by a series of criteria (e.g. lack of potholes, lack of debris, clean ditches, etc.). These prices are assessed by the road authority and the contract awarded. During the term of the contract, when the road condition is at or above the performance standard, the contractor is paid a contracted and normally constant amount. Reductions may be applied for poor quality works or works not meeting required service-level criteria. Contracts can be let for a single road or a package of roads for 2 years or more up to over 10 years. Performance contracts suit only experienced contractors who are able to properly determine the activities to meet required standards and related risks, innovate to reduce costs, and bid with great care. Equally, the road authority must be experienced enough to set and enforce compliance with appropriate performance standards (Bull et al. 2014).

Performance-Based maintenance contracts can be only utilised effectively for roads in a maintainable condition. For roads in poor condition, initial rehabilitation must be specified using agreed rates and measured quantities, to bring the road into a maintainable condition. The hybrid approach of combining rehabilitation and maintenance in this way can encourage the contractor to produce high-quality work. PBCs are gaining increased popularity, particularly on MDB-funded programmes. There are potential issues that may require clarification on the levels of climate impact risk to be carried by the parties involved.

Contracting entities

Conventional contractors: Private contractors can range from large national entities to Small and Medium Enterprises (SMEs). Although larger private contractors can participate in bidding for management and maintenance of provincial and district roads, they are unlikely to work on village access roads, unless as part of wider package. When participating in the management and maintenance of provincial and district roads, contractors may be required in some LMICs to subcontract Community Micro Enterprises (CMEs) or local workers groups or social organisations for simple maintenance work.

Micro-enterprises usually consist of 15–20 members or more recruited from a local community. The leader must have relevant skills in contract management, planning, financial and human resources management. CME members should normally have had some practical training on maintenance works.

Community contracts: Community or local group contracts enable organisations which have grown from within the local community to carry out work under a transparent contracting arrangement (World Bank 2015). They can be for a variety of services from the supply of sub-contracted labour to the maintenance and rehabilitation of a section of road. They can be paid on the basis of measurement or performance, although in most situations the former is more appropriate. The contracts should be drawn up according to the required services and the roles of the various parties. Costs will normally be negotiated between the contractor group and the client, although if two community contractors are interested, the technical proposals may also be assessed. Contract documentation should be transparent and understandable by those unfamiliar with contracting. Monthly measurements take place as the contract is being completed.

Table 21.5 Comparison of Community and Contractor Models

Issue	Community group (CG)	Community micro-enterprise (CME)	Small medium contractor/SME
Scope of maintenance work	Routine I	Routine I and some routine 2	Routine I and 2; periodic.
Extent of direct works	Roads within community boundaries.	Village/farms road or packages of these roads.	Roads package within district/ province
Sub-contracting	Sub-contracted to SME or CME for routine I activities in province/district roads.	Sub-contracted to SME for routine I and some routine 2 activities in province/district roads.	Employ CGs and CMEs as sub-contractors.
Contracting options	Standardised contract document.	Simplified bidding process.	Competitive (NCB) bidding.
Use of performance-based options	Not generally recommended. Possible in very simplified version.	Training essential. Performance criteria to be appropriate to Road Environment.	Appropriate performance criteria required.
Payment period	Fixed monthly or perhaps quarterly lump sums adjusted for performance.	Fixed monthly or perhaps quarterly lump sums adjusted for performance.	Fixed quarterly Lumps Sums adjusted for performance.
Contract duration	I–2 years	I–3 years	2–5 years

The involvement of local communities and local organisations is increasingly being seen as a way forward for the sustainability of the lower priority local roads (Table 21.5); this can comprise three distinct but related options:

1. Local associations operating as, or leading, Road Maintenance Groups (RMGs).
2. Commercially minded locally based micro-enterprises.
3. Small to medium maintenance contractors (SMEs).

The lengthworker option: In this model a lengthworker (male or female) is expected to keep his or her allocated length of road (normally between 1 and 2 km) in good condition using routine maintenance activities. A lengthworker is often paid a monthly sum based on a specified attendance time (for example, 10 days per month), in effect to a day-works contract. This can lead to problems if standards are not achieved. The situation may be confused by a loose definition of 'good condition', although there is an option to include service-level criteria, making it, in effect, a very simple form of PBC. The lengthworker's reliance upon supplies of tools and materials from other parties is another downside to this option.

As an example of the lengthworker approach, the Bangladesh Second Rural Infrastructure Improvement Project specifically targeted poor rural women for employment creation and income generation, by providing them with training and pilot contracts to plant, nurse, and protect roadside greenery along specified sections of the road during and after the road construction period. As women were keen to continue their employment and had demonstrated their on-the-job abilities, the Local Government Engineering Department (LGED) absorbed them as ongoing roadside maintenance workers. In fact, this approach of hiring local women for ongoing and regular roadside maintenance proved so effective and successful that LGED subsequently replicated this approach nationwide in all their road improvement projects (ADB 2013).

IMPLEMENTATION CHALLENGES

Training requirements

Training of maintenance personnel used to be satisfactorily accommodated in the Force Account systems of many road authorities by in-house training

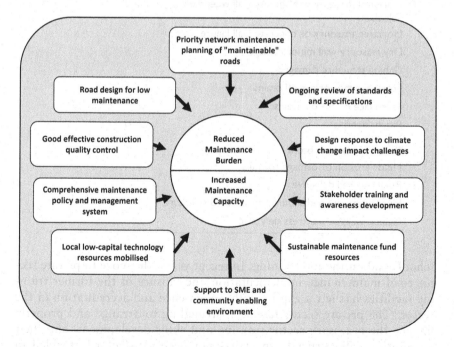

Figure 22.13 Improving maintenance performance by reducing burden and improving capacity.

Table 21.6 Essential Climate Resilience Maintenance Requirements

Defect	Routine	Periodic
Road surface		
Road surface cross-fall eroded	√	
Road surface eroded and impinging on access		√
Grass on shoulder	√	
Shoulder badly eroded		√
Drainage		
Ditch silted	√	
Ditch eroded (minor)	√	
Ditch eroded (major)		√
Turnouts blocked	√	
Culvert blocked	√	
Culvert damaged		√
Earthwork		
Minor earth slip	√	
Slope eroded (major)		√
Mortared masonry or concrete wall weep holes blocked	√	
Mortared masonry or concrete wall failure		√
Dry masonry wall minor damage	√	
Gabion structure damaged		√
Erosion in face or cut-off drains	√	
Failed face or cut-off drains		√
Bioengineering	√	
River or stream bank		
Minor erosion or damage to walls	√	
Retaining wall defective		√
River or stream bed scoured adjacent to structure		√
Gabion walls or mattress defective		√

schools and on-the-job training. In recent years, the move to private sector road maintenance implementation and closure of the former training facilities has left a gap in skills, experience and accreditation in the sector. The private sector has been struggling to arrange and properly finance the necessary sector training and skills development, and their recognition. This crucial aspect has not been satisfactorily tackled in many LMICs.

Skills development requires the academic components to be catered for. However, road maintenance is a very practically orientated activity and demonstration of good practice is an essential part of skills development.

Reducing the maintenance burden

Road authorities can adopt strategies that combines reducing the maintenance burden as well as increasing their maintenance capacity. A number of initiatives can be taken to narrow the gap between needs and capacity, as indicated in Figure 21.13.

Climate change adaptation

Maintenance is in the front-line as regards mitigating the impacts of climate change on rural road networks (World Bank 2017). Impacts, in general, are reviewed and discussed in Chapter 17. Table 21.6 focuses on maintenance requirements of specific importance to combatting climate impacts.

REFERENCES

ADB. 2011a. *Community-Based Routine Maintenance of Roads by Women's Groups. Manual for Maintenance Groups*. Asian Development Bank Metro Manila, Philippines. https://www.adb.org/publications/community-based-routine-maintenance-roads-womens-groups-manual-maintenance-groups.

ADB. 2011b. *Performance-Based Routine Maintenance of Rural Roads by Maintenance Groups. Manual for Maintenance Groups*. Asian Development Bank Metro Manila, Philippines. https://www.adb.org/publications/performance-based-routine-maintenance-rural-roads-maintenance-groups-manual-maintenance-groups.

ADB 2013. *Gender Tool Kit: Transport Maximizing the Benefits of Improved Mobility for All*. Asian Development Bank Metro Manila, Philippines. https://www.adb.org/sites/default/files/institutional-document/33901/files/gender-tool-kit-transport.pdf

ADB. 2018. *Compendium of Best Practices in Road Asset Management*. Asian Development Bank Metro Manila, Philippines. https://www.adb.org/publications/compendium-best-practices-road-asset-management.

Bull, M., R. Brekelmans and P. Wilson. 2014. Lessons learned in output and performance-based road maintenance contracts, World Bank PPIAF. https://ppiaf.org/documents/2065/download.

Ethiopian Roads Authority (ERA). 2016. Manual for Low Volume Roads. Part G, Road Maintenance. https://www.gov.uk/research-for-development-outputs/manual-for-low-volume-roads-part-a-introduction-to-low-volume-road-design.

Larcher P., R.C. Petts and R. Spence. 2010. *Small Structures for Rural Roads. Volume 2, Detailed Design, Construction and Maintenance*. https://www.gov.uk/research-for-development-outputs/small-structures-for-rural-roads-a-practical-planning-design-construction-maintenance-guide.

PIARC. 1994–2006. International road maintenance handbooks (4 Volumes). https://www.piarc.org/en/order-library/4369-en-Road%20Maintenance%20 Handbook%20-%20Practical%20Guidelines%20for%20Rural%20 Road%20Maintenance.

TRL. 1985. *Overseas Road Note 2 – Maintenance Techniques for District Engineers.* 2nd edition. https://www.gov.uk/research-for-development-outputs/ orn2-maintenance-techniques-for-district-engineers.

TRL. 1988. *Overseas Road Note 7 – Volume 1, A Guide to Bridge Inspection and Data Systems for District Engineers, Volume 2, Bridge Inspectors Handbook.* https://www.gov.uk/research-for-development-outputs/orn7-volume-1-a-guide-to-bridge-inspection-and-data-systems-for-district-engineers.

TRL. 1999. *Overseas Road Note 18. A Guide to Pavement Evaluation and Maintenance of Bitumen-Surfaced Roads in Tropical and Sub-tropical Countries.* Crowthorne: Transport Research Laboratory. https://www.gov.uk/ research-for-development-outputs/a-guide-to-the-pavement-evaluation-and-maintenance-of-bitumen-surfaced-roads-in-tropical-and-sub-tropical-countries-overseas-road-note-18.

TRL. 2003. *Overseas Road Note 20. Management of Rural Road Networks.* TRL for DFID. https://www.gov.uk/research-for-development-outputs/management-of-rural-road-networks-overseas-road-note-20.

World Bank. 2015. Empowering local women to build a more equitable future in Vietnam. World Bank blog. http://blogs.worldbank.org/transport/allaboutfinance.

World Bank. 2017. *Integrating climate change into road asset management.* https:// documents1.worldbank.org/curated/en/981831493278252684/pdf/114641-WP-ClimateAdaptationandAMSSFinal-PUBLIC.pdf.

Chapter 22

Appropriate technology

INTRODUCTION

Early development of road infrastructure was based on the use of labour, while animal power could be used to assist in tasks such as materials haulage. Durable all-weather surface types such as stone slab or cobblestone paving were ideally suited to labour construction and repair, and still are. Over time, road makers developed labour saving and ever more powerful tools and equipment and new paving materials and techniques to expand the range of options available to them. Road works can now be carried out through a range of methods and materials using various combinations and levels of labour and equipment. These methods can be broadly grouped and defined by technology options for economic, social, policy and works planning purposes (Petts 2012). Technologies can be broadly categorised as heavy equipment, intermediate equipment and Labour Based (using hand tools). Many operations involve combinations of technology; however, it is not widely appreciated that there is a range of technology options for nearly every road sector construction or maintenance activity (Petts 2022).

This chapter provides commentary and guidance on the technology options available for constructing and maintaining rural roads.

KEY DEFINITIONS

Local resources

These can include human resources such as local government bodies, private and community institutions, NGOs, and local entrepreneurs such as contractors, consultants, industrialists and artisans. Local physical resources can include locally made, or assembled, intermediate equipment along with local materials such as timber, stone, bricks and natural construction materials. These latter could be marginal in terms of international standards but acceptable in terms of locally appropriate road designs. There are obvious

DOI: 10.1201/9780429173271-22

economic, social and sustainability advantages to giving preference to local resource use over imported materials, equipment, and enterprises.

Labour-based operations

These are operations carried out principally by manual methods, using hand tools (preferably of construction, rather than agricultural quality). They may be supported by intermediate or more sophisticated equipment for activities not ideally suited to labour methods, e.g. medium-long distance haulage or heavy compaction. Labourers usually walk or cycle to work each day from their homes to avoid significant accommodation and other costs and socio-economic issues relating to 'imported labour'. The use of local labour and equipment resources on local roads has been shown, through development of ownership and stakeholder links, to have add-on benefits in terms of quality supervision and maintenance awareness (ILO 1998).

Sometimes the term *labour-intensive* is used. This is usually in relation to road works programmes designed to create employment in locations lacking such opportunities or in post-crisis or post-natural-disaster circumstances. Unfortunately, some labour-intensive initiatives have paid insufficient attention to appropriate technical standards or Quality Control with consequences for durability of the resulting works (O'Neill et al. 2010).

Intermediate equipment

Modern wheeled agricultural tractors are a low-cost (capital and operating) mobile power source and with various attachments can be used to substitute for heavy equipment for a proven range of tasks in the road sector, often with substantial cost savings and flexibility of applications. There are also lower carbon footprint advantages (Petts 2012). This simple or intermediate equipment is typically designed for ease of maintenance and repair in the conditions typical of a limited-resource environment. It is preferable if the equipment can also be manufactured or fabricated locally.

Heavy equipment based or capital intensive

Sophisticated civil engineering equipment is typically designed for, and generally manufactured in, high-wage, low-investment-charge economies. It is expected to operate with close support and high annual utilisation; usually designed for limited function or task with high-efficiency operation. Typical limited-function, high-cost heavy equipment would include:

- Heavy bulldozers (D8–D10).
- Motor graders.
- Motor scrapers.
- Articulated dump trucks.

- Heavy front-end shovels.
- Large back-actors.

ROADWORKS TECHNOLOGY BACKGROUND

Development of labour-based approaches

Widespread interest in the possibility of labour substitution for capital-intensive construction techniques in the transport sector in the 1970s stimulated the research and implementation activities by the World Bank, resulting in the 1978 report on the topic (World Bank 1986). The socio-economic, technical, economic and financial issues were particularly relevant in the face of rising unemployment and under-employment in many developing countries, and the increasing cost of acquiring and operating capital-intensive equipment for civil engineering works. There were also relevant issues around saving on foreign exchange expenditures on imported equipment and its in-service support.

Many programmes and initiatives were developed in the following decades that sought to substitute labour for heavy equipment (ILO 1998, 2002). However, the necessary support and sustainability issues were not properly addressed in many cases (O'Neill et al. 2010), and few initiatives investigated the 'middle way' of the adoption of intermediate equipment techniques.

Choosing an appropriate technology

The choice of technology has important cost implications for the road maker, manager or client. There are also social and economic implications for the local population. Consideration of technology choice in developing and emerging economies should be driven by four fundamental factors in the financial and economic costs of the works:

- Availability of labour.
- Cost of local labour.
- Availability of suitable equipment.
- Cost of finance or opportunity cost of capital.

Other influential factors are:

- Policies on technology, local employment and local capacity development.
- Appropriate legal and implementation framework.
- Taxation regime for technology options.
- Technologies permitted by local contract arrangements.
- Local awareness and support of the technology.
- Local human resources capacity for the technology.
- Availability of local training in the technologies.

- Local budgeting and cash flow regime.
- Local market and mechanical support for equipment.
- Local availability of construction quality hand tools.

The viable technology options should each be properly and realistically costed to ensure that the most technically compliant and cost-effective application of available resources is used. It is important to use realistic data on finance/opportunity and depreciation/amortisation costs in relation to typical equipment lives and annual utilisation. These factors are complex and generally poorly understood, and often lead to poor decisions on the most appropriate choice of technology (Petts 2002). Government and enterprise policy on socio-economic issues should also influence the final decisions on choice of technology. Finally, Whole Life or Life Cycle costing environmental and sustainability considerations should be included.

TECHNOLOGY CHOICE REVIEW

Key issues

The cost of labour and finance are two key inputs into the consideration of the most appropriate technology in any circumstances. If we take the USA as the prime example of heavy equipment road works approaches, the median hourly wage of a plant operator is about US$23 (2021 US Bureau of Labor Statistics data). Bank base rates in the USA are less than 1% per annum (2021) and have been for most of the last decade, although now rising. The average interest rate on a conventional small business loan is around 3%–8% per annum. Such figures are common to most well-developed countries. By contrast, in most developing economies the wage earnings are a small fraction of those in the advanced economies (typically US$0.5–2 equivalent/hour) and commercial credit is both difficult to obtain for SMEs, and the real rates of interest are typically 15%–40% per annum. Developing country interest rates and inflation are also now rising (2023), and the differentials with the advanced economies remains, or is increasing.

These resource cost differential factors of up to more than ten times suggest that use of capital-intensive technologies should be minimised in medium- and low-income countries, and the potential, needs and constraints of low-capital and flexible intermediate equipment and local labour methods should be rigorously assessed and implemented where appropriate.

Heavy equipment use for unpaved rural roads

The rural road networks in most LMICs are still largely to earth or gravel standard, and the traditional maintenance and construction techniques for roads in these countries in recent decades have normally been based on the use of heavy equipment. Motor graders, for example, are commonly seen as the default maintenance option for the essential maintenance of the camber and side drainage system (as shown in Figure 22.1). These approaches are expensive and in a low-budget environment contribute significantly to the known serious funding and operational constraints on sustainable network management.

Many problems encountered in the road sector in emerging and developing countries can be attributed to the application of inappropriate heavy equipment technology. Heavy equipment-based methods are not physically suited to the geometry of most rural roads, which are relatively narrow, compared to main roads. However, the principal disadvantage is the large capital investment costs of heavy equipment operations, which can include the use of low-bed transporters to move equipment between work sites. They also require specialist operational and support skills and resources. Key issues are summarised in Table 22.1.

With relatively low fleet numbers in many LMICs, there may be serious support and spares stocking and supply issues. With most unpaved road maintenance and rehabilitation equipment tasks fundamentally requiring no more than 100 hp (75 kW), the use of over-powered heavy equipment is an extremely expensive, inflexible and an avoidable luxury in the LMIC context.

For equipment owners, equipment life and annual utilisation are key considerations regarding profitability of its deployment. Even in advanced economies, the general advice is that if owners cannot achieve upwards of 1,500 hours of productive utilisation per year, then they should not own but rather hire in equipment items when required. The problem for LMICs is that active and competitive plant hire options rarely exist. Finally, the carbon footprint of heavy equipment LVRR works is approximately twice as high

Figure 22.1 Heavy motor-grader. Maintenance of an unpaved rural road.

Table 22.1 Summary of Potential Heavy Equipment Problems

	Issues	Concerns
1	Operational	a. Dedicated function (can only be used for one operation).
		b. Inter-dependence, that is, dozer, loader, trucks, motor-grader, bowser, roller all required for gravelling and the fleet could be idle if one link in the chain breaks down.
		c. Lack of continuity of workload for plant items with a dedicated function.
		d. Usually based at locations remote from worksites, plant transporters required, and long mobilisation/demobilisation distances could be involved.
2	Technical	a. Sophisticated mechanisms with high-pressure hydraulics.
		b. Effective plant maintenance unlikely at a local level.
		c. Disposable components, difficult to repair or refurbish.
3	Local support and equipment maintenance	a. Limited local market for equipment sales of each model.
		b. Specialist repair and maintenance skills, tools and facilities required (often only available in the capital or regional centre).
		c. Few dealers able to provide the necessary close support for rural road projects.
		d. Long spares supply lines and delivery times.
		e. Frequent model 'improvements' causing spares stocking and procurement problems with 'planned' obsolescence.
4	Cost	a. Equipment and spares most likely to be imported, consuming scarce foreign exchange.
		b. High capital and finance costs.
		c. High costs of stocking and provision of spares.
5	Climate	a. Carbon footprint much higher than intermediate equipment or labour based alternatives.

or more than the proven intermediate equipment or labour based technology alternatives (Sturges and Petts, 2010 and Petts and Gongera, 2021).

Labour-based road works technology

The 1978 World Bank Study identified the differences in Labour Based (LB) technology purpose between relief, employment generation and income distribution initiatives on the one hand, and durable asset generation programmes on the other. The more durable and sustainable outcomes of the latter require more attention to the planning, technical standards, management and operational environment. Government commitment (genuine and not merely rhetorical), policy support and local resource mobilisation are essential for a positive long-term outcome. Labour-based approaches were seen as the first step in developing a healthy and meaningful local construction industry through local Micro Small and Medium Enterprises (MSMEs).

In the 1990s, the UKDFID-funded Management of Appropriate Roadworks Technology (MART) initiative held international workshops on

Table 22.2 Labour-Based Challenges

Issue	Challenge
Technical capacity	Previous focus on unsealed roads with consequent challenges in terms of lack of knowledge and effective maintenance costs on other surfaces.
Central government support	Lack of local high-level policy support for local resource-based (LRB) methods.
Implementation strategies	Frequent lack of strategies to effectively mainstream LRB approaches into the Project Cycle: planning and design, social impact assessment, approval, bidding, work implementation, Quality Assurance and auditing.
Central Road Authority management	Can lead to a failure to effectively involve and develop key stakeholders for locally based infrastructure implementation. Lack of understanding of the LRB private sector.
Workload	Lack of workload continuity for local contractors or local enterprises after project closures.
Appropriate support framework	Lack of contract frameworks for small-scale contractors. Lack of affordable finance for contractor development, allied to common delays in making contract payments. LRB approaches not included in most civil engineering training curricula. Lack of training and accreditation for local organisations. Lack of forums for local contractor representation.
Standards and cost norms	Frequently not adapted for LRB road works.
Mentoring of LRB	Required to develop next generation of local road engineers.
Financial and technical auditing	Auditing required to counter the frequently opaque and potentially fraudulent practices within local road management.

Labour Based contracting to set out the framework for a facilitating environment for the establishment of cost-effective and sustainable rural road works using the private sector, local resource-based (LRB) approaches. However, many of the essential pre-requisites identified by these initiatives have not been mainstreamed into projects and programmes, so that a sustainable 'enabling environment' has usually not been created with several challenges yet to be addressed (O'Neill et al. 2010). These are summarised in Table 22.2.

It is clear from the foregoing that more attention needs to be paid to the complex range of issues surrounding the labour-based technological approach to rural road works. However, this approach, remains an important option for low-wage, high unemployment environments, or where the logistics of equipment deployment are challenging (for example in remote, mountainous or island communities). Guidelines and techniques are well developed and proven (e.g. MoW Tanzania 1997). However, the creation of an appropriate 'enabling' environment (as with any change in technology practice) is essential for all stakeholders to benefit from its implementation.

Intermediate technology equipment

Fortunately for LMICs, there are proven, low-capital intermediate equipment solutions available, which vitally do not suffer from the adverse equipment issues highlighted previously. Currently, a range of low-cost but powerful, premium brand tractors (as shown in Figure 22.2) are now more widely available to the agricultural sector in many LMICs. The brands are established and have good backup support with future spares availability for the substantial national fleets assured. It is estimated that there are significant numbers of underutilised tractors in many of these countries, mostly deployed to the agricultural sector for commercial and small-scale crop farming. However, with applications restricted only to the agricultural sector, their commercial viability can sometimes be lacking due to low overall utilisation (Hancox and Petts 1999).

The global manufacturing production of agricultural tractors is approximately 100 times that of motor graders. The availability of support, spares and skills for the simpler and cheaper tractor-related equipment means that there is now the prospect to address many of the road sector sustainability issues by utilising proven tractor technology for road rehabilitation, Spot Improvement and maintenance techniques at a much lower cost than traditional heavy equipment approaches.

The tractors can be used with a range of low-cost, simple attachments for the various road construction and maintenance activities, in addition to other rural sector activities. Table 22.3 illustrates the versatility of the

Figure 22.2 100 hp, 4WD agricultural tractor.

Table 22.3 Potential Agricultural Tractor Applications in the Rural Economy

Sector	Operations
Agriculture	Ploughing, harrowing, rotovating, haulage, land clearance and levelling, root removal, planting, seed drilling, fertiliser/pesticide, harvesting, pond/dam construction fencing (post hole boring), access roads.
Forestry	Winching, loading, hauling, poling, sawing, access roads.
Roads	Gravel/materials haulage, water haulage and distribution, personnel transport, fuel and plant haulage (low loader trailer or semi-trailer), towed grading, dragging, towed compaction (rubber tyred/steel roller), earthworks excavation and haulage (towed scraper), excavation, loading (front shovel), grass and bush control, spreading materials, bitumen Sealing (towed bitumen/emulsion heater/sprayer), stone crushing (towed crusher and screens), recycling pavement (milling attachment), brushing/sweeping, mixing (disc harrow).
Agro-processing	Threshing, hulling, milling, haulage, trenching.
Water sector (non-road)	Pipeline excavation, pipe laying, cranage, loading, earth dam construction, irrigation channel construction, water pumping, water haulage, borehole drilling.
Building contracting	Materials haulage, loading.
Mining/quarrying	Stone crushing (from power take off), loading, access roads, materials haulage.
Transport services	Loading, short haulage: goods, materials and personnel.
Plant hire	Hire for all the applications in this table.
Academic/technical institutions	Demonstration, training.

wheeled agricultural tractor to provide low-cost services to the various rural sectors (Petts 2012). Thus, rural tractor utilisation could be raised, benefiting both road and agricultural sectors with lower unit costs. Consequently, road maintenance will become cheaper, more affordable and more sustainable, and agricultural production and rural transport costs will be reduced.

The tractor attachments, such as towed graders, loaders, trailers and bowsers, could be manufactured locally, potentially contributing to the support of the local commercial economy. Light towed graders have been manufactured for over 50 years in Zimbabwe (Figure 22.3). Heavier towed graders (4–5 tonnes) suitable for heavy, rehabilitation grading are manufactured in Kenya and South Africa (Figure 22.4). For the investment costs of one motor-grader between four and six tractor and towed grader units may be purchased and work unit rate costs savings for many tasks are expected to be more than 50%. The capital investment is thus more affordable for MSMEs in an environment where credit/capital is scarce and expensive.

As regards the key road maintenance activities of grading, the power requirement for light (Routine) grading is about 70 hp (52 kW) in conjunction with a 2-tonne towed grader. For heavy (Periodic) grading, which

Figure 22.3 Two tonne towed grader manufactured in Zimbabwe for more than 50 years.

Figure 22.4 Five tonne towed grader manufactured in South Africa.

involves recovery of lost camber on earth and gravel roads, or reshaping of coarse material (such as stone macadam surfaces), the power requirement is about 100 hp (75 kW) with a towed grader of 4–5 tonnes weight.

The case for introduction of tractor technology into the rural road sector in LMICs is powerfully demonstrated. The utilisation of the low-capital investment equipment could be raised to commercially viable levels in the typical high credit charge (and scarce credit availability) economies of the developing world, benefiting both the agricultural and rural transport sectors with lower unit costs (Cook et al. 2017).

CASE STUDIES IN APPROPRIATE TECHNOLOGY FOR ROADWORKS

Kenya

Agricultural tractors were successfully used for the haulage of surfacing gravel in the national Rural Access Roads Programme (RARP) and the Minor Roads Programme (MRP) in Kenya (de Veen 1980; Intech 1991; World Bank 1991; MoPW Kenya1993). Local community labour was paid to construct the roads to engineered natural surface (earth road) standard prior to gravelling. These techniques were applied on rural road networks eventually totalling more than 11,000 km over a period of 16 years.

The RARP used International model 444 (two-wheel drive) tractors of rating 45 hp (34 kW) with naturally aspirated engines. They were used to successfully haul natural gravel for surfacing using 3 cubic metre trailers, deploying one tractor to work with two trailers. One trailer was filled at the quarry by hand, while the other was being hauled to the deposition site. Haul distances were typically up to 10 km. The tractors were standard specification but fitted with heavy duty pick-up hitches.

A Kenya Classified Road network maintenance study (Intech 1991) identified the potential to use agricultural tractors and labour techniques to rehabilitate the more extensive earth- and gravel-classified road networks. The resulting pilot project co-funded by Sida, Danida, KfW and SDC used the civil service Force Account operational structure plus casual labour employment in two medium-rainfall districts (Kericho and South Nyanza). Half of the classified road networks in those districts were rehabilitated (700 km) and brought under effective maintenance in 18 months. The equipment used was 4 No. Ford model 7810 100 hp (75 kW) naturally aspirated 4WD tractors and 5 tonne towed graders manufactured in Nairobi. Camber and drainage systems were restored, new/upgraded culverts were provided where required, and spot surfacing improvements were made on problem sections. This was achieved without any formal design documentation, merely with effective training, operational manuals and good direction and supervision.

This pilot project demonstrated that earth and gravel roads could be successfully and inexpensively rehabilitated using agricultural tractor and local support labour techniques, and brought under affordable maintenance. The various tractor attachments were successfully designed and manufactured locally. The Ford 4WD tractors were adequate for all rehabilitation and maintenance operations when fitted with heavy duty fixed drawbars with pin coupling. The locally made 5 tonne towed grader could achieve heavy regrading, even without a scarifier attachment, although scarifier/ripper attachments were added to later production models. Earth and gravel road rehabilitation rates were between 0.6 and 1.13 km/tractor-grader/day. Direct costs (equipment, labour, hand tools, transport and equipment support) of camber, side and turnout drainage were less than US$700/km (1993 prices). Finance, overhead and Spot Improvement costs were additional.

Zimbabwe

A Tractor Based Road Maintenance System was developed in Zimbabwe to manage the road assets that had been constructed under the German–Zimbabwe cooperation agreement (Gongera and Petts 2003). The programme involved construction of 25,000 km of roads and developing institutional arrangements to maintain the investment. The maintenance model used fixed maintenance areas, each one covering between 150 and 200 km lengths of road. Through practice, this length of road was established to be the optimum length a single unit could effectively maintain.

A single operational unit comprises:

- 1 Agricultural tractor (MF275, 70 hp: 53 kW).
- 1 Towed grader (2 tonne).
- 1 Tractor drawn general purpose trailer.
- 1 Tractor drawn water bowser (5,000 L).
- 20 Tyre drags.

Service levels that can be achieved by using the tractor-based road maintenance system to keep the road in its original state after improvement and minimise deterioration. The work cycles can be split into two:

- Labour-based activities.
- Equipment-based activities.

Equipment-based activities comprise cycles of towed grading using the tractor drawn grader, as shown in Table 22.4. This is a wet season activity as the road surface material will be moist, allowing the grader to re-shape and restore camber on the road without compaction. The traffic adequately re-consolidates the material after reshaping. The light towed grader is heavy enough to scrape about 20–30 mm of gravel or soil on the surface during

Table 22.4 Equipment-Based Road Cycles

Activity	High ≤ 30 vpd	Medium ≥ 10–30 vpd	Low ≤ 10 vpd
Towed grading	1 cycle/month	1 cycle/2 months	1 cycle per year
Tyre dragging	1 cycle/week	1 cycle/2 weeks	1 cycle/month
		Based on a rainy season November to March	

Table 22.5 Labour-Based Road Cycles

Activity	No. of cycles per year
Verge clearing	2
Cleaning culverts	2
Cleaning drains and repairing erosion damages	2
Patch gravelling	2
Road furniture maintenance	1

reshaping. This process does not damage the consolidated and established camber with good structural integrity. Similarly, tyre drags are drawn by the tractor during the dry season to minimise establishment of ruts and corrugations and to re-distribute loose gravel on the surface. Tyre drags are located at strategic locations about every network 20 km, to be hitched to the tractor for the local network. These equipment-based activities are programmed according to terrain, climate and traffic on each road by the unit teams themselves.

Labour cycles are not dependent on traffic but are more responsive to terrain and climate, Table 22.5. With more hilly terrain and high rainfall, the more the road will benefit from higher level of labour input, especially for drain clearing and for vegetation encroaching on to the road and around curves. The tractor and trailer are used to transport the labour gang to their work location each day, before continuing with the equipment-based activities.

The system proved to be effective in providing regular and timely response to road maintenance needs throughout the year. This was made possible by the appropriate size of maintenance areas allocated per tractor together with personnel and a dedicated budget. Funding of routine maintenance was given priority over construction, rehabilitation and periodic (re-gravelling) operations.

The use of the tractor as the only motorised equipment meant that activities were undertaken by changing the accompanying attachment to the tractor. The repair and maintenance of the tractor and attachments was relatively simple, and trained artisans responsible for the routine checks on the tractor ensured that it was well kept and serviced. All these arrangements resulted in long equipment life (typically 10,000 hours per tractor) and a reduced cost for road maintenance.

The total costs of the system, including overheads, capital depreciation and finance costs were assessed to be US$260/km/year in 1997 prices for a typical network of 150 km (Gongera and Petts 2003). More recent (2021) costs would be in the order of US$600/km/year (Gongera and Petts 2015).

Vietnam: Intermediate technology in rural road construction

Concerns regarding the climate resilience of Gravel Wearing Course (GWC) as a Low Volume Rural Road (LVRR) surfacing in Vietnam were raised by the Government of Vietnam in the early 2000s. Subsequently, UKAID-DFID-funded projects under the South East Asian Community Access Programme (SEACAP), with additional support from the World Bank, were initiated to investigate this and identify alternative pavement options (Intech-TRL 2007; World Bank 2014).

The road trials comprised an extensive programme of 156 km of short-section construction and monitoring of various bituminous and non-bituminous surfacing options in a representative selection of geological, terrain and climatic environments. While the principal focus of the project was on the performance of specifically constructed trial sections, a key supporting issue was the use and assessment of appropriate technology involving available local labour in combination with agricultural equipment in pavement construction.

Construction of the trial sections was undertaken through commercial contracts involving the use of small local contractors, who sub-contracted to varying degrees labour and/or equipment from local communities. The principal uses of local labour alongside equipment-based operations on the Vietnam trials programme were as follows:

- Site clearance.
- Construction of road formation.
- Construction drainage system.
- Excavation of construction materials.
- On-site screening and sorting of aggregates.
- Spreading of stone aggregate and bitumen (Chip Seal surfacing).
- Sand spreading (Sand Seal surfacing).
- Concrete Block/Brick placement.
- Cobble Stone placement.
- Hand Packed Stone construction.
- On-site cold-mix preparation.
- Concrete slab preparation.

Table 22.6 summarises the principal applications of local equipment, as also illustrated in Figure 22.5.

Table 22.6 Local Agricultural Equipment Used in Low Volume Rural
Road Construction

Local equipment	Activity
Locally built small trucks/trailers (Figure 22.5a)	Construction materials transport.
Agricultural tiller tractors (Figure 22.5b)	Construction materials transport. Labour transport. Mobile power source for rotavators.
Small conventional tractors (Figure 22.5c)	Towing trailers – materials. Towing rotavators – soil stabilisations.
Rice-field rotavators (Figure 22.5d)	Subgrade scarification. In situ mixing – soil stabilisation.

(a)   (b)

(c)   (d)

Figure 22.5 Local equipment used in low volume rural road construction. (a) Trailer to
be used with an agricultural tractor for hauling construction materials. (b)
Agricultural tiller tractor used for hauling construction sand. (c) Light agricultural tractor with a rotavator attachment, bitumen stabilisation of sand.
(d) Rice-field rotavator used to mix-in lime for base stabilisation.

The following general lessons can be drawn from the appropriate technology mix of labour, local equipment and heavier construction plant used in the Vietnam trial road pavement construction programme:

1. In general, the use of local labour within an appropriate mixed technology environment proved to be beneficial in terms of local social-economic issues as well as pavement construction.
2. In general, on-site, hands-on training proved to be an effective means of introducing new techniques and more effectively utilising labour in mixed technology construction.
3. The cost-effectiveness of labour-based activities is variable depending on the seasonal availability of labour; this proved an issue in some areas of Vietnam where two or three rice harvests per year restricts labour availability.
4. Use of mixed, appropriate, technology allowed for significant gender diversification in low volume road construction.
5. The link between labour and the local community served by the roads being constructed provided an additional motive for improved Quality Control, and left a residual local understanding of village access road features and function.
6. Mentoring of contractors in new techniques and good cost awareness helped to achieve high-quality, low-cost outcomes.
7. There are potentially issues to be addressed regarding site safety in construction; in particular, the lack of personal safety equipment always has to be addressed.

CREATING THE ENABLING ENVIRONMENT FOR TECHNOLOGY CHANGE

Key issues

The experiences over recent decades are that introduction of new technologies or approaches in a limited-resource environment requires very careful planning to create the necessary 'enabling environment'. The concept and terminology 'Technology Transfer' needs to be used with great care and direct transfer between High Income and Low-Income economies usually proves to be inappropriate and unsuccessful. Even transfer between LMIC regions or countries can bring significant challenges.

Initiatives will usually require piloted project initiatives to successfully develop the methods in the local environment. Careful consideration needs to be given to the technical, management, institutional, financial, economic and social issues, to ensure successful 'roll out' into cost-effective uptake and sustainable embedment.

The following issues/actions should be addressed in the process of introduction of a new or modified technology approach in a limited-resource environment:

1. Review experiences of the technology elsewhere, adapt and pilot if necessary.
2. 'Buy in' by political and senior management cadres; especially persons responsible for national standards, designs, specifications and sector documentation.
3. Awareness creation concerning benefits, costs, availability and performance (leaders–agencies–media–private sector).
4. The pros and cons of implementation options of private sector, Force Account or other approaches.
5. Real cost awareness regarding all equipment (particularly intermediate equipment) in the local operating environment.
6. Possibility of developing and sustaining benchmark costing systems.
7. Stable, specific and diversified market for local MSMEs.
8. Intermediate equipment designs and specifications for procurement.
9. Equipment procurement guidelines (usable by civil engineering orientated managers).
10. Equipment economics and commercial management and support guidelines.
11. Training-demonstration and accreditation of courses in the technical and management aspects of the technology.
12. Availability of finance for MSMEs for the technology.
13. Availability of equipment for hire on the local market.
14. Dissemination and mainstreaming of the technology knowledge.

Stakeholder roles

The roles of stakeholders in developing innovative technology approaches and the necessary 'enabling environment' should be shaped by the actions summarised in Table 22.7.

SUMMARY

Construction enterprises in most Emerging Economies use heavy equipment technology for main road construction because:

- It is powerful (high productivity when working).
- It is fast (work completed quickly).
- Large road projects must be completed quickly to achieve the intended benefits for the large investments.

Table 22.7 Actions to Support Technology Change

Actions to support and encourage an enabling environment for technology change	Government	Road authority	Funding agencies	Training establishments	NGOs	Community groups
Regularly review research and technology developments in similar socio-economic and operational environments.	√	√	√	√	√	
Explore technology for local change potential together with pilot schemes.	√	√	√			
Encourage local manufacture/ assembly of equipment.		√	√		√	
Provide policy, legal and taxation support.	√					
Ensure stable funding for micro-enterprise roadworks.		√	√			
Ensure cross sector support.	√	√	√		√	
Encourage media promotion.	√		√		√	
Encourage LB and Intermediate options in contracts & specifications.	√	√				
Develop specific contract and assured payment arrangements.	√	√	√		√	
Support Government appropriate technology initiatives.		√	√	√	√	√
Ensure effective mechanical support for all equipment in use.	√			√	√	√
Provide credit for intermediate equipment and SMEs.	√	√				
Facilitate knowledge dissemination.	√	√	√		√	
Facilitate training and demonstration in technology options.	√	√	√	√	√	√
Support the Government appropriate technology initiatives.	√	√	√	√	√	√

However, main road construction works are based around larger budgets than low volume rural roads, and the case for using heavy equipment on these lower standard rural roads for construction and maintenance is much weaker.

For rural roads in LMICs, the sector management aim should be to use the most suitable mixture of labour and equipment in a given social, technical and economic context—the approach known as 'appropriate technology'. In developing regions, the two principal inputs of labour/skills and credit/finance have very different costs and implications than in developed economies. Technologies for designing, constructing and maintaining rural roads in LMICs should:

1. Employ appropriate design standards and specifications.
2. Utilise intermediate equipment technology options and local labour and skills resources to reduce reliance on heavy equipment imports where feasible and cost-effective.
3. Promote road construction and maintenance technologies that create local employment and enterprise opportunities.
4. Use types of contracts that support the development of domestic contractors and consultants.
5. Be robust to the vagaries of climate and recognise potential impacts of a changing climate.
6. Be affordable, manageable and sustainable.

REFERENCES

Cook, J. R., R. C. Petts, C. Visser and A. You. 2017. The contribution of rural transport to achieve the sustainable development goals. Research Community for Access Partnership (ReCAP) Paper, ref. KMN2089A, for UKAID-DFID. https://www.research4cap.org/index.php/resources/rural-access-library.

de Veen, J. J. 1980. The Rural Access Roads Programme, appropriate technology in Kenya. SBN 10: 9221022048 ISBN 13: 9789221022046. International Labour Office. https://trid.trb.org/view.aspx?id=165302.

Gongera, K. and R. C. Petts. 2003. A tractor and labour based routine maintenance system for rural roads, Institution of Agricultural Engineers, LCS working paper no 5, DFID. https://gtkp.com/assets/uploads/20100102-224547-7081-LCS%20Working%20Paper5.pdf.

Gongera, K. and R. C. Petts. 2015. Agricultural tractor-based solutions for rural access and development. In *7th Africa Transportation Technology Transfer Conference*, Zimbabwe. https://www.academia.edu/34006415/Agricultural_Tractor_Based_Solutions_for_Rural_Access_and_Development.

Hancox, W. and R. C. Petts. 1999. Guidelines for the development of small-scale tractor-based enterprises in the Rural and Transport Sectors.

ILO. 1998. Employment-intensive infrastructure programmes: Labour Policies and Practices. https://www.ilo.org/emppolicy/pubs/WCMS_114940/lang--en/index.htm.

ILO. 2002. *The Labour-Based Technology Source Book. A Catalogue of Key Publications.* 6th (revised) edition. https://www.ilo.org/emppolicy/pubs/WCMS_ASIST_8370/lang--en/index.htm.

Intech Associates. 1991. Roads 2000: A programme of labour and tractor-based maintenance of the classified road network. Project Document for Government of Kenya. https://www.gtkp.com/assets/uploads/20100103-142337-7115-Roads%202000%20pilot%20project%20brochure.pdf.

Intech-TRL. 2007. Rural road surfacing trials: Construction guidelines. SEACAP 1 Final report appendix A. Report for DFID and Ministry of Transport, Vietnam. https://www.research4cap.org/index.php/resources/rural-access-library.

MoPW, Kenya. 1993. Roads 2000: A programme of labour and tractor-based maintenance of the classified road network. Pilot project final report, prepared by Intech Associates.

Ministry of Works, Tanzania. 1997. Labour-based roadworks manuals, prepared by Intech Associates.

O'Neill, P., R. C. Petts and A. Beusch. 2010. Improved asset management, climbing out of the road sector pothole! World Road Association (PIARC) international seminar on sustainable maintenance of rural roads, Hyderabad, Andhra Pradesh, India.

Petts, R. C. 2002. Low-Cost Road Surfacing (LCS) project. LCS working paper 3. Costing of roadworks. https://www.gtkp.com/assets/uploads/20100103-220522-2539-LCS%20Working%20Paper3-aa.pdf.

Petts, R. C. 2012. *Handbook of Intermediate Equipment for Road Works in Emerging Economies.* UKAID-DFID. https://www.gov.uk/research-for-development-outputs/handbook-of-intermediate-equipment-for-road-works-in-emerging-economies.

Petts, R. C. 2022. Technology Information Note (TIN) TE.1. Technology choice for rural roads.

Petts, R. C. and K. Gongera. 2020. Establishment of tractor-based road works Demonstration-Training Unit in Zambia. AFCAP Project Reference Number: ZAM2059B. https://www.research4cap.org/ral/ZAM2059B-Zambia-DTU-Close-Down-Report-Volume-1-10-09-2020.pdf

Sturges M. and R. C. Petts R. 2020. Developing an approach for assessing the carbon impact of rural road infrastructure provision in developing countries. A proposed methodology and preliminary calculations, for gTKP. Home - IRF gTKP - global Transport Knowledge Practice

World Bank. 1986. Study of labor and capital substitution in civil engineering construction. Project completion report. https://documents1.worldbank.org/curated/en/232611468771663997/pdf/432390WP0NO0PR1Box327355B01Public10.pdf#:~:text=In%201971%2C%20the%20World%20Bank%20launched%20the%20%22Study, socge%20economic%20enviromnt%20of%20laborabundant%20and%20capital-scarce%20countries.

World Bank. 1991. Project performance audit report, Kenya, rural access roads project. https://documents1.worldbank.org/curated/en/807531468914093925/pdf/multi-page.pdf.

World Bank. 2014. *Improving Vietnam's Sustainability: Key Priorities for 2014 and Beyond. Helping Vietnam to Achieve Success as a Middle-Income Country.* Washington, DC. http://documents.worldbank.org/curated/en/752531467986246866/Improving-Vietnam-s-sustainability-keypriorities-for-2014-and-beyond.

Appendix A
Standard data collection sheets

A1 Typical Traffic Count Form for Low Volume Rural Traffic

Province						
District					SURVEYOR	
					LOCATION	
Daily 12 hour counts DATE						
Traffic Class	0600-0900hrs	0900-1200hrs	1200-1500hrs	1500-1800hrs	Option for additional Hours	Daily Average
MOTORCYCLE						
CAR, 4WD, PICKUP						
Tractor						
LIGHT TRUCK =< 5 TONS GWW						
TRUCK > 5 T (2 axle) GVW						
TRUCK > 5 T (3 axle +) GVW						
Mini-bus/Bus						
PEDESTRIAN, WALKER						
ANIMAL/HAND CART						
BICYCLE						
TOTALS						
Rain This Period?						
Daily Survey Period: 6.00 hours to 18.00 hours					GVW = Gross Vehicle Weight	

A2 Typical Axle Load Form

AXLE LOAD SURVEY

Road Name/Number:

Province:

Survey Date:

Time of survey:

Surveyor:

No	AXIAL CONFIG	AXIAL LOADS (TONNES)				Comment on vehicle type	Type of loading
		1	2	3	4		
1							
2							
3							
4							
5							
6							
7							
8							
9							
10							
11							
12							
13							
14							
15							
16							
17							
18							
19							
20							
21							

A3 Standard DCP Field Data Sheet

SITE/ROAD				DATE				
				TEST NO				
SECTION NO/CHAINAGE				DCP ZERO READING mm				
DIRECTION				TEST STARTED AT				
WHEEL PATH								
No OF BLOWS	TOTAL BLOWS	READING mm	No OF BLOWS	TOTAL BLOWS	READING mm	No OF BLOWS	TOTAL BLOWS	READING mm

A4 ROAD CONDITION & VULNERABILITY ASSESSMENT FORM

PROVINCE:　　　ROAD NAME:
DISTRICT:　　　ROAD CODE:
INSPECTOR:　　DATE:　　TRAFFIC:　　PAGE: 01

STRIP MAP

Village / urban area:　Stream
Rock Outcrop:　River
Road:　Gully
Path/Track:　Culvert:
Quarry　Bridge:
Gravel pit:　Outlet:

Strip Map

Chainage (m) +200 +400 +600 +800 ... +200 +400 +600 +800 ... +200 +400 +600 +800 7. PROFILE

1. WAYPOINT NUMBER
2. PICTURE REFERENCE No
3. TOPOGRAPHY (F / R / H / M)
4. LANDUSE / VEGETATION (IRR / DRY / FOR / SHR / RES)
5. ROAD SURFACE (CL / SA / GR / LA / BIT / CO)
6. ROAD GRADIENT
7. PROFILE
8. CARRIAGEWAY PROBLEMS
9. ROAD CARRIAGEWAY WIDTH (M)
10. OVERALL ROAD DRAINAGE CONDITION
11. MISSING SIDE DRAIN(S) IF REQUIRED ? (L, R, LnR)
12. MISSING OR DAMAGED CULVERT (MC/DC)
13. CUT (C) OR FILL (F) SLIPS, [S,M,L]
14. FLOOD VULNERABILITY (1-5)

3. F: FLAT   R: ROLLING   H: HILLY   M: MOUNTAIN

4. IRR: IRRIGATED LAND   DRY: DRY FARMLAND   FOR: FOREST   SHR: SHRUB   RES: RESIDENTIAL

5. RO: ROCK   ST: STONE   GR: GRAVEL   BO: BOG / PEAT   CL: CLAY   SA: SANDY

6. A 0-2%   B 2-6%   C 6-8%   D 8-10%   E 10-15%   F >15%

8. Carriageway Problems
1 Poor shape   a Minor
2 Rutting   b Moderate
3 Erosion   c Severe
4 Potholes
5 Low alignment

10. Road Drainage Condition
DR 1: "Good": No water on the road during rain, no erosion, no silt, good working drainage system
DR 2: "Fair": Some water on the road during rain, some erosion in side drains or half silted
DR 3: "Poor": Much water on the road during rain, severe erosion/siltation of side drains
DR 4: "Bad": Non existing / non functioning drainage system

13. Cut or Fill Slips
C/F 1 No slips
C/F 2 Minor slips <3Cu M
C/F 3 Moderate; 3-10 CuM up to 25% carriageway
C/F 4 Significant, up to 50% carriageway impacted
C/F 5 Major 50-100% carriageway impact

14. Flood Vulnerability
1. No risk to access
2. Slight risk, little impact on access
3. Moderate risk
4. High risk of access being compromised
5. Very high risk - severe engineering issue

Glossary of technical terms

AADT

Annual average daily traffic, commonly taken to refer to motorised vehicles only.

Aggregate (for construction)

A broad category of particulate material including sand, gravel, crushed stone, slag and recycled material that forms a component of road pavement layers and composite materials such as concrete and pre-mix asphalt.

Apron

The flat invert of the culvert inlet or outlet.

Asphalt

Used as an alternative term for *Bitumen* in some regions.

Asphaltic concrete

A mixture of inert mineral matter, such as aggregate, mineral filler (if required) and bituminous binder in predetermined proportions (sometimes referred to as Asphaltic Concrete or Asphalt Concrete). Usually pre-mixed in a plant before transport to site to be laid and compacted.

Atterberg limits

Basic measures of the nature of fine-grained soils which identify the boundaries between the solid, semi- solid, plastic and liquid states.

Basin

A structure at a culvert inlet or outlet to contain turbulence and prevent erosion.

Berm
a. Platform on a fill or cut slope that has been constructed as a series of steps rather than as a continuous slope.
b. Stabilising embankment at toe of an earthwork slope.

Binder, bituminous
Any bitumen based material used in road construction to bind together or to seal aggregate or soil particles.

Binder, modified
Bitumen based material modified by the addition of compounds to enhance performance. Examples of modifiers are polymers, such as PVC, and natural or synthetic rubbers.

Bitumen
A non-crystalline solid or viscous mixture of complex hydrocarbons that possesses characteristic agglomerating properties, softens gradually when heated, is substantially soluble in trichlorethylene and is usually obtained from crude petroleum by refining processes. Referred to as Asphalt in some regions.

Bitumen, cutback
A liquid bitumen product obtained by blending penetration-grade bitumen with a volatile solvent to produce rapid curing (RC) or medium curing (MC) cutbacks, depending on the volatility of the solvent used. After evaporation of the solvent, the properties of the original penetration-grade bitumen become operative.

Bitumen, penetration grade
That fraction of the crude petroleum remaining after the refining processes, which is solid or near solid at normal air temperature and which has been blended or further processed to products of varying hardness or viscosity.

Bitumen emulsion
A mixture of bitumen and water with the addition of an emulsifier or emulsifying agent to ensure stability.

Bitumen emulsion, anionic
An emulsion where the emulsifier is an alkaline organic salt. The bitumen globules carry a negative electrostatic charge.

Bitumen emulsion, cationic
An emulsion where the emulsifier is an acidic organic salt. The bitumen globules carry a positive electrostatic charge.

Black cotton soil
A smectite-rich soil that has the characteristic behaviour of excessive volume change in response to changes in moisture content, expanding during rainy season, and contracting and cracking during the dry season. Commonly derived from the chemical weathering of basic igneous rocks.

Blinding
An application lean concrete or fine material, e.g. sand, to fill voids in the surface of a pavement, structural foundation or earthworks layer.

Brick (fired clay)
A hard, durable block of clay material formed from burning (firing) clay at high temperature in a kiln.

Bridge
A structure providing a means of crossing above water, a railway or another obstruction, whether natural or artificial. A bridge consists of abutments, deck, and sometimes wingwalls and piers.

Camber
The slope of the road surface that is shaped to fall away from the centre line to either side, necessary to shed rainwater. The slope of the camber is called the crossfall. On sharp bends the road surface should fall directly from the outside of the bend to the inside (superelevation).

Cape seal
A multiple bituminous surface treatment that consists of a single application of binder and stone aggregate followed by one or two applications of slurry.

Carriageway
The road pavement or bridge deck surface on which vehicles travel.

Cascade
A drainage channel with a series of steps, sometimes with intermediate silt traps or ponds, to take water down a steep slope without erosion impact.

Catchpit
A manhole or open structure with a sump to collect silt.

Catchwater drain
See Cut-off.

Causeway
A low-level structure constructed across a stream or river that may become submerged in flood conditions.

Cement (for construction)
A dry powder that hardens and sets to bind aggregates together to produce concrete on the addition of water (and sometimes other additives). Cement can also be used to stabilise certain types of soil.

Chippings
Clean, strong, durable pieces of stone aggregate made by crushing or napping rock. The chippings are usually screened to obtain material in a small size ranges.

Chip seal, single
An application of bituminous binder followed by a layer of normally single-sized stone aggregate or clean sand.

Chip seal, double
An application of bituminous binder and stone followed by a second application of binder and stone or sand. The second seal usually uses a smaller aggregate size to help key the layers together.

Chute
An inclined pipe, drain or channel constructed in or on a slope.

Cobble stone (dressed stone)
Cubic pieces of stone larger than setts, usually shaped by hand and placed to form a road surface layer or surface protection.

Collapsible Soil
Soil usually comprised of a sensitive fabric that undergoes a significant, sudden and irreversible decrease in volume upon soaking and/or the application of a load or compaction.

Compaction
Increase in density from loose material to compacted material. The compression of soil or other granular material into a dense state through the application of weight or mechanical manipulation.

Complimentary interventions
Additional actions or initiatives that are implemented through a roads project that are targeted towards the communities that lie within the influence corridor of the road and are intended to optimise the benefits brought by

the road and to extend the positive and mitigate the negative impacts of the project.

Concrete
A construction material composed of cement (most commonly Portland cement), a coarse aggregate such as gravel or crushed stone, plus a fine aggregate such as sand and water. Sometimes chemical admixtures are added to improve performance or for special requirements, such as rapid setting or for submerged placement).

Concrete block paving
A wearing course of interlocking or rectangular solid concrete blocks placed on a suitable base course and bedded and normally jointed with sand.

Consolidation
Settlement of loaded soil that occurs over a period of time due to dissipation of pore pressure as water, or air and water are expelled from the voids in the soil.

Cribwall
Timber or reinforced concrete beams laid in an interlocking grid, and filled with soil to form a retaining wall.

Crossfall
See Camber

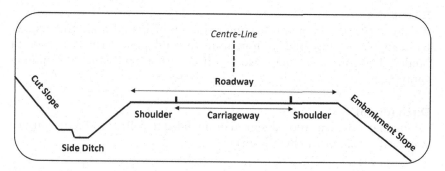

Crushed stone
A form of construction aggregate that is typically produced by extracting a suitable rock deposit and breaking the removed rock down to the desired size using mechanical crushers or manually using hammers.

Curing
The process of keeping freshly laid/placed concrete moist to prevent excessive evaporation with attendant risk of loss of strength or cracking. Similarly, the measures to minimise moisture loss during the initial period of strength development with cement or lime stabilised layers.

Cut-off/catchwater drain
A ditch constructed uphill from a cutting face to intercept surface water flowing towards the road.

Debris grill
Grill, grid or post structure located near a culvert entrance to hold back floating debris too large to pass through the culvert.

Deck
The part of a bridge that spans between abutments or pier supports and carries the road traffic.

Design speed
The assessed maximum safe speed that can be maintained over a specified section of road when design features of the road govern the speed.

Dispersive soil
Soil in which the clay particles detach from each other and form the soil structure in the presence of water and go into suspension.

Distributor
A vehicle or towed apparatus comprising an insulated tank and a spray bar capable of applying a thin, uniform and predetermined layer of binder. The equipment may also be fitted with a hand lance for manual spraying.

Ditch (drain)
A long, narrow excavation designed or intended to collect and drain surface water off a road or earthwork surface.

Drainage
Interception and removal of ground water and surface water by artificial or natural means.

Drainage pipe
An underground pipe to carry water.

Dressed stone
See Cobble Stone

Drift
A stream or river crossing at bed level over which the stream or river water can flow.

Dry-bound macadam
A pavement layer constructed where the voids in a large, single-sized stone skeleton are filled with a finer material and vibrated-in with suitable compaction equipment.

Earth road
See ENS.

Embankment
Constructed earthworks below the pavement raising the road above the surrounding natural ground level.

ENS (engineered natural surface)
A road carriageway built on soil or weathered rock in place at the road location, and provided with a camber and drainage system.

Expansive soil
See Black Cotton Soil.

Fog spray/seal
A light application of diluted bitumen emulsion to the final layer of stone of a reseal or chip seal, or to an existing bituminous surfacing as a rejuvenating maintenance treatment.

Ford
See Drift.

Formation
a. The shaped surface of the earthworks, or subgrade, before constructing the pavement layers.
b. A geological term to describe a stratigraphic unit between Group and Member in size (thickness).

Faulting
Rupture within rock strata due to tectonic stresses as a result of which displacement takes place.

Formation Width
Total required width for construction of carriageway, shoulders and side drains.

Gabion basket/mattress
Stone-filled wire or steel mesh cages. Gabions are often used as retaining walls or river bank/bed scour protection structures.

Geocells
A concrete pavement option, typically cellular confinement systems made from ultrasonically welded high-density polyethylene (HDPE) expanded on-site to form a honeycomb-like structure and then filled with concrete. When filled with rock or sand also used for erosion control, soil stabilisation on flat ground and steep slopes, channel protection, and structural reinforcement for load support and earth retention.

Geosynthetics
Polymer-based products including geotextiles, geogrids and geomembranes. Filter fabrics for drainage control fall into this category.

Geotextile
A permeable woven or non-woven needle-punched or heat-bonded synthetic fabric that is used in civil engineering to filter, reinforce, protect and drain a soil.

Gravel (construction material)
A naturally-occurring, weathered or naturally transported rock material within a coarse particle size envelope. Gravel is typically used as a pavement layer in its natural or modified condition, or as a road surface wearing course. Suitable gravel may also be used in a graded gravel seal in appropriate circumstances.

Gravel (soil group)
In geological terms, a granular material with a precise grain size larger than sand (2 or 4 mm) and smaller than cobble (63 mm).

Hand Packed Stone
A layer of large, angular broken stones laid by hand with smaller stones or gravel rammed into the spaces between stones to form a road surface layer.

Incremental Paving
Road surface comprising small blocks such as shaped stone (setts) blocks or bricks, jointed with sand or mortar.

Intermediate equipment
Simple or intermediate equipment, designed for low initial and operating costs, durability and ease of maintenance, and repair in the conditions typical of a limited-resource environment, rather than for high theoretical efficiency.

Invert
The lowest point of the internal cross-section of a ditch, pipe or culvert.

Labour-based construction
Economically efficient employment of as great a proportion of labour as is technically feasible throughout a construction or maintenance process to produce the standard of construction or service as demanded by the specification.

Labour-intensive construction
Works using large numbers of labourers with an objective of creating temporary or permanent employment, as well as achieving sustainable and durable infrastructure.

Layby
An area adjacent to the road for the temporary parking of vehicles.

Liquid Limit
The water content of a soil at which the soil starts to behave as a liquid.

Low volume road
Roads carrying less than 300 motor vehicles per day and up to approximately 1–3 million equivalent standard axles over their design life.

Macadam
Layer of coarse, graded, angular mineral aggregate with a filler of fine aggregate, interlocked by compaction. Bitumen macadam uses a bituminous binder to hold the material together.

Manhole
Accessible pit with a cover forming part of the drainage system and permitting inspection and maintenance of underground drainage pipes.

Margins
The right of way or land area maintained or owned by the road authority or owner.

Marginal material
A potential construction material that while may be unacceptably outside normal technical specifications may be deemed suitable under certain proven circumstances.

Mitre drain (turn out drain)
A drain that leads water away from the side drains to the adjoining land.

Net Present Value (NPV)
Measures the excess or shortfall of cash or benefits inflows against the present value of cash outflows over a selected period of time, once financing charges are met.

Plasticity
The range of water contents over which a soil exhibits plastic properties.

Plastic limit
The water content of a soil at which the soil starts to exhibit plastic properties.

Pore water pressure
Pressure exerted by water contained within the pore spaces of soil or rock. Positive pore water pressure acts to reduce effective stress, while negative pore water pressure creates soil suctions, thus increasing apparent soil cohesion.

Pre-split blasting
A technique by which blasting is used to create a linear shear in a rock mass prior to bulk blasting to allow a smoother, more easily supported face.

Otta seal
A carpet of graded natural gravel or crushed rock aggregate spread over a freshly sprayed bituminous 'soft' (low viscosity) binder and rolled in with heavy pneumatic tyred roller.

Outfall
Discharge end of a ditch or culvert.

Parapet
The protective edge, barrier, wall or railing at the edge of a bridge deck.

Pavé
See Sett.

Paved road
A paved road is a road with a stone, bituminous, brick or concrete surfacing.

Pavement
The constructed layers of the road on which vehicles travel.

Penetration macadam
A pavement layer made from one or more applications of coarse, open-graded aggregate (crushed stone, slag, or gravel) followed by the spray application of bituminous binder. Usually comprising two or three applications of stone, each of decreasing particle size, each grouted into the previous application before compaction of the completed layer.

Permeable soils
Soils through which water will drain easily, e.g. sandy soils. Clays are generally relatively impermeable except when cracked or fissured.

Prime coat
A coat of suitable bituminous binder applied to a non-bituminous granular pavement layer as a preliminary treatment before the application of a bituminous base or surfacing. While adhesion between this layer and the bituminous base or surfacing may be promoted, the primary function of the prime coat is to assist in sealing the surface voids and bind the aggregate near the surface of the layer.

Quarry
Open or surface mine for the extraction of aggregate, minerals, or rock.

Reinforced concrete
Concrete that has a metal grid or framework, or fibres of synthetic material, in it to withstand greater stress and strain.

Reseal
A surface treatment applied to an existing bituminous surface.

Residual soil
In situ soil remaining after full chemical weathering decomposition of the parent rock mass.

Rejuvenator
A material that may range from a soft bitumen to petroleum, which, when applied to reclaimed asphalt or to existing bituminous surfacing, has the ability to soften aged, hard, brittle binders.

Ripping (excavation)
To dislodge or fracture soil and rock by pulling through it a ripper tooth or teeth (tynes) mounted on a bulldozer or similar equipment.

Riprap
Rough stones of various sizes, from about 5 to 150 kg placed irregularly and compactly to protect a sea-wall or the banks or bed of a river or watercourse from scour.

Road base and sub-base
Pavement courses between surfacing and subgrade.

Road maintenance
Suitable regular and occasional activities to keep pavement, shoulders, slopes, drainage facilities and all other structures and property within the road margins as near as possible to their constructed or renewed condition.

Roadway
The portion within the road margins, including shoulders, for vehicular use.

Route corridor
The linear area or zone within which road alignment is designed to run.

Scarifying
The systematic disruption and loosening of the top of a road or layer surface by mechanical or other means.

Scour -defect
Erosion of a channel bed area by water in motion, producing a deepening or widening of the channel.

Scour checks
Small check dams in a ditch or drain to decrease water velocity and reduce the possibility of erosion. May be of concrete, stone or bioengineered construction.

Seal
A term frequently used instead of "reseal" or "surface treatment". Also used in the context of "double seal", and "sand seal" where sand is used instead of stone.

Selected fill
Pavement layers of imported selected gravel or soil materials used to bring the fill or subgrade support properties up to the required structural standard for placing the sub-base or road base layer.

Sett (pavé)
A small piece of hard stone trimmed by hand to a size of about 10 cm cube used as a paving unit.

Shoulder
Paved or unpaved part of the roadway next to the outer edge of the pavement. The shoulder provides side support for the pavement, provides access for pedestrian traffic, and allows vehicles to stop or pass in an emergency.

Site investigation
Collection of essential information on the soil and rock characteristics, topography, land use, natural environment and socio-political environment necessary for the defined location, design and construction of a road.

Slope
A natural or artificially constructed surface at an angle to the horizontal.

Slurry Seal
A mix of suitably graded fine aggregate, cement or hydrated lime, bitumen emulsion and water, used for filling the voids in the final layer of stone of a new surface treatment or as a maintenance treatment.

Slurry-bound macadam
A surfacing or pavement layer constructed where the voids in single-sized stone skeleton are filled using bituminous slurry.

Soffit
The highest point in the internal cross-section of a culvert, tunnel or the underside of a bridge deck.

Spray lance
Apparatus permitting hand application of bituminous binder at a desired rate of spread through a nozzle.

Squeegee
A small wooden or metal board with a handle for spreading bituminous mixtures by hand.

Sub-base
See Road Base.

Subgrade
The native or imported material layer forming the foundation underneath a constructed road pavement.

Sub-soil drainage
See Underdrainage.

Surface dressing
A sprayed or hand applied film of bitumen followed by the application of a layer of stone chippings, which is then lightly rolled.

Surface treatment
A general term incorporating chip seals, Slurry Seals, micro surfacing, or fog sprays.

Surfacing
The road layer with which traffic makes direct contact. It consists of wearing course, and sometimes a base course or binder course.

Tack coat
A coat of bituminous binder applied to an existing bituminous surface as a preliminary treatment to promote adhesion between the existing surface and a subsequently applied bituminous layer.

Template
A thin board or timber pattern used to check the shape of an excavation or a placed pavement layer surface (camber board).

Traffic lane
The portion of the carriageway usually defined by road markings for the movement of a single line of vehicles.

Transverse joint
Joint normal to, or at an angle to, the road centre line.

Turf
A grass turf is formed by excavating an area of live grass, and lifting the grass complete with about 5 cm of topsoil and roots still attached.

Turn out drain
Discharge point from a side drain, intermediate between culverts to accommodate side drain surcharge. See Mitre Drain.

Underdrainage (sub-soil drainage)
System of pervious pipes or free draining material, designed to collect and carry water in the ground.

Unpaved road
A road with a soil or gravel surface.

Vented drift
See Causeway.

Waterbound macadam
A pavement layer constructed where the voids in a large, single-sized stone skeleton are filled with a fine aggregate, washed in by the application of water and then compacted, usually by a deadweight or vibrating roller.

Water table
The upper level of the zone of groundwater saturation in soils and permeable rocks.

Wearing course
The upper layer of a road pavement on which the traffic runs and is expected to wear under the action of traffic. This applies to gravel and bituminous surfaces.

Weephole
Opening provided in retaining walls or bridge abutments to permit drainage of water in the filter layer or soil layer behind the structure. Weepholes prevent water pressure building up behind the structure.

Windrow
A ridge of material formed by the spillage from the end of the machine blade or continuous heap of material formed by labour.

Wingwall
Retaining wall at a bridge abutment to retain and protect the embankment fills behind the abutment.

Index

Printed in the United States
by Baker & Taylor Publisher Services